Developmental Processes in Higher Vertebrates

Developmental Processes in Higher Vertebrates

RUTH BELLAIRS

*Department of Anatomy and Embryology,
University College, University of London*

UNIVERSITY OF MIAMI PRESS
Coral Gables, Florida

Library of Congress Catalog Number 70–80928
ISBN 0–87024–204–0

Manufactured in Great Britain

CONTENTS

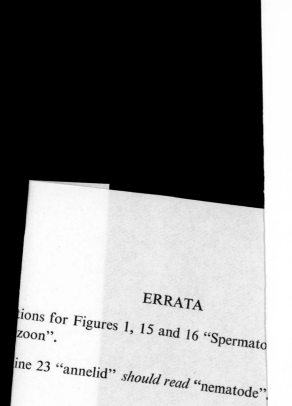

ERRATA

tions for Figures 1, 15 and 16 "Spermato
zoon".

ine 23 "annelid" *should read* "nematode".

PREFACE

MANY of the most important advances in embryology have been made on animals other than the higher vertebrates. One of the purposes of this book is to try to assess how relevant this information is to our understanding of the development of reptiles, birds and mammals.

Embryology, like other sciences, has exhibited phases in its own development. In the last century it passed through a morphological stage, in the early part of this century it became largely an experimental science, and recently molecular biology has been in the ascendant. The emphasis has come to be upon the cells and on the nature and function of the sub-cellular elements. Sometimes it may appear as if we have gone too far away from the problems of the embryo, proper, though in fact this is probably untrue; on the contrary, we may now be ready for a shift of interest back to the whole organism. This book therefore covers many of the general aspects of embryonic life, knowledge of which may provide a springboard for further advances.

The selection of topics that are discussed and the examples that have been chosen have been governed largely by what has especially interested me or has seemed suitable for illustrating the points I have tried to make. I am only too conscious of the omissions and limitations. In mitigation of these offences, I can only plead that I have tried to provide the reader with a wide selection of references, so that he may be able to find his way easily to the literature of those topics that especially interest him.

In preparing this book I have had reason to appreciate the kindness and help of my friends. I owe the greatest debt to Professor A. d'A. Bellairs, Dr. E. M. Deuchar and Professor D. R. Newth who have each painstakingly read the entire manuscript and made many invaluable suggestions. I have also been greatly assisted by others who have either read parts of the book or have spared time for helpful discussions. They are: Professor P. Curzen, Dr. D. A. Ede. Mrs. M. E. England, Dr. M. J. Evans, Dr. J. E. M. Heaysman, Dr. F. Lefford, Dr. V. B. Morris, Dr. R. Niedergerke, Dr. A. K. Tarkowski, Dr. R. A. Weiss, Mr. C. C. Wylie and Professor J. Z. Young, FRS.

Figures 1 and 75 were drawn by Mrs. J. Astafiev, Figures 30 and 57 by Professor A. d'A Bellairs, and the remainder by Mrs. K. Longstaff. I am sincerely grateful for the care and patience they have displayed and for their intelligent approach to the problems of presentation that have arisen. I am also indebted to Mr. R. F. Moss who skilfully prepared the plates, to Dr. A. M. C. Burgess and Mrs. M. E. England, who helped with the arduous task of checking the bibliography, and to Miss Vivien Bellairs for her assistance in the preparation of the indices. Finally I wish to thank Miss Alison Anderson of Logos Press, who so efficiently guided this book through all stages of its preparation.

University College London   Ruth Bellairs
September 1970

ACKNOWLEDGEMENTS

I AM especially grateful to the authors, editors and publishers who have given permission for drawings to be copied. The name of the author is cited under the appropriate drawing, and the original source is quoted in the References. The following journals and publishers have kindly granted permission:

Academic Press; Anatomical Record; Cambridge University Press; The Carnegie Institute; Elsevier Publishing Co. (Amsterdam); Journal of Cell Biology; Little, Brown and Co.; Macmillan, New York; Masson et Cie, Paris; McGraw-Hill Inc.; Saunders; Scientific American; Springer-Verlag, Berlin; John Wiley and Sons, Inc., New York; Zoological Society of London.

CHAPTER 1

INTRODUCTION AND CONCEPTS

PERHAPS one should start a book of this nature with an explanation of its title. The term 'higher vertebrates' will be used to mean the reptiles, birds and mammals, a coherent group with affinities for each other greater than for the amphibians or fishes. These three classes of vertebrates are conveniently grouped together by embryologists and taxonomists and known as the *amniotes*, since they are all primarily terrestrial and possess an amnion in their embryonic stages. Because this group includes man and his domestic animals one might expect its embryos to be most thoroughly studied, but (apart from certain aspects of descriptive morphogenesis and foetal physiology in mammals) probably far less is known about them than about the development of amphibian embryos. Indeed, much of our knowledge of developmental processes is even derived from the study of non-vertebrate forms. This book is an attempt to assess what we know about amniote development, and to discuss how far the processes involved correspond with those in other animals.

The term 'developmental processes' is in a sense self-explanatory, for it means those processes that go on during development. By this is implied not just the events of the embryonic period and the preparation for them that takes place in the germ cells of the parents, though as an embryologist I shall place the main emphasis here, but it includes many activities, such as production of hair and nails, that continue throughout adult life.

It is often useful to consider development as consisting partly of growth, an increase in the mass of tissue, and partly as an increase in complexity. This increasing complexity is given the special term 'differentiation'. To some extent the two processes are separable, and are often dealt with separately, but it is important to remember that during normal development both take place and affect one another. For example, in many tissues differentiation is often preceded by a burst of cell division (Ebert and Kaighn, 1966). Traditionally, the term 'differentiation' has been applied to visible changes in complexity, but is now generally taken to imply related changes in the chemistry of the tissues. Consequently, an exact definition is not always possible. Recent ideas on the meaning of differentiation and its relation to cell division are discussed in Chapter 12.

A. CONTROL OF DEVELOPMENT

We can look at development in many ways, and each of these is reflected in the questions we ask. We may, for example, study the embryo in relation to its environment. This is perhaps the oldest approach and it involves asking 'common sense' questions, such as, 'What happens to the chick embryo if the temperature of the incubator is at say 25°C instead of the normal one of 38·5°C?', or 'Do reptiles ever brood their eggs?' or 'What hormones/drugs/viruses can cross the human placenta from the mother and how do they affect the foetus?'. These are important questions from the point of view of the natural historian or the teratologist, and many of them will be discussed. But to most modern embryologists, or developmental biologists as they now often prefer to be called, these are not the most fundamental queries, for they do little to help us formulate general laws about development. To do this we must ask a different type of question, such as: 'Why does an animal ever stop growing?', 'Why does it undergo differentiation instead of just forming a massive ball of cells?' or 'How does it happen that the parts of the body are in suitable proportion to one another and that the organs lie in the correct relationship?' Re-expressed, in all these questions we are asking 'How is development in all its aspects controlled; how are the processes started, integrated and stopped?'

There is, of course, not just one type of control mechanism in the embryo but many, and these operate on nuclear, cytoplasmic, cellular, tissue and organ levels. Furthermore, the embryo exists in an environment which must be favourable if development is to proceed.

During the last 20 years we have begun to understand much more about some of the types of control involved. In considering them, however, some of the basic concepts of development must also be discussed, for they are described in terminology which will be used in this book. Most of these terms have been introduced in the past as attempts to explain the results of experiments, and as such they suffer from a great disadvantage: they are somewhat vague. As embryologists, one of our tasks is to see how well our older terminology and concepts can be linked with the new ideas on control mechanisms.

B. GENETIC CONTROL

One way of looking at development is to regard it as a programme of protein synthesis. As differentiation of the tissues takes place new proteins are needed, and as the embryo grows the amount of proteins present must be increased. If the embryo is to develop normally then it must control the way in which these proteins are formed so that they appear only in the right tissues at the right time (p. 179). The science of molecular biology tells us something about these controls, for its major preoccupation is with the way in which protein synthesis is supervised by the genes.

Several years ago it would have been in order to have started with the basic idea that all cells in the body possess an identical set of genes, and hence the same information. In other words, we would have assumed there to be no loss or alteration of genes during development. It is important to note, however, that this concept, which has achieved the status of dogma is not really proven and is now being widely criticised. (The problem is discussed in Chapter 9.)

Meanwhile, whether or not some of the genes are lost or modified in differentiating cells, it is evident that they cannot all function all of the time. Thus we may conclude that other factors affect which genes are active at a specific moment. The most widely-accepted scheme that has been put forward to explain how the activity of the genes is controlled is that of Jacob and Monod (1961), who have proposed that in a bacterium, *Escherichia coli*, a series of controller genes, the operators and regulators, become active when conditions in the cell change in some way. These controller genes then switch on or off the genes responsible for protein synthesis (structural genes) (but see discussion in Chapter 9).

It is possible to criticise the Jacob-Monod hypothesis in various ways. Many authors now believe that much greater control may lie in the cytoplasm than has hitherto been suggested, and that to regard differentiation as the result of a switching of genes on and off is probably too naive a concept (see, for example, Harris, 1968). Despite the great amount of work devoted to this subject in recent years, it has not been possible to demonstrate conclusively that the Jacob-Monod type of control of gene activity really occurs in multicellular organisms, let alone in higher vertebrates. One of the nearest analogies to the system worked out by Jacob and Monod on *E. coli* is that presented for the 'bithorax loci' in *Drosophila* where a sequential blocking of certain genes appears to be under the control of other genes (Lewis, 1964).

In a complex system like a developing embryo many interacting genes are involved and these must be continually affected by the feedback of information from the cytoplasm. It is inevitable, therefore, that interest in the cytoplasmic control of gene activity should have recently increased.

It is perhaps unfortunate that the only proof we have of whether a gene is functional or not is if we can recognise the end-product of its actions. This might be, say, an enzyme, or a pigment, or perhaps a particular type of fibril. However, the converse is not necessarily true; if we cannot find the product this does not prove that the gene is inactive. Our failure may be because the end product has not yet been formed, the chain of events leading to protein synthesis having become temporarily blocked. We shall see (p. 59) that such a block intervenes to prevent protein synthesis in the unfertilized egg.

In some situations an indication of whether or not certain genes are active can be gained from inspecting the chromosomes. Where regions are seen to be compacted in cytological preparations, this may be correlated with inactivation of those parts of the chromosome, precursors of messenger* RNA being physically prevented from having access to them. This is thought by some authorities to be the situation in female mammals, where one of the X-chromosomes becomes compacted at gastrulation and is subsequently genetically inactive because it is not physically possible for the RNA precursors to attach themselves to the DNA template (Lyon, 1968; Schultz, 1965). Conversely, the extension out into loops of certain regions of the chromosomes, as in the lampbrush chromosomes of oöcytes (p. 32), occurs at times of great synthetic activity, and is thought to be correlated with greater physical access to the genes.

Although our knowledge is still fairly rudimentary about the way in which the activities of the genes or their products are controlled, we can yet feel confident that control of gene activity does occur. Our present task is now to consider how we can apply this idea of genetic control to some of the basic concepts of embryology.

C. SOME BASIC CONCEPTS OF EMBRYOLOGY

The concepts of *preformation* and *epigenesis* have been exhaustively considered by others (e.g. Needham, 1959; Meyer, 1939; Oppenheimer, 1967). The idea of preformation, current in the 17th and 18th Centuries, was that a complete though miniature embryo was present in each egg or sperm (Fig. 1). This charming idea was not abandoned until the microscopical investigations of the 19th Century showed it to be untrue. The concept of epigenesis, which supposed that the embryo became morphologically more complex as it developed, was then universally accepted

---

* For rôle of messenger RNA in protein synthesis, see p. 31.

and embryology became concerned with trying to explain how increasing complexity (differentiation) is brought about. At first it was believed that each part of the fertilized egg was completely set to develop into a separate organ or tissue. Each region was thought to be able to form more or less independently on its own and the development of the entire embryo was thus regarded as being the sum of the development of the individual parts, in much the same way as a bunch of grapes forms by the development of many separate grapes. This type of development became known as *mosaic* development.

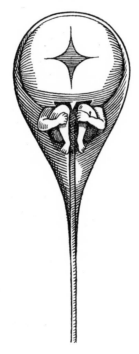

Figure 1. *Preformation. 17th Century idea of a human spermatozoan, which was supposed to contain a minute embryo. (From Needham, 1959.)*

This idea of mosaic development is one that we now know to be untrue for most animals that have been studied, including the vertebrates, because experiments of many types have not led to the results that we should expect if mosaicism were a reality. It was neatly tested in some simple classical experiments by Driesch (1891) who took a 2-celled sea-urchin embryo and shook it in sea water so vigorously that the cells separated from one another. One cell was destroyed, and the other cell subsequently developed into a single, well-formed, though half-sized, embryo (Fig. 2). If this cell had developed in a mosaic way, it would have formed exactly half an embryo. Conversely, if two eggs were fused together a single giant embryo developed. Any process of this type by which the cells rearrange themselves so that they give rise to a normal part or individual is given the name of *regulation*. Until recently there was doubt as to whether mammalian embryos were capable of regulating, and it is of special interest to note that similar experiments to those of Driesch have now established that they are (see p. 135).

Figure 2.   *Evidence against mosaic development. A sea-urchin embryo at the two-celled stage is separated into two cells. One cell is destroyed and the other cell develops into a normal, though half-sized, embryo. In the diagram, the dead cell is shown on the left. The other cell has now divided into a ball of cells which have begun to gastrulate. (After Driesch, 1891.)*

The time comes in the life of every embryo when regulation of the parts to form a complete individual is no longer possible, even though the reorganization of a part of the body can still be achieved. Waddington (1946) has compared this gradual loss of regulative ability to the path taken by a train as it passes through a railway sorting yard: as the train approaches each set of points it has the choice of moving on to different tracks. Gradually the number of possible pathways becomes more and more restricted as the train passes each series of points until eventually it has no more choice but must move along one particular track. If an embryo (or part of an embryo) has passed the stages at which it can modify its development, it is said to be *determined*. Thus, a group of cells may originally be capable of differentiating into epidermal or neural tissue. However, eventually it will become determined so that it becomes capable of forming only one or the other.

The state of determination that an embryo has reached at any particular time is something that is revealed to us only through our experiments. This means that with new and better techniques we are sometimes able to achieve regulation in embryonic tissues at an older stage than expected. Indeed, this has proved to be so in two important situations.

The first has been in relation to the so-called mosaic eggs. As we have seen above, experiments of the early embryologists of this century made it clear that the eggs and early embryos of most animals were capable of extensive regulation, but side by side with this idea persisted another one, that the eggs of protochordates and of certain invertebrates were determined from the time of fertilization (i.e., they had mosaic development). To some extent this was based on the classical morphological descriptions of Conklin (1905) on the ascidian, *Styela*, who showed that the variously coloured pigments of the fertilized egg became distributed in an orderly way. It seemed as if each part of the egg must always give rise to the same tissue. At the two-celled stage, however, complete regulation can be obtained in *Styela* if the two cells are separated (Dalcq, 1932). Experimental analyses have also shown us that even at the 8-celled stage a variety of operations can lead to regulation. Thus, nowadays, we tend to regard the differences between the so-called mosaic and the regulation eggs as being ones of degree, rather than of a more radical nature. That is, some eggs (e.g. *Styela*) lose their ability to regulate during the cleavage stages whilst others (e.g. the chick) retain it until the end of gastrulation.

The second situation in which cells have been found to display regulative ability at an unusually late stage has been when they have been removed from the body and grown in abnormal situations such as in tissue culture. Many cells growing in tissue culture undergo morphological changes and become one or other of two cell types, fibroblasts or macrophages, so-named because superficially they resemble the fibroblasts and macrophages of the body proper. These changes are especially interesting for they can take place in cells taken from adult organisms or from embryos. Furthermore, if cells that have undergone these modifications are now grafted back into the body they can often regain their normal appearance. This process by which cells can change their morphology when they change their environment is known as *modulation* (Weiss, 1939). It appears, therefore, that even an apparently determined cell may still have some potentialities remaining which can be expressed under appropriate conditions.

These potentialities are not, however, unlimited, for the cells that undergo modulation can develop only into cells of the same general type as themselves. A classic example is provided by the work of Fell and Mellanby (1953) on the modulation of types of epithelium growing in tissue culture. With excess vitamin A the epithelium became of the mucous variety; with reduced vitamin A it became keratinous.

The situation is, however, even more complex, for it appears that there are two different types of determination (Hadorn, 1965). In one type, determination leads to immediate differentiation of the cells, whereas in the other type the cellular differentiation is postponed, and instead some change in state of the cell is passed on to its daughter cells. Although Hadorn's examples are based mainly on insect material he points out that these two types of determination are also found in vertebrates. For example, the embryonic nerve cells proper, the neuroblasts, having become determined, soon become differentiated and do not divide further either in embryonic or post-embryonic life. By contrast, some cell systems become determined but remain undifferentiated throughout life; this happens in the spermatogonia in the testes. These cells become determined early in life to form only spermatogonia, but they do not actually differentiate into them until the animal is sexually mature. Thus, we have a type of perpetually undifferentiated or *stem* cell, which gives rise to daughter cells that undergo differentiation. We are ignorant, however, about the mechanisms that permit differentiation to be postponed in one type of cell and to be carried through straight away in another. Nevertheless, these ideas are of special interest in emphasising the interaction of environmental and genetic effects during differentiation and determination.

In the next section we shall consider further how the concepts of determination and regulation fit into the scheme of modern molecular biology.

D. REGULATION AND GENE CONTROL

For our present purposes it is useful to think of the fertilized egg as possessing a programme for development which is encoded in its genes. If the embryo were to undergo completely mosaic development, the genes would be switched on and off in a predetermined way at appropriate times and there would be no possibility of altering this programme.

Generally, however, most embryos are capable of regulation at least in the early stages. In other words, they have a flexible programme. To take an extreme case, if part of the embryo is damaged in some way, certain feedback mechanisms appear to inform the neighbouring cells of this event and these respond by changing their development so that they come to replace the affected region. For instance, if the region of the chick embryo which will normally give rise to the notochord is completely removed, a new notochord will sometimes develop from the adjacent cells (Fig. 3). These cells, which would normally have given rise to somites, have therefore undergone a change in their programme of development and formed notochordal proteins instead of somitic ones.

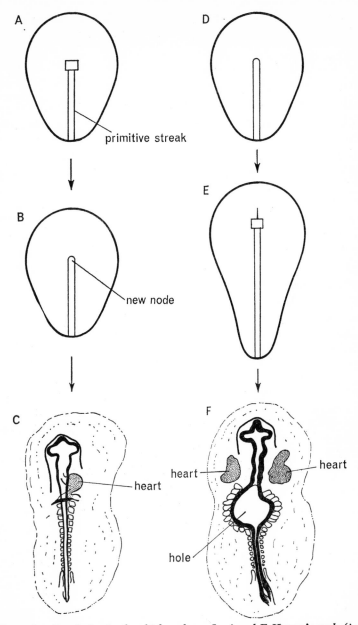

Figure 3. *Regulation in the chick embryo. In A and E Hensen's node (i.e. the presumptive notochord) has been removed, leaving a square hole. If the extirpation is performed at the primitive streak stage (as in A), the wound is sometimes repaired and a new node regulates (B), and a normal embryo forms (C). If the extirpation is performed not at the primitive streak stage (D), but at the head process stage (E), the wound is seldom repaired and the embryo is marred by a hole (F).*

It seems probable that changes of this type involve a whole series of co-ordinated genes.

Regulation can, therefore, be considered as the process whereby genes, probably through the intervention of 'controller' genes, respond to the changes in other parts of the embryo. Conversely, determination may be thought of as the stage when these adjustments are no longer possible.

In the past, regulation has often been considered as an activity brought into play by the embryo only when it needed to replace damaged or dead tissues. It seems more probable, however, that regulation is taking place all the time in the early stages of vertebrate development and that *continual adjustments* are made by the controller genes (or by some other controlling mechanism) to the protein synthesis that goes on in each cell. Only in this way is an orderly integrated embryo likely to form. If this is so, it follows that each part of the embryo must be constantly 'aware' of what is going on in other parts. This idea is one of the central themes of this book and I shall return to it later (p. 14, 62).

E. EFFECT OF CYTOPLASM ON THE NUCLEUS

Meanwhile it is appropriate to discuss the ways in which the cytoplasm can affect the nucleus during the early stages of development. Much of our knowledge has been gained from experiments carried out largely in amphibians in which nuclei and cytoplasm of different origin have been combined. For instance, nuclei from cleavage or gastrula cells

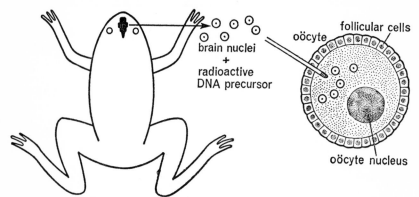

Figure 4. *Effect of cytoplasm on nucleus. Nuclei from adult brain cells are injected into an amphibian oöcyte (as shown). Radioactive precursors for nucleic acids are injected simultaneously. The pattern of synthesis by the injected nuclei changes to that of the oöcyte. (After Gurdon, 1968a.)*

or even from adult brain neurons have been injected into oöcytes or ferti-lized eggs (Fig. 4). Gurdon and Woodland (1968) who have recently reviewed this subject have concluded 'wherever nuclei are introduced experimentally into the cytoplasm of another cell, they very quickly assume, in nearly every respect, the nuclear activity characteristic of the host cell'. They suggest that in unoperated (i.e. normal) cells also the nucleus is influenced by its own cytoplasm so that it comes to function in a particular way. They also propose that during mitosis, when the nuclear membrane has broken down, chromosomes may become associated with cytoplasmic proteins in some way which results in the reprogramming of gene activity. This idea is especially fascinating for it suggests yet another feedback mechanism, and one which possesses the attraction of occurring with great regularity. Unfortunately, the cells of amniote embryos are smaller and technically more difficult to manipulate than those of amphibian embryos, so that no nuclear transplants have yet been performed with them. We do, however, have clues from another type of experiment which suggest that a similar control of the nucleus by its environment may take place in amniote cells. It is possible to fuse together two cells of different histological type to form one single cell with two nuclei (a heterokaryon). For example, Harris et al. (1966) fused together malignant and normal cells. They found that if a cell which synthesised DNA or RNA (such as a HeLa cell or an Ehrlich ascites cell) was fused with one that did not (such as a rabbit macrophage or a hen's erythrocyte), the active cell initiated the synthesis of the nucleic acid by its partner's nucleus.

F. SUPPLEMENTARY CONTROL MECHANISMS IN DIFFERENTIATION

Recently a number of theoretical papers have appeared arguing the case for control mechanisms in the cell supplementary to those in the genes. These will be discussed in this section.

We have already seen that cells with what we assume to be the same genetic make-up (i.e. the same genome) develop differently in different regions of the body, forming such varied tissues as muscle, nerve and kidney. Thus, it is generally believed, different parts of the genome become functional in different regions of the body. This suggests that each part of the embryo actually controls which genes become functional within its cells. In a sense, therefore, differentiation may be looked on as being the result of interaction between environment and the genes.

Most cells, once they have become differentiated, will, when they divide, give rise to cells of the same type. A liver cell, when it divides, forms two liver cells directly. We say that there is a stability of the phenotype. That is, the phenotype is retained after each cell division. How is this stability maintained?

One possibility is that it is due to a persisting and continual interaction between the genetic and environmental factors. This implies that if we were to alter one of these factors the cell would change. To some extent this happens during the process known as modulation (p. 7) in which a cell that has been put into a different environment undergoes morphological change and yet when placed back in its usual environment reverts to its old form. This cannot be the whole story, however, for, as Abercrombie (1967) has pointed out, the following experiment shows that other factors are involved.

If two quite different cell types are taken from the same animal (i.e. the genotype is the same but the phenotypes are different) and placed in the same environment (say in tissue culture) they do not necessarily come to resemble each other. It is true that each may undergo some alteration in its morphology or physiology as a result of the new environment, but this is not important. For instance, groups of epithelial cells and of muscle cells taken from the same animal and grown in tissue culture remain distinguishable from one another. The fact is that each has remained relatively true to type even though both have the same complement of genes and environment. Thus, something has been handed on at each cell division in the new environment that is specific for each cell type; and this implies that once cells have undergone differentiation of some sort there is another, supplementary, control operating in the cell.

Nanney (1958) has proposed that this type of control be known as an *epigenetic system* to distinguish it from the genetic system, though Abercrombie (1967) has suggested that *epigenotype* might be more suitable. Whatever we call it, however, we are still remarkably ignorant about how such systems operate. Nanney suggested that at least some of these controls are located in the nucleus; it is possible that certain regions of the DNA may be permanently altered in differentiated cells (p. 180). Another possibility is that the controls are located in the cytoplasm for, as we have already seen, there is evidence that the cytoplasm can influence the nucleus (p. 10). It is even possible that these controls should not be considered as being initiated by the individual cells but rather by the interactions that go on between the cells in a population. Grobstein (1966) has drawn particular attention to the importance of regarding the cells in relation to their neighbours. He has pointed out the importance for a differentiating tissue of both its cellular arrangements and its mass. There are now many examples in the literature to indicate that a certain minimum mass of cells is often necessary if differentiation is to take place properly. For example, he found that if epithelium from embryonic mouse pancreas is grown in tissue culture it will differentiate only if a sufficiently large mass of cells is present (Grobstein, 1966, 1967). The situation is, however, a complex one for if the mass of undifferentiated pancreatic cells is artificially increased this does not lead to precocious differentiation (Wessels, 1968).

In many organs differentiation proceeds first as a laying down of a morphologically recognisable organ, and the differentiation of the cells into specific cell types follows only secondarily. Grobstein (1966) suggests, therefore, that it is only as a result of group action within the cell population that cellular differentiation proper becomes possible.

Ideas of this sort, which stress the interaction between a cell and its neighbours, are of especial importance, since there has for a long time been considerable and mounting interest in the idea of extra-nuclear inheritance. One of the earliest to envisage such a possibility was Goldschmidt (1924), though, as Oppenheimer (1967) pointed out, it was not until the mid-forties that such an idea was acceptable to embryologists.

One of the most promising examples is the possibility that the cell surface may possess properties which are inherited directly from one generation of animals to the next without the intervention of the genes. Curtis (1965) reported that if the grey crescent region of *Xenopus* eggs was damaged this tended to cause an arrest of development. In those embryos that survived, reached adulthood and bred, the arrest was revealed in their children and in their grandchildren. Curtis concluded that these results implied self-replicating properties of the grey crescent cortex. It will be particularly interesting to see if similar results can be obtained for other species.

G. PATTERN FORMATION

So far we have considered the cells as separate entities in the developing organism, and if we are to think of embryology solely as a branch of molecular biology it is not easy to go further; for molecular biology aims essentially at the relations between nucleus and cytoplasm, almost as a closed system, though conceding that 'signals' (i.e. other factors influencing the cells) may also be received from outside the cell. As yet, molecular biology hardly begins to help us in understanding why individual cells in a mass (that all possess the same genetic make-up) should develop not only along different but also along co-ordinated lines. For instance, why do the highly specialised cells of the retina develop in a regular pattern within that organ? Further, how does it happen that the different organs of the body form in appropriate proportions and relationships to one another?

This process of co-ordinated development is often known as *pattern formation*, though some authorities prefer the older term, *individuation* which has a more restricted meaning: that of the formation of the embryo as a whole or of any organ as a whole. Our task is now to consider how the cells may interact in the process of pattern formation.

(a) The interaction of the cells

We have already seen that each part of the embryo is probably constantly 'aware' of what is going on in other regions of the body. This is an idea that arose from experimental embryology but has been re-expressed recently by a number of authors. For instance, Abercrombie (1958) said, 'Directly or indirectly, it seems that all cells must influence each other, so that each is informed of its relative position in the system'. A similar view has been expressed by Smith (1960) who has pointed out that just, as cells must be 'aware' of their positions, so must groups of cells (such as the group of mesoderm cells that form each somite.)

Bonner (1965) has suggested that each cell in the developing embryo is constantly carrying out tests to discover such things as whether it lies close to any other cell, and, if so, whether it is in contact with other cells over its whole surface, what types of cell lie around it or what foodstuffs are available. He states (his p. 142), 'The progress of development requires that cells take appropriate, informed action based upon the outcome of tests just as is done in carrying out any other complex task'. Thus, we can consider development as a series of choices selected by the embryo as a result of the tests it continually carries out. In other words, the embryo may be thought of as continually scanning itself and adjusting its development in much the same way as an automatically controlled machine scans itself and adjusts its activities. A similar conclusion has been reached by Apter (1966). Wolpert (1969), in an important theoretical paper has developed these ideas and suggested that each cell may have its position specified with respect to one or more points in the developing system of cells. He calls this 'positional information' and argues that it is the positional information of each cell, which, together with its genetic constitution and developmental history, determines the nature of its molecular differentiation.

The idea of the continual test implies a certain rhythm in development, so that testing is carried out at regular time intervals. We have little knowledge of what these intervals consist. Gurdon and Woodland (1968) have suggested that the nucleus might be reprogrammed at mitosis (p. 11); Goodwin and Cohen (1969) in an interesting theoretical paper put forward the idea that within any differentiating tissue there is a periodically recurring event which synchronises the cells and so provides a time base for development. They say, 'It is as though every cell has access to, and can read, a clock and a map'.

The manner in which information is passed from one cell to another is obscure, though recently we have gained a little insight into one of the ways in which it may happen.

It is possible to correlate certain types of cell behaviour with the development of those regions of close attachment now so familiar in electron micrographs and which are known as attachment zones, desmosomes, or *maculae adhaerentes*. It has often been noticed that these structures are absent from the cells of the embryo at cleavage and gastrula stages, periods during which a continual shifting and rearrangement of cells is occurring, but that they appear once this stage is over (Overton, 1962; Balinsky and Walther, 1961). For this reason it has been suggested

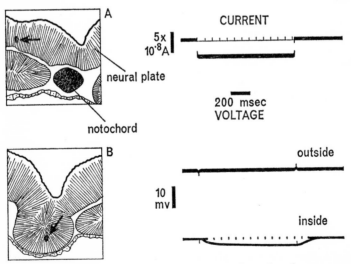

Figure 5. *Intercellular communication. Electrical coupling between neural plate cell (arrowed in A) and developing notochord cell (arrowed in B) in chick embryo. The two cells were 91μ apart. Electrical records show (at top) the current supplied to the neural plate cell, and (bottom) the voltage recorded just outside and just inside the notochord cell. (From Sheridan 1966.)*

that these and similar structures are regions of adhesion between adjacent cells. The cells of the area pellucida during gastrulation, however, possess other types of morphological specialisations which appear to be simpler versions of the desmosomes, and Trelstad, Hay and Revel (1967) have suggested that their presence can be correlated with certain aspects of cell behaviour. As the cells migrate during gastrulation they are constantly making and breaking connections with each other in much the same way as cells do in tissue culture.

The nature of these temporary contacts that these cells make is of special interest for there is some evidence that intercellular communication may take place at points of contact. This has been demonstrated in other types of cells by experiments in which micro-electrodes have been inserted into individual cells (Fig. 5). These have shown that there is a passage of ions from one cell to another at the surfaces where the cell membranes are in contact (Loewenstein, 1966; Sheridan, 1966; Furshpan and Potter, 1969). The significance of these findings, if they can be substantiated in the living embryo, is that we may now have evidence of what as embryologists we have so often supposed: that is, the cells, jolting each other as they move, making and breaking contact, are in a state of continual communication. If this is so, it is not surprising that such complex processes as regulation can occur where all the cells co-operate to act to produce a 'whole' organ or even a whole embryo.

It should be emphasised that this is probably not the only way in which different parts of the embryo may communicate with one another. Indeed, they may influence each other also by direct action of gene products, such as hormonal secretions (p. 272) or perhaps by chemical inducing agents (p. 119, 243).

(b) Polarity and gradients

One of the most tantalising aspects of this problem of cell interaction is that of polarity or orientation, which must be established early in all embryos. In all the vertebrates, except perhaps the mammals, a dorso-ventral polarity is present from the time of fertilization. In mammals it is probably not developed until the end of cleavage (but see discussion in Chapter 7). However, whether it is established at fertilization or not may be unimportant so long as it is laid down by the time of gastrulation, for it is during gastrulation that the antero-posterior axis of the body is fixed. Inevitably, as the embryo achieves these two axes it acquires the third, that of the right/left symmetry.

The establishment of the dorso-ventral polarity in large-yolked eggs is easily understood, for it is based on the position of individual cells in relation to the inside or outside of the embryo as a whole, or on differences between one side of the egg and the other. Thus, for instance, the early chick embryo lies as a flat disc of cells on the surface of the yolk; that part that lies furthest from the yolk is the future dorsal side of the embryo and the part that lies nearest it is the future ventral side. This once more implies an 'awareness' by each cell of its own position.

The establishment of the antero-posterior axis would appear to be a much more complex and formidable undertaking. The first visible indication is at the beginning of gastrulation when the primitive streak starts to develop (p. 85), but the orientation of the streak itself is presumably controlled by the entire area pellucida (p. 114), and this implies that the factors responsible for the establishment of the antero-posterior axis are present even at cleavage; that is, some sort of differential must be set up to start off the gastrulation movements in one place rather than another. Thus, although gastrulation movements reinforce and establish the antero-posterior organisation of the embryo, they are not in themselves responsible for it but are rather the first results of it. Indeed, in the amphibian egg an overriding factor is that of the cortical influence. Such an influence, long suspected, was elegantly demonstrated by Curtis (1960) who transplanted pieces of cortex from one part to another of the fertilized egg of *Xenopus*. Grafts of the cortex of the grey crescent area from uncleaved, fertilized eggs induced a secondary dorsal lip, and a subsequent secondary axis, from the host tissue around them. Curtis concluded that the cortex possesses 'morphogenetic properties'. (The term 'cortex' is difficult to define, but may be taken as the outermost region(s) of the egg, including the plasma membrane.)

Another way of explaining how polarity of the embryo may arise is to suppose that gradients are present in the early embryo. Thus, if there is a gradient in the embryo, then there is inevitably polarity. A gradient has been defined as 'a spatial distribution of a physical or chemical quantity, whereby the value of this quantity gradually changes from point to point' (Raven, 1954). In other words, if the gradient is a chemical one, the concentration of this material is especially high in one region but decreases gradually as we move away from this point. The most obvious gradients are the visible ones, such as the distribution of yolk in most eggs. We have seen that in birds and reptiles this yolk-gradient corresponds with the dorso-ventral polarity of the embryo, and is brought about by the fact that the yolk which has a high specific gravity sinks to the bottom of the egg, whereas the cytoplasm rises to the uppermost side. It used to be thought that this yolk gradient was solely responsible for establishing the dorso-ventral polarity. However, this view has now been largely abandoned in favour of one which supposes that, in amphibians at least, there is an interaction between the yolk gradient and the so-called cortex of the egg (Pasteels, 1964). This theory is based largely on the results of experiments in which amphibian eggs were rotated into abnormal positions so that both yolk and cortex shifted relative to each other. The position in which the blastopore developed was determined by an interaction between an area of white yolk and some influence from that region of the cortex known as the grey crescent (Pasteels, 1964).

The idea of interacting gradient systems has also been used by many workers to explain various aspects of morphogenesis. For example, Saxén and Toivonen have used it in their theory of how neural induction takes place (p. 123).

The concept of physiological gradients in the embryo is usually attributed to Child (1928), though Oppenheimer (1967) gives priority to Boveri (1910). Like so many theories, it arose not as an entirely new product, the fruit of one man's ideas, but as the synthesis and elaboration of what had gone before. It is possible to demonstrate that certain simple physiological gradients do exist: for instance, most embryos when treated with poisons show a gradient of susceptibility of their cells. In the chick, Hensen's node (p. 84, 93) is generally the high point of the gradient and the effect becomes less well-defined as the distance from it increases. A striking example of this can be seen when the chick is treated with hydrogen cyanide (Hyman, 1927). The deleterious effects are highest in the node, and gradually become less pronounced further from the node.

Doubts have, however, often been expressed as to whether or not such reactions have any specific meaning for the development of the embryo. More relevant, therefore, is the fact that both in sea urchins (Runnström, 1966) and in amphibian embryos (Brachet, 1967) the animal-vegetal pole gradients appear to correspond with the distribution of ribonucleic acids. In amphibians, this gradient is characterised by a basophilia which can be correlated not only with a gradient of mRNA (messenger type ribonucleic acid) but also with the distribution of ribosomes seen in electron micrographs (Fig. 6). It appears also that the effects of many chemical substances on the early amphibian embryo are correlated with their effects on the RNA gradients (Brachet, 1967). During cleavage, mRNA is scarcely synthesised (p. 64), but during gastrulation its synthesis takes place preferentially on the dorsal side. In the unfertilized egg, some functional polyribosomes are distributed along this gradient (Brachet, 1967).

Closely related to the idea of gradients is that of *fields*, another term which, because of its vagueness, has been used in a variety of ways.

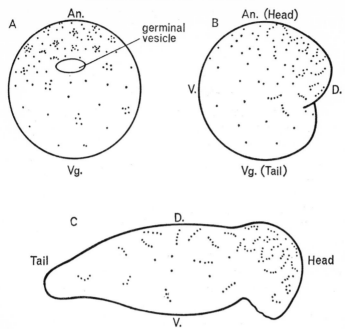

Figure 6. *Gradients of rRNA. A. In the unfertilized amphibian egg the ribosomes are more numerous toward the animal pole (An) than toward the vegetal pole (Vg). Some ribosomes are clumped together to form polysomes. B. In the gastrula stage ribosomes are more numerous toward the region of the dorsal lip (shown at D). Many more have become aggregated to form polysomes. C. At the neurula stage, the ribosomes are most highly concentrated in the head region. Most ribosomes are in the form of polysomes. (After Brachet, 1967.)*

Perhaps it is best to restrict its usage to that of Waddington (1956), who says a field is 'a number of processes that go on in a geographically-definable area of the embryo'. It will be apparent that since we are seldom certain of what exactly does go on in each area, we can seldom speak with precision about it. For this reason the term 'field' is not at present in great favour, though it is still valuable as a concept, for it emphasises that embryonic development is not just due to a collection of individual cells being assembled in a fairly haphazard way.

In a recent discussion of the subject, Waddington (1966) restates his point by emphasising that from the embryological point of view it is most profitable to consider fields as 'individuation fields', individuation meaning the organisation of cells into tissues and organs. He stresses that it is not possible to employ the 'field' concept usefully unless we also

B

take notice of the temporal aspect of development. That is, since an embryo is constantly changing, the boundaries and the nature of the fields within it must also be continually altering.

(c) Morphogenetic movements

In considering the interactions that take place between the cells of the developing embryo, we find that cell movements play an important rôle, perhaps not just by bringing cells into place, but by the actual stimulus of the locomotion itself (p. 103). The courses of many of these complex morphogenetic movements have been painstakingly mapped out for individual species (p. 86, 150, 169, 220). To be aware of the routes they take in the normal embryo or even of the ways in which they can be deflected is not, however, to understand the physiology of these movements nor the ways in which they are controlled and co-ordinated. It is only within recent years that we have begun to gather some clues, and these are derived mainly from model systems of cells growing in tissue culture. Our problem is now to consider what meaning concepts obtained from such studies may have for the living embryo.

The study of cells in culture has become an important branch of scholarship in itself, but the problems that are pursued are often ones that are not directly related to those of the living embryo; for instance, the ability of cells to move over such abnormal substrates as glass or millipore filters. For this reason, much of the tissue culture work, though of interest in itself, has often appeared to have limited value for embryologists. This view is reinforced by the fact that the methods by which the cells are prepared for culture, such as dissociation from one another with trypsin, never occur in nature and have been shown to lead to changes at the surfaces of these cells (Overton, 1968). Nevertheless, we should not forget that the *raison d'etre* of tissue culture is to help us to understand the processes occurring in the living organism. Indeed, we shall lean heavily on *in vitro* experiments throughout this book. Similarly, a number of articles by leading workers have recently tried to correlate the *in vitro* and *in vivo* findings (for instance, Trinkaus, 1965, 1966; Abercrombie, 1965, 1966, 1967; Curtis, 1967; L. Weiss, 1967).

Perhaps the two main ways in which cell movements may be controlled in tissue culture are those known as *contact guidance* and *contact inhibition*. The former was originally suggested by P. Weiss (1934) who found that fibroblasts in culture move along grooves or along lines of stress in the substratum. It is possible to think of many situations within the embryo where one tissue may use another to guide itself. For instance,

DeHaan (1963), studying the migratory patterns of the presumptive heart mesoderm of the chick concluded that they were guided by the morphology of the underlying endoderm (p. 222).

Contact guidance also operates in the chick blastoderm which, as it expands over the surface of the yolk, is constantly attached by its periphery to the inner surface of the vitelline membrane. In an elegant series of experiments New (1959) demonstrated that these peripheral cells possessed some specificity for the inner side of the vitelline membrane, for they would not spread over the outer side of it, nor over certain other substrates. This conclusion was supported by the fact that the inner layer of the vitelline membrane was found to differ chemically and physically from the outer (Plate I) (Bellairs et al., 1963; Bellairs et al., 1969).

The importance of tension as a factor in contact guidance has been stressed by certain workers, primarily with cells in culture (e.g. L. Weiss, 1967). It is of interest that a similar tension is necessary in the spreading of the chick blastoderm (New, 1959; Bellairs et al., 1967).

The term *contact inhibition*, originally put forward by Abercrombie and Heaysman (1953, 1954) to explain certain findings on cells in tissue culture, has been defined by Abercrombie (1966) as 'an inhibition of the locomotion of a cell in a direction that would take it across the surface of another cell, the inhibition occurring when contact is made between the two cells'. In other words, when a cell that is migrating in tissue culture comes into contact with one or more other cells, its speed of movement is usually decreased, and when it becomes completely surrounded by cells it ceases to move at all. Most cells in tissue culture move by means of what Trinkaus (1965) calls a gliding movement and this involves production of ruffled membranes, whether migrating singly (Abercrombie and Ambrose, 1958) or as an epithelium (Vaughan and Trinkaus, 1966). A characteristic of this movement is that a thin, fan-like membrane which appears at the leading edge periodically folds and ruffles on its upper surface; this ruffled membrane activity becomes reduced or even ceases with contact inhibition. The term 'contact' is used to mean contact that is visible by light microscopy and it is not yet clear what meaning this may have in electron microscopical or biochemical terms. The possible mechanisms involved in the process have been discussed by Curtis (1967) who suggests that there is an actual increase in adhesiveness at the surface of the two cells when they come into contact.

The question now arises as to how far the phenomenon of contact inhibition can be applied to cells that are migrating in the embryo and not just *in vitro*. One of the difficulties is that our knowledge has been derived largely from cells which, because they are in tissue culture, tend to lie in a single layer. In the living animal they are usually present in a three-dimensional arrangement. Nevertheless, certain situations do exist in the embryo where the cells are arranged in flattened sheets, and contact inhibition appears to play a rôle in their cell movement. For example, contact inhibition seems to take place when epithelial sheets meet during wound closure (Lash, 1955). Similarly the edge of the chick blastoderm as it migrates (see p. 21) has been shown by time-lapse cinematography to move along by means of ruffled membranes comparable to those seen in cells in tissue culture (Bellairs *et al.*, 1969), and like those other cells will, if it collides with another advancing blastoderm, submit to contact inhibition and halt (Bellairs and New, 1962).

Abercrombie (1966) in discussing the problem points out that those cells which do not become contact-inhibited *in vitro* are the ones that migrate in the body and infiltrate other tissues, i.e. certain types of tumour cells. He suggests that contact inhibition may help to maintain order in the living organism by trapping the moving cells so that they fill up empty spaces. That is, a cell will migrate if it is not completely surrounded by other cells, but once it has reached a position where it is completely surrounded its movement ceases.

A third way by which the cells may possibly be guided in the embryo during cell migration is by *chemotaxis* (discussed by Trinkaus, 1965; Abercrombie, 1967; Curtis, 1967). Few examples of this type have so far been established, but there is some possibility that it may play a rôle in attracting the primordial germ cells to their ultimate destination (p. 234).

We have seen that contact guidance, contact inhibition and chemotaxis may perhaps be of importance in controlling the direction of cell movements in embryos. What rôle do these and other events play in initiating and stopping cell migration?

Inevitably, this is a question that has been studied mainly in tissue culture. It is generally said that most embryonic and indeed adult cells are capable of migrating if placed in a suitable environment. We have little knowledge, however, of the subtle changes in the early embryo that set the stage for the onset of gastrulation, though it is possible that a change in the adhesiveness of the cells to one another is an important factor. In a classical experiment, Holtfreter (1943) found that different germ layers (i.e. ectoderm, mesoderm and endoderm) in amphibian embryos had different degrees of adhesiveness to each other and was able to demonstrate a sequence of changes in these cells as they passed through successive phases of gastrulation. Since then, a number of workers have been able to show that as cells enter new phases during development their adhesive properties change (DeHaan, 1964). The importance of the adhesiveness of cells to one another has been widely studied in tissue culture. It has been confirmed that different types of cells have their own specific adhesiveness (Roth and Weston, 1967; Steinberg, 1964) and that with a few exceptions (such as the neural crest), cells will adhere to their own kind rather than to another. These investigations have largely involved experiments in which two or more tissues have been disaggregated by treating with substances such as trypsin (p. 20) or EDTA (ethylene diamine tetra-acetic acid), which disrupt the intercellular matrix, after which the disaggregated cells are mixed together and allowed to develop. The random mixture of cells then sorts itself out in a characteristic and striking way, usually in such a manner that the cells from one tissue reaggregate as a mixture in the centre of the clump of cells, or sometimes as several small masses, whilst the cells from the other tissue become arranged around the periphery (Fig. 7). Cells will even aggregate with similar cells from another species, in preference to cells from a different tissue of their own species (p. 243).

Most authors now agree that such phenomena are best explained as being due basically to differential adhesions between cells (Abercrombie, 1965; Steinberg, 1964; Trinkaus, 1965) though details of interpretation differ (Curtis, 1967).

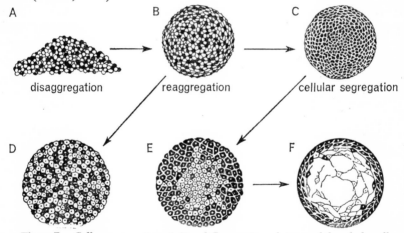

Figure 7. *Cell reaggregation. A, B and C are external views of the whole cell mass; D, E and F are sections through the mass. A. Epidermis (black) and mesenchyme (white) from an amphibian embryo have been disaggregated with alkali; the cells lie in a heap. B & D. The cells round up into a ball but are still randomly arranged. C & E. Eventually, the epidermal cells come to occupy the outside of the mass, and the mesenchyme cells move into the inner part. Differentiation into tissues takes place after the cells have reached their new situation. (After Townes and Holtfreter, 1955.)*

Once again we are confronted with the idea that, in some way or other, cells are able to test their neighbours and to recognise each other. The actual mechanism of adhesion is not well understood. Moscona and Moscona (1963) have put forward the view that aggregation depends on the cells synthesising an adhesive, intercellular cement; electron microscopical observations support this idea, for extra-cellular material lies between explanted cells and their substrate (Flaxman *et al.*, 1968; England, 1969). Similarly, a fibrillar intercellular material that stains heavily with lanthanum has been demonstrated in reaggregates of chick embryo limb bud cells (Overton, 1969). Further, epithelia growing *in vitro* are invariably united by a substrate of basement membrane.

Moscona and Moscona suggested that failure of disaggregated cells to reaggregate was due to failure to synthesise this substance. By contrast, Curtis and Greaves (1965) have produced strong evidence that the aggregation of cells, in tissue culture at least, depends on the condition of the serum used in the culture medium, since fresh serum was found to contain a protein capable of inhibiting reaggregation. It would be interesting to find out whether this protein, when active, inhibits the production of Moscona's exudate.

The structural changes occurring at the surfaces of reaggregating cells are not clear. Lesseps (1963), who has studied reaggregating heart and retina cells of the chick embryo, has shown that when cells are dissociated undulations and microvilli appear on their surfaces, and that as the cells reaggregate the initial contact is made at these places. He suggested that the first adhesions form in these regions. Such a phenomenon is not universal, however, for in an electron microscopical study of reaggregating sea urchin cells Millonig and Giudice (1967) found that microvilli were lost on disaggregation of the cells and not reformed during reaggregation.

Most of these disaggregation experiments have been carried out in tissues taken from embryos in the later stages of development (e.g. mesonephros, limb bud) when cell migrations, though still important, are of less significance than during gastrulation. It is of interest, therefore, that Zwilling (1960b) has been able to show that disaggregated cells of the area pellucida of the chick blastoderm will, if allowed to develop, sort themselves into two types of differing adhesiveness: the cells of the primitive streak which adhere closely, and those of the extra-embryonic area pellucida which are less adhesive and spread out.

Although we have no proof that the behaviour of the cells in this type of experiment exactly mirrors their precise behaviour in the normal embryo, these results fit in well with our knowledge of the chick embryo at this stage. Thus, one type of cell has a low adhesiveness in tissue culture and spreads outwards, just as it does in the embryo when forming the extra-embryonic tissues. The other type of cell has a higher level of adhesion and remains a fairly compact mass just as it does when it forms tissues of the embryo proper.

Contact guidance, contact inhibition and selective adhesion are by no means the only findings of tissue culture, but they are perhaps the most readily applicable to the problems of the embryo, and at this stage of ignorance they give us some confidence that concepts of tissue culture already derived and still to be formulated have some meaning for us as embryologists. Furthermore, since they are all processes that are concerned in the interactions between cells they are likely to play a rôle in the basic requirement that each cell is able to know its position.

(d) Cytoplasmic bridges

In certain specialised situations, cytoplasmic bridges exist between cells. For instance, developing oöcytes of the rat (Franchi and Mandl, 1962) and of the golden hamster (Weakly, 1966), are united in small clusters by intercellular bridges. Similarly developing spermatocytes are also joined to one another in this way (Dietert, 1966), and a certain synchrony of developmental stage seems to be a characteristic of such cells.

(e) Suppression of cell activities

Finally, we should perhaps remind ourselves that cells and tissues may influence each other by suppressing, as well as promoting, certain activities. We have already seen that the pathways along which development can take place become restricted during development, and that this restriction is an essential part of the process of determination. This problem of control through suppression has recently been discussed by Moscona and Garber (1968).

H. SUMMARY

In this chapter I have emphasised that the embryo largely controls its own development, both on a molecular level within the individual cell, and on a cellular and tissue level by collaborations and interactions between cells. These interactions are of various kinds, such as by morphogenetic movements or by the exchange of materials. The individual cells collaborate with each other in forming the embryo, and each cell appears to behave appropriately for the position it occupies. Thus, in some way each cell is 'aware' of its place in the organism as a whole.

CHAPTER 2

PREPARATION FOR EMBRYONIC DEVELOPMENT

FERTILIZATION is one of the key points in development. This is not only because it marks the moment when the individual acquires its full complement of chromosomes, but it is also the time when the egg dramatically switches into a new physiological phase. To appreciate what happens at fertilization itself, and the events that come after it, it is essential to know as much as possible about the gametes. In particular, if we are to understand something of the way that the genes carry out their task at fertilization and during cleavage, we must begin our enquiry with the events of oögenesis.

A. OÖGENESIS

The adult vertebrate ovary contains many oöcytes, each of which is surrounded by a capsule of follicle cells (Fig. 8). All these oöcytes are derived from the primordial germ cells (p. 229) which colonised the

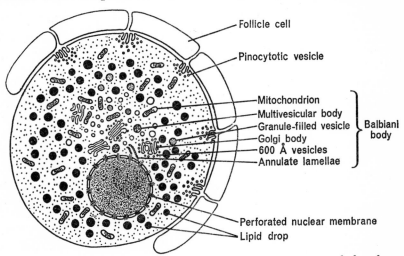

Figure 8. *Oöcyte of hen's ovary. An early stage of differentiation before the yolk has begun to form. (After Bellairs, 1967.)*

ovary during the embryonic development of the mother. Soon after it enters the ovary each primordial germ cell undergoes a series of mitotic divisions so that a large number of new cells, the oögonia, are eventually formed. When this phase of multiplication is over, each oögonium becomes known as an oöcyte.

As the oöcyte develops, a number of events take place which prepare it for fertilization and the early stages of embryogenesis. These are: meiosis, synthesis of nucleic acids and of ribosomes, a certain amount of protein synthesis, and the formation of yolk. In addition, various membranes and protective coverings are secreted around the egg. Shortly before ovulation the nucleus and cytoplasm move to one side of the egg, this being especially apparent in oöcytes that possess a large amount of yolk.

These events will now be considered separately. In our discussion, however, we shall be obliged to lean heavily on knowledge which is derived from animals other than the amniotes, for much of this work is so recent that it has been carried out only on echinoderms and amphibians and it is not always clear to what extent it can be applied to higher vertebrates.

(a) Meiosis

The events involved in meiosis (or reduction division, as it is also called) are described in detail in most text-books of genetics or histology and will only be discussed briefly here. More elaborate descriptions of meiotic chromosomes at the level of electron microscopy will be found in the review by Moses and Coleman (1964). The most important event of meiosis is that the diploid number of chromosomes (2N) becomes reduced to the haploid number (1N) (Fig. 9).

In oöcytes, the process begins to take place in the infancy of the female. For instance, in the newborn mouse the oöcytes are in leptotene and early zygotene, but by four days of age they have reached the stage of pachytene or early diplotene (Peters et al., 1962). Meiosis is not usually completed until about the time of ovulation or even of fertilization. The exact stage at which the first and second polar bodies are thrown off varies in different species, though probably in most amniotes it occurs at about the time of fertilization (p. 58). This delay in the completion of meiosis, which in long-lived species, such as man, may be spread over twenty or thirty years, suggests that a block to development is imposed on the oöcyte during the late embryonic stage or the infancy of the mother, which is not removed until fertilization. This blocking mechanism will be discussed in the next chapter.

PROPHASE {	LEPTOTENE	CHROMOSOMES VISIBLE	2N	OÖGONIUM
	ZYGOTENE	ASSOCIATE IN PAIRS	2N	
	PACHYTENE	LONGITUDINAL DIVISION OF EACH CHROMOSOME INTO 2 CHROMATIDS	2N(4)	
	EARLY DIPLOTENE	CROSSING-OVER OR SYNAPSIS	2N(4)	
	MID DIPLOTENE	NUCLEOLI VISIBLE LAMPBRUSH STAGE	2N(4)	
	LATE DIPLOTENE	GROWTH OF OÖCYTE PAUSE IN DEVELOPMENT	2N(4)	PRIMARY OÖCYTE
	METAPHASE	CHROMOSOMES ON SPINDLE	2N(4)	
	ANAPHASE	CHROMOSOMES MOVE TO POLES OF SPINDLE	2N(4)	
	TELOPHASE	DIVISION OF CELL INTO OÖCYTE & 1st POLAR BODY	1N(2)	
	PROPHASE		1N(2)	SECONDARY OÖCYTE
	METAPHASE	EACH CHROMATID BECOMES A CHROMOSOME	2N	
	ANAPHASE		2N	
	TELOPHASE	DIVISION OF CELL INTO OVUM & 2nd POLAR BODY	1N	OVUM

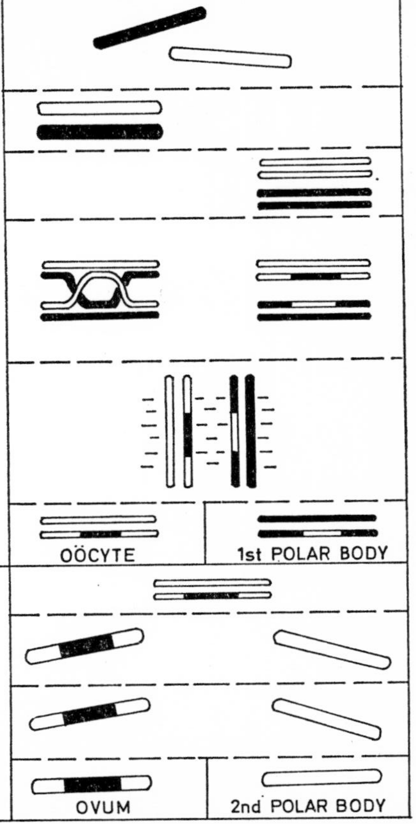

FIRST MEIOTIC DIVISION

SECOND MEIOTIC DIVISION

Figure 9. Meiosis. Schematic representation of the stages. For simplicity, only one pair of chromosomes is shown. The initial pair is depicted with one black and one white chromosome. In the column showing chromosomes per cell, the number of chromatids is indicated in parentheses.

Perhaps one of the most consistent morphological events of meiosis is the rapid growth of the nucleus, which becomes distended by 'nuclear sap' and is then known as the 'germinal vesicle'. At the first meiotic division the nuclear membrane surrounding the germinal vesicle breaks down, so that the nuclear contents mingle with the cytoplasm. This appears to be important for cleavage (p. 62). With completion of this division, one set of chromosomes passes into the first polar body and the other acquires a new nuclear membrane and remains within the oöcyte (Fig. 9).

(b) Nucleic acid synthesis during oögenesis

In this section I shall be merely relating a few of the basic facts as we know them about nucleic acid synthesis in oögenesis. Their significance is in relation to fertilization and early embryogenesis which will not be discussed until Chapter 3.

It is advantageous to begin with a diagram illustrating in simplified form the interrelationships of nucleic acids in animal cells. Further information is available in many recent text-books (e.g. Watson (1965)).

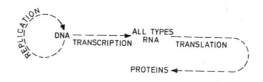

It is now generally accepted that the genes are each composed of DNA (deoxyribonucleic acid), and that apart from such exceptions as cells undergoing mitosis or meiosis, the same amount of DNA is present in each cell (p. 180). This means that as the cells multiply, so the amount of DNA must also increase. The formation of new DNA is known as *replication*.

The DNA in each gene contains a 'blueprint' or programme of instructions, which is used by the cell when it is synthesising a particular protein. The details of this programme depend on the way the bases (i.e. adenine, guanine, cytosine, thymine) are arranged in the DNA. The possible number of different arrangements is immense and this enables every cell to possess many different genes, each of which is able to direct the cell to form a specific protein. Before the cell can carry out the instructions of the gene, a copy of the information in the DNA must be sent from the gene to the cytoplasm. The copy is made with a molecule of RNA (ribonucleic acid). This copying is called *transcription* and it means that the RNA possesses the complementary arrangement of bases to the DNA it is copying. When the programme has been carried to the cytoplasm it is *translated* into protein by the ribosomes. The ribosomes are small particles in the cytoplasm, each about 150 Å in diameter. When protein is not being synthesised by the cell, the ribosomes lie separate from one another. When protein synthesis is taking place, the ribosomes become assembled in strings or clusters. The main types of RNA are:

messenger (mRNA) (or informational RNA) which carries the message from the gene to the ribosome;

soluble (sRNA) (or transfer RNA) which attaches the amino acids to the ribosomes and also activates them;

ribosomal (rRNA) which forms part of the ribosomes where the mRNA and the sRNA-amino acid combinations are brought together and the message is translated in to protein. At least three types of rRNA are known: 28S, 18S and 5S RNA.

Some authorities avoid the use of the terms *messenger* or *informational* RNA in preference for the more biochemical term, DNA-like RNA (dRNA). This is used to describe RNA which has the same base sequences as DNA. It should be noted however that not all the dRNA is necessarily the same as mRNA, for it is possible that some of the dRNA remains within the nucleus.

For our present discussion we will ignore the fact that the situation is more complex than I have indicated, (for instance, there are many different types of sRNA) and will concentrate on the question of which synthetic processes are known to take place in the oöcyte. The evidence is derived from experiments reviewed by Monroy (1965), Tyler (1967) and Davidson (1968).

There are two sites of nucleic acid synthesis in the oöcyte nuclei, that is, the chromosomes and the nucleoli. DNA replication is probably confined to the chromosomes. The possible rôle of cytoplasmic DNA is discussed on p. 35 and 41. It is thought that after replication some of the chromosomal DNA is shifted to the nucleoli.

All oöcytes are rich in RNA and although synthesis takes place throughout development much of it occurs in the earliest stages (Monroy and Tyler, 1967). Brown and Gurdon (1966) showed that in *Xenopus*, at least, no further rRNA synthesis occurred even after fertilization, right up to the time of gastrulation.

(*i*) CHROMOSOMAL ACTIVITY

Much of our knowledge of the activity of chromosomes in the oöcyte has been derived from the study of lampbrush chromosomes (Fig. 10). These are found in the oöcytes of a wide variety of species, and are present at the diplotene stage of meiosis, that is, when the chromosomes are beginning to split apart (Fig. 9). Most of the work on lampbrush chromosomes has been carried out on amphibian oöcytes. They are so-called because in superficial appearance they resemble old-fashioned lamp brushes, or bottle brushes, as each has loops of material projecting out laterally to its main axis. Chromosomes of this type are associated with a period of great ribonucleic acid synthesis and it is significant that before and after this period normal-sized chromosomes are present in the oöcyte (Izawa *et al.*, 1963). It should be noted that in some species they are never visible, the chromosomes apparently taking up a different configuration (Raven, 1961).

The doubling of the DNA in preparation for meiosis occurs before the lampbrush chromosomes have formed, the amount of chromosomal DNA in the prematuration oöcyte nucleus being twice that in the somatic cells. The loops of the lampbrush chromosomes are mainly strands of

A   B

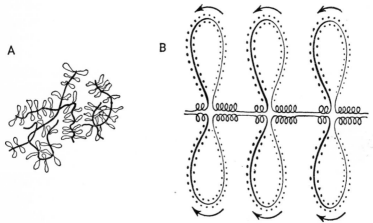

Figure 10. *Lampbrush chromosomes of amphibian oöcyte. A. Appearance of typical lampbrush chromosomes showing projecting loops of material. B. Diagram illustrating the possible mechanism of action of a small section of a lampbrush chromosome. The two lateral loops are thought to be formed by the bilateral unwinding of a section of the chromosome. The region which became unwound first is thicker than the rest because it has been synthesising and so accumulating RNA for a longer time. (Diagram B after Wischnitzer, 1967.)*

DNA that have become partially unwound; they can be largely digested with DNAse. The unwinding of the strands does not appear to be a random process but apparently occurs in definite regions (Callan and Lloyd, 1960). The loops become surrounded by material which can be digested away with RNAse, and which is believed to be RNA which is being synthesised. Indeed, in *Triturus* the base ratio of this RNA appears to be the same as that of the chromosomal DNA (Edstrom and Gall, 1963). This synthesis of RNA apparently occurs along all the extended regions of the lampbrush chromosomes (Davidson and Mirsky, 1965).

Most of this RNA formed on the lampbrush chromosomes is thought to be ribosomal in type, though the question does not appear to have been fully settled. Davidson and Mirsky (1965), who extracted RNA from oöcytes in amphibians, found that about 97% of the RNA synthesised in the nucleus at this stage of the lampbrush chromosomes is of the ribosomal type, but that much of this synthesis took place in the nucleoli (see below). In view of this high proportion it seems possible that some of the RNA of the lampbrush chromosomes is also ribosomal.

Some soluble RNA is synthesised during oögenesis but this is rapidly used (Brachet, 1967), presumably in the formation of the cellular structures of the oöcyte, and does not accumulate.

About 1% of the RNA synthesised in *Xenopus* during oögenesis is of the messenger type (Davidson *et al.*, 1966) and this is also formed on the chromosomes. This mRNA is of especial interest for the understanding of the early post-fertilization stages of development in this species, for, as we shall see, it is at least partially stored by the oöcyte and plays an important rôle in cleavage (p. 59, 64).

Lampbrush chromosomes are not always visible in amniotes, though it seems likely that the mid-diplotene chromosomes carry out a similar function without undergoing such extensive looping. 'Synaptinemal chromosome complexes' are, however, common and their fine structure has been described for both spermatocytes (Coleman and Moses, 1964), and oöcytes of the hatchling (Greenfield, 1966), of the domestic fowl (Plate II), though it is not quite clear what relationship the structures seen by electron microscopy bear to the chromosome components of light microscopy. Using staining techniques, Coleman and Moses who examined spermatocytes of the rooster, concluded that DNA was present in the two dense lateral elements of the synaptinemal complex but not of the central one. Lampbrush chromosomes are present in human oöcytes (Baker and Franchi, 1966) but apparently not in those of rats (Franchi and Mandl, 1962).

(ii) NUCLEOLAR ACTIVITY

Another site of RNA synthesis is the nucleoli, which may be present as a large single mass, or as a number of small patches, the number depending both on the species and on the stage of development. In *Triturus*, for example, it has been estimated that more than 1,000 nucleoli are present in the oöcyte (Gall, 1954, 1958).

The nucleolar synthesis of RNA occurs at many times the rate of chromosomal synthesis at this stage. As we have seen, about 97% of all the RNA synthesised at this period of development occurs in the nucleolus (Davidson and Mirsky, 1965). The base ratios of nucleoli isolated from both starfishes (Edström and Kawiak, 1961) and spiders (Edström, 1960) resemble those of the rRNA, so that it seems that nucleoli contain rRNA. Further, if oöcytes are presented with radio-actively labelled precursors of nucleic acids, these become taken up by the nucleoli and it is believed that an active synthesis of RNA either occurs there or is possibly stored there after synthesis on the chromosomes. Davidson *et al.* (1964) suggested that, since rRNA is DNA-dependent and the nucleolar RNA is also DNA-dependent, it may be that a nucleolar-organiser region on the lampbrush chromosome produces the DNA which is then parcelled out into the rRNA-synthesising units known as nucleoli. In *Triturus*, at least, there is much more DNA in the nucleus of the oöcyte than in that of a somatic cell, and there is nearly as much DNA in the nucleoli as there is in the chromosomes (Izawa *et al.*, 1963). Further evidence for this 'amplification of the genes' by the multiplication of sites suitable for RNA synthesis, is discussed by Davidson (1968).

At least some of the rRNA produced in the nucleoli passes out into the cytoplasm; there is abundant cytological (Raven, 1961) and electron microscopical evidence that nucleoli are extruded into the cytoplasm during oögenesis. For example, this happens in the rat egg (Szollosi and Ris, 1963) as well as in certain reptiles (Terio, 1962). One of the characteristic features of young oöcytes is that the nuclear membrane is studded with pores and it is probably through them that RNA passes into the cytoplasm.

Closely associated with the nuclei are stacks of perforated membranes lying in the cytoplasm and known as 'annulate lamellae'. They are especially common in oöcytes, though not restricted to them, and have been reported in a wide range of somatic cells, both embryonic and adult (see review by Kessel, 1968). Their function is not known but it has frequently been suggested that they are derived from the nuclear mem-

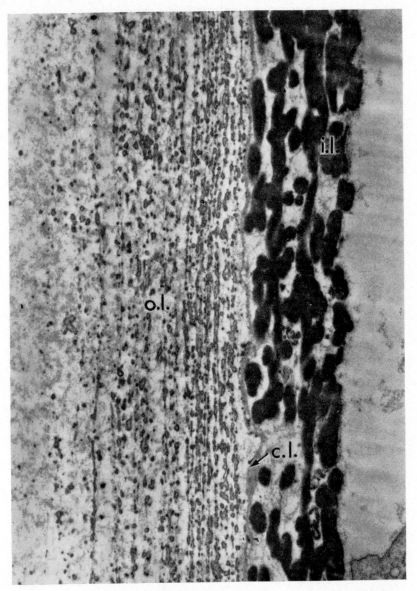

I. Electron micrograph of a section through the vitelline membrane of a hen's egg. The inner layer (i.l.) is laid down before ovulation. The narrow continuous layer (c.l.) and the outer layer (o.l.) are added whilst the egg passes down the oviduct (× 10,000).

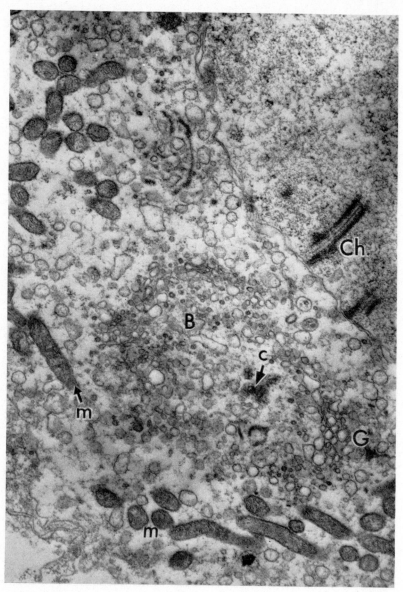

II. Electron micrograph of a section through an oöcyte of a one-day-old chick hatchling. The Balbiani body (B) contains centrioles (c) surrounded by Golgi vesicles (G) and mitochondria (m). Chromosome filaments (Ch) are present in the nucleus (× 25,000). (By kind permission of Dr. M. L. Greenfield.)

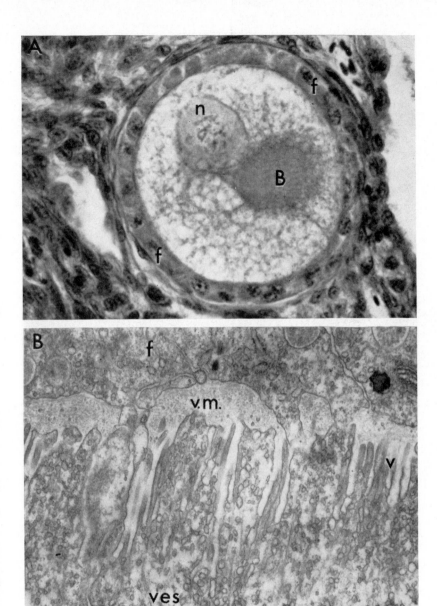

III. **A.** Section through a hen's oöcyte of 70 μ diameter. Fixed in Carnoy's fluid and stained with haematoxylin and eosin. The oöcyte contains a nucleus (n) in which chromosomes can be seen, and a Balbiani body (B). A capsule of follicle cells (f) surrounds the oöcyte ($\times$ 850). **B.** Electron micrograph of a section through the periphery of a hen's oöcyte of 2·5 mm diameter. Villi (v) project from the surface of the oöcyte and many pinocytotic vesicles (ves) are present. The vitelline membrane (v.m.) is beginning to form in the gap between the follicle cells (f) and the oöcyte ($\times$ 20,000).

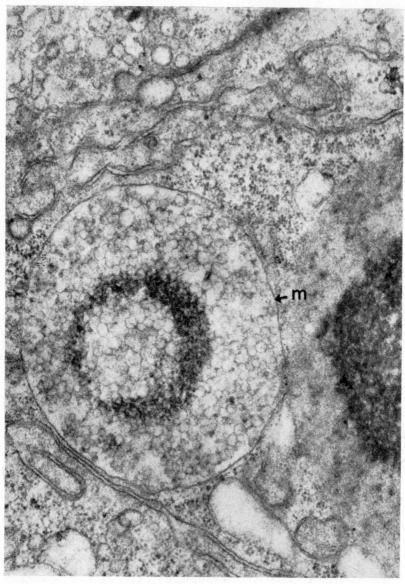

IV. Electron micrograph of a section through an intracellular yolk drop in a chick embryo. Taken from the area pellucida after 36 hrs of incubation. The yolk drop is partially digested but is still surrounded by a membrane (m) ($\times$ 45,000).

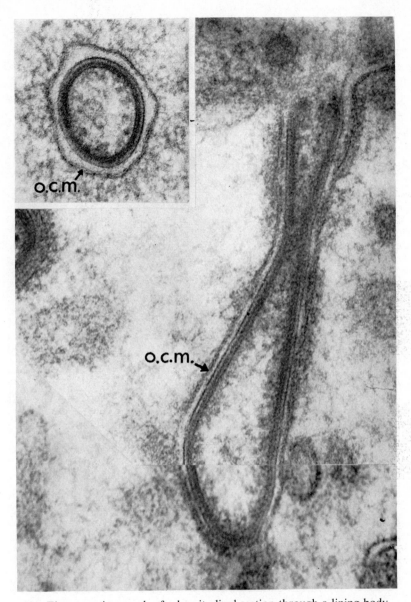

V. Electron micrograph of a longitudinal section through a lining body. These structures are produced by the ovarian follicle cells of the hen and indent the cell membrane of the oöcyte. There is some evidence that they provide a mechanism by which material of maternal origin may be injected into the oöcyte ($\times$ 130,000). Inset: transverse section across a lining body. o.c.m.: oöcyte cell membrane ($\times$ 130,000).

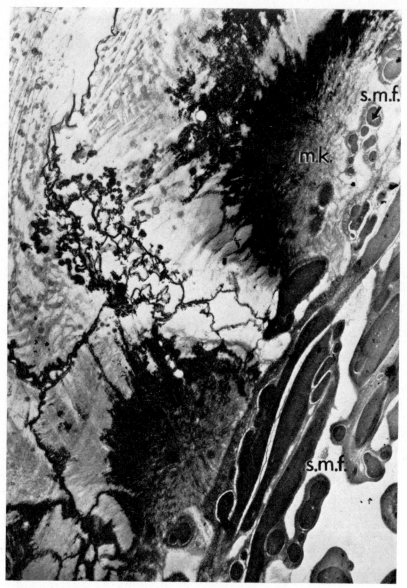

VI. Electron micrograph of a section passing through the inner surface of an experimentally decalcified shell of a hen's egg. Shell membrane fibres (s.m.f.) penetrate into the mammillary knobs (m.k.) of the shell (× 20,000).

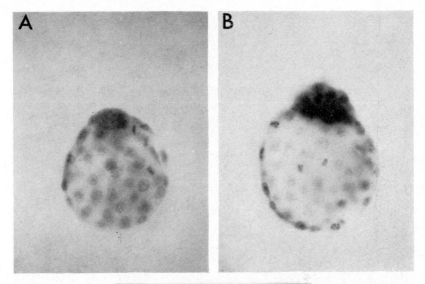

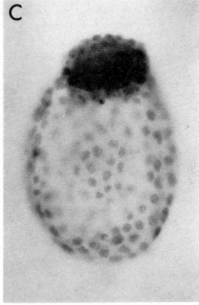

VII. Experimentally produced mouse chimaeras. **A**: Normal single embryo. **B** and **C**: Chimaeras produced by the fusion of two and four embryos, respectively. The chimaeras have become single individuals, even though larger than normal ($\times$ 400). (By kind permission of Dr. A. K. Tarkowski.)

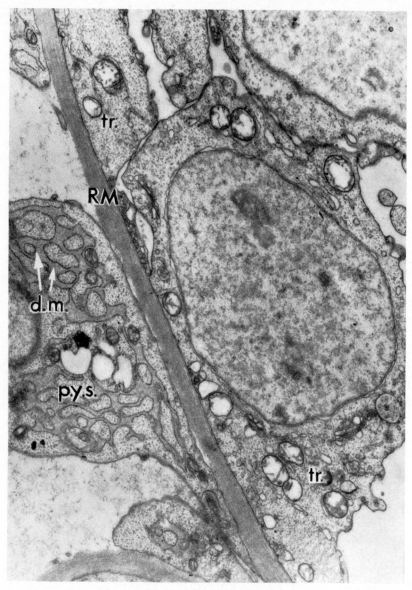

VIII. Electron micrograph of a section through the yolk sac of a mouse embryo to show Reichert's membrane (RM). This structure is probably secreted by the cells of the parietal yolk sac (p.y.s.). Note the densely staining material (d.m.) within the smooth-surfaced endoplasmic reticulum of these cells. This material is not present in the trophoblast cells (tr.). (By kind permission of Dr. F. S. Goldby.)

brane and may act as templates when the nuclear membrane reforms after mitosis, or that they play some rôle in the transfer of genetic information from the nucleus to the cytoplasm, especially since RNA appears to be associated with them in some animals (Kessel, 1968).

(iii) EXTRA-NUCLEAR DNA

Cytoplasmic DNA is found not only in the mitochondria but, in certain amphibians and the chick, it is also associated with the yolk (p. 41). At various times it has been suggested that a further source of DNA for the oöcyte might be provided by the follicle cells but this has not been substantiated in a recent enquiry using the mouse (Petersen *et al.*, 1967).

(iv) PROTEIN SYNTHESIS IN THE OÖCYTE

Nucleic acids are synthesized by cells largely as a prerequisite for the formation of proteins. It is not surprising, therefore, that the time of greatest protein synthesis in oögenesis is in the earliest phase when nucleic acid synthesis is also at its maximum. During the later stages the formation of both the nucleic acids and the new proteins is greatly depressed and only rises again on fertilization; we have seen that some of the mRNA is stored in the oöcyte against this time. The proteins that are synthesised in the young oöcyte, are used in the formation and replication of the nuclear and cell membranes and of various cytoplasmic structures such as the mitochondria. As we shall see, however, they do not play a direct rôle in the formation of the yolk in amniotes (though they may possibly do so in some classes of invertebrates) but they are indirectly important for the laying down of the yolk, since this process can only occur in an oöcyte that has already passed through the earliest phase of oögenesis. Furthermore, it is possible that the enzymes necessary for the formation of yolk are included among the proteins formed at the early stage, though if this is so they are probably stored in an inactive form since they may not be required for many years until the individual reaches the reproductive age.

(c) Yolk formation and other cytoplasmic events

The cytoplasm of the oöcyte undergoes a number of important changes during development. These are associated not only with the laying down of the yolk reserves in the oöcyte, in the final stages of oögenesis, but probably they also reflect the physiological adjustments necessary in the normal metabolism of the cell itself.

(*i*) VARIATION IN YOLK

The amount of yolk present in amniote eggs is very variable, although it tends to be large in birds as compared with that in amphibians, a fact that can be correlated with the necessity for a prolonged embryonic life. This prolongation enables the animal to reach a stage of development by the time it hatches which is sufficiently advanced to allow it to survive in a terrestrial environment and avoid desiccation. The 'typical' amniote egg is known as a cleidoic egg and is exemplified by that of the domestic fowl. It contains sufficient yolk to serve it during development and usually possesses an additional watery component, the egg white, though this may be present in only small amounts in certain reptiles (see p. 43). Finally, it is surrounded and protected by shell membranes and a shell (Fig. 11). As development proceeds, certain membranes, the amnion (Figs. 51, 58, 65) and the chorio-allantois, are formed to cope with special problems of the terrestrial embryo. None of these components are exclusive to the eggs of amniotes. For example, the eggs of fishes and amphibians and of many invertebrates possess yolk and protective membranes. A cleidoic-type egg has also been developed by certain insects for, like birds and reptiles, they usually lay their eggs on dry land.

The eggs of eutherian mammals possess little, if any, yolk and so are an exception to the normal amniote plan. There can be little doubt that yolk has been lost during evolution, though there is some difficulty in deciding whether certain fatty drops or crystalloid inclusions in eutherian ova should be considered as yolk or not (Hadek, 1965), since there is no satisfactory definition of yolk. There are, for instance, considerable differences between amphibian and avian yolk, the protein component of the former having an orderly, crystalline structure (Ward, 1962 a, b) that is totally absent in the latter (Bellairs, 1961a) (Plate IV). There are also some differences in the chemical composition, though both consist mainly of proteins and lipids (Williams, 1965; Bellairs, 1961a).

(*ii*) THE SURFACE OF THE OÖCYTE

Formerly, it was thought that the raw materials for the yolk were manufactured in the oöcyte, though this hypothesis has now been abandoned. By the time the yolk begins to form, however, important changes have already taken place in the oöcyte and it seems likely that some, at least, may be concerned with preparations for yolk formation.

Shortly after birth or hatching, at about the time of the disappearance

A

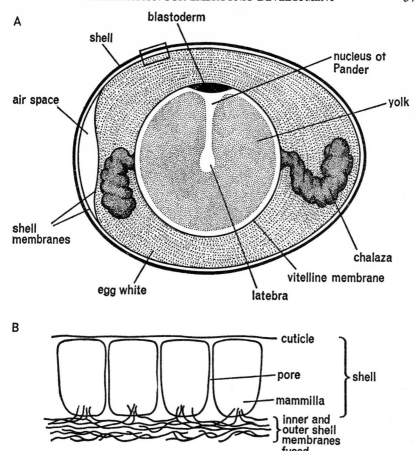

B

Figure 11. *Cleidoic egg. A. Sagittal section of the hen's egg. The latebra and the nucleus of Pander are regions of white yolk. The blastoderm lies over the nucleus of Pander and beneath the vitelline membrane. The chalazae are balancers which support the yolk. The two shell membranes are separated by an air space at the blunt end of the shell. B. Enlargement of the region of the shell shown in A.*

of the meiotic chromosomes (Greenfield, 1966), each oöcyte becomes surrounded by a capsule of follicle cells and this in its turn is embedded in connective tissue. Thus, from now on, any materials that enter the oöcyte from the blood vessels of the ovary must first pass through the covering of maternal cells. It is perhaps, not surprising, therefore, that the cell membrane of the oöcytes in many species is extended out into villous processes (Fig. 8) and that these become more elaborate as the egg develops. The increase in surface area presumably reflects the enhanced passage of materials through it. The follicle cells show a similar increase in surface area (Bellairs, 1963a).

The growth in size of the oöcyte does not proceed at a regular pace but passes through fairly well-defined phases. This is especially obvious in the large-yolked species such as the domestic fowl. Here, growth is slow until the oöcyte is about 6 mm in diameter (Riddle, 1911), but after that it takes place with great speed and in 8–10 days reaches its full size. During this latter period the diameter increases by about 2·5 mm per day, largely because of the deposition of yolk.

(iii) THE BALBIANI BODY

The structure of the early stages of the oöcytes of many species has been studied. In almost all animals there is a specialised region near the nucleus which was formerly believed to be concerned with the formation of yolk, and for this reason it was known as the 'yolk nucleus' or the 'yolk body of Balbiani' or the 'mitochondrial yolk body' (Plates II and IIIA). There has been considerable confusion as to the nature of the Balbiani body, and even when the discussion is restricted to birds its composition is not clear. Recent electron microscopical studies of a wide range of species have shown that it tends to consist of paired centrioles surrounded by mitochondria and Golgi bodies. Fat drops may sometimes be associated (Guraya, 1962, 1964), a fact that may account for the old idea that yolk was formed by the Balbiani body. The real function of the Balbiani body is not understood but its position, adjacent to the nucleus (Fig. 8), is a strategic one for intercepting materials that pass from nucleus to cytoplasm.

Elsewhere in the early oöcyte, mitochondria and Golgi bodies are relatively sparse (Raven, 1961; Bellairs, 1965) but large numbers of vesicles of smooth-surfaced endoplasmic reticulum, as well as quantities of free ribosomes, are present. Rough-surfaced endoplasmic reticulum and polysomes are relatively uncommon, a fact that suggests that not very much protein synthesis is taking place at this stage. The free ribosomes, however, are subsequently used by the early embryo for protein synthesis in some, if not all, species (p. 65).

In the bird's oöcyte, the Balbiani body disappears before the yolk begins to be laid down. The mitochondria and Golgi bodies migrate to sub-cortical portions. Thus, they have exchanged one strategic position for another and are now well placed to intercept and influence raw materials entering the oöcyte from the mother. These raw materials are of two kinds, those that will form yolk and those that will help to maintain the growing cell.

(*iv*) THE FORMATION OF YOLK

It is only within the last thirty years that it has been possible to understand where the raw materials for the yolk are synthesised. There is now general agreement that in vertebrates they are formed in the liver and pass from there in the blood plasma to the ovary. This explanation is supported by several lines of evidence. Firstly, the rate of deposition of the yolk in birds' eggs is too great for it to have been manufactured in the oöcyte or in the follicle cells. Riddle (1911) calculated that to maintain this rate during the last phase of deposition of the hen's egg, each follicle cell would have to secrete a quantity of yolk equal to its own volume every half hour.

Secondly, the blood serum and liver of laying hens has been shown to contain large amounts of lipids and proteins with properties closely resembling those of similar material found in yolk, and these substances cannot be found in large quantities in the liver and serum of male birds or of non-laying females. Immunological comparisons have demonstrated that these substances in blood and yolk are frequently identical. Finally, radioactively-labelled serum proteins which have been injected into the blood of the mother have also been found subsequently in the yolk (see Bellairs, 1964, for references).

The substances that pass through the follicle cells and into the oöcyte are the definitive proteins and lipids from which yolk is formed. Thus, in vertebrates at least, there is no real synthesis of yolk in the oöcyte but merely a morphological rearrangement of materials.

The raw materials from which the yolk forms in reptiles are supplied to the oöcyte by the liver in much the same way as in birds. However, some reptiles possess in addition to the liver a 'fat body' organ. In the lizard, *Uta stansburiana*, the fat body apparently enables the female to reproduce early in the season by providing a store of fat which passes to the liver where it is rapidly utilised to make yolk materials (Hahn and Tinkle, 1965). These authors found that although the fat body itself was not essential for reproduction, its extirpation resulted in belated oviposition. The significance of the fat body in males is, however, not clear.

The way in which materials pass into the oöcyte is not well understood. In the early stages of oögenesis there are visible at the surface of the oöcyte many small vesicles of the type usually considered to be pinocytotic. They are present not only in the oöcytes of animals that produce yolky eggs (such as birds), but are also a conspicuous feature of mammalian oöcytes. It seems possible that they are concerned more with transporting materials for the general nourishment of the developing cell than with the laying down of yolk. The absence of similar vesicles in the adjoining region of the follicle cells suggests that the process is a one-way one. As the oöcyte grows, its surface becomes extended out in villi, which presumably increases its absorptive area (Plate IIIB). In the later stages when yolk is being laid down rapidly, certain modifications occur in the follicle cells in birds. The cells no longer lie so close to one another, being separated by a gap, and Wyburn et al. (1966) have suggested that the raw materials for the yolk pass through the channels between the cells. The oöcyte is capable of exerting some selectivity (Bellairs, 1964, 1965) and it probably eliminates some of the materials it takes in. Indeed, Grau and Wilson (1964) have suggested that at this stage the bird oöcyte loses its identity as a cell and that the functional physiological unit should now be considered as the follicle capsule plus its contained oöcyte. This concept is supported by the fact that in the ultimate stage of yolk deposition, the oöcyte cell membrane becomes discontinuous and can be seen as a series of vesicles lying beneath the partly-formed vitelline membrane (Bellairs, 1967). It seems not unreasonable to suppose that the functions of the cell membrane in delimiting and supporting the oöcyte contents have been taken over by the vitelline membrane.

Furthermore, we may suppose that once the cell membrane has broken up in this way, the speed of flow of raw materials will be enhanced. No such break-up appears to occur in the non-yolky eggs of eutherian mammals. Nevertheless, there is some evidence, obtained from studying human ova growing in vitro, that the ovum is nourished by the cells of the corona radiata (Shettles, 1958). The corona radiata is a layer of follicle cells that adhere to the oöcyte (ovum) after it has been extruded from the ovary.

In the past, some authors have suggested that in birds the yolk might form by a direct transformation of mitochondria. Such a situation may possibly occur in amphibia where yolk seems to develop in an intimate association with mitochondria, lumps of crystalline yolk protein being visible by electron microscopy within individual mitochondria (Lanzavecchia and LaCoultre, 1958; Ward, 1962). But there is no evidence for such a conversion in avian or reptilian yolk (Bellairs, 1959b, 1962). It does seem possible that both mitochondria and Golgi bodies may play a rôle in yolk formation by influencing the raw materials as they enter the oöcyte. Wallace (1964b) has shown that in Rana pipiens a mitochondrial enzyme, a protein kinase, plays an important rôle in yolk formation, and it is possible that similar enzymes are produced by the mitochondria of the hen's oöcyte.

Oöcytes are generally rich in cytoplasmic DNA most of which is located in the mitochondria which is the usual, though not the only, site of it in embryonic cells (Nass and Nass, 1963). In amphibian oöcytes DNA is also associated with the yolk platelets (Brachet, 1967). Some of the cytoplasmic DNA differs from nuclear DNA in having a circular form, but experiments have shown that it is comparable biochemically in that it is able to prime RNA. Its possible function in development is discussed by Brachet (1967) and Tyler (1967). It is probably not used in cleavage as an immediate precursor for nuclear DNA, since the nuclei readily utilise such low molecular weight DNA precursors as thymidine.

(v) THE LINING BODIES

Other structures have also been described that enter the developing oöcytes of birds. The most striking are specialised regions of follicle cell walls that have been termed *lining bodies* (Bellairs, 1963a) (Plate V), though they have also been given various other names (Schjeide *et al.*, 1963, 1964; Wyburn *et al.*, 1965). It has been suggested by Schjeide *et al.* that these are developing mitochondria, a suggestion which has been criticised in detail elsewhere (Bellairs, 1964). There is some indication that some of the lining bodies may become nipped off and engulfed by the oöcyte (Bellairs, 1965; Greenfield, 1966). Structures of this type have been seen as much as 1μ deep within the oöcyte and here they appear to become broken down. Their eventual fate is not clear but it is tempting to consider that it may be connected with the production of highly yolked eggs. Lining bodies have not been described in mammalian oöcytes, nor in the smaller though yolky eggs of teleosts (Droller and Roth, 1966). However, light microscope studies indicate that something comparable may be present in the tortoise, *Testudo graeca* (Bhattacharya *et al.*, 1929). Little electron microscopical work has been done so far on the relationship between follicle and oöcyte in reptiles, though it must be admitted that in preliminary studies on the tortoise, *Testudo hermanni* (Zahnd and Porte, 1963) and on *Lacerta s. sicula* (Ghiara and Taddei, 1966) no structures strictly comparable to the 'lining bodies' of birds have been described. Both groups of authors have, however, illustrated specialised, elongated, cisterna-like structures in the oöcyte which may be homologous. Alternatively, it is possible that maternal RNA is being injected into the oöcyte cytoplasm, perhaps in a masked form for use during cleavage (see p. 59 for discussion of masked mRNA).

The yolk drops of amniotes are spheres floating in a watery fluid. The fine structure of the yolk spheres of the hen and of certain lizards and snakes have many features in common (Bellairs, 1959a, 1961a), and differ markedly from the yolk platelets of amphibians (Ward, 1962 b). Similarly, the yolk spheres of the turtle *Geoclemys reevesi*, when studied by light microscopy, appear to have a similar structure and osmotic behaviour to the yolk spheres of the hen (Grodzinski, 1951, 1953).

(d) The egg membranes

(i) VITELLINE MEMBRANE AND THE EGG WHITE

The innermost egg membrane is laid down around the cell membrane of the oöcyte whilst it is in the ovary, and so is inserted between the oöcyte and its follicular capsule (Plate IIIB). In birds' eggs this is the inner layer of the vitelline membrane (Bellairs *et al.*, 1963) and in mammalian eggs it is called the zona pellucida. These membranes are thus homologous to some extent and are thought to be secreted by the follicle cells (Merkar, 1961; Bellairs, 1965), and indeed in its earliest stage of development the vitelline membrane of birds possesses the same granular appearance as the zona pellucida of mammals. As the bird's oöcyte grows, however, a complex system of fibres becomes laid down in the vitelline membrane, and perhaps by this means extra mechanical strength

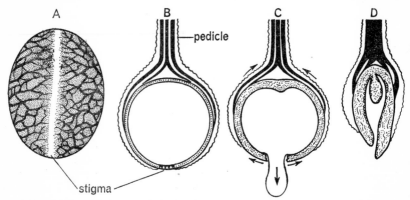

Figure 12. *Ovulation of the hen's egg. A. Surface view illustrating the vascular network that covers the mature ovarian follicle except in the region known as the stigma. (After Warren, 1949.) B, C and D. Sections through the ovum at the three stages during ovulation. Each ovum is suspended from the main body of the ovary by a pedicle containing longitudinally running bands of smooth muscle (shown in solid black). Contraction of these muscles leads to rupture of the stigma and to escape of the ovum. (After Guzsal, 1966.)*

s provided to support the large-yolked ovum. So far the vitelline membranes of reptiles have not been studied by electron microscopy. A further 'membrane', the zona radiata, was often formerly described by light microscopists as a radially striated structure lying at the immediate surface of the oöcyte (Raven, 1961). It is now thought that it is the villi on the surface of the oöcyte (p. 37) that give the impression of a striated band by light microscopy (Bellairs, 1965). The eggs are shed into the peritoneal cavity but, in mammals at least, are then engulfed by the fimbriated ends of the Fallopian tubes. Ovulation in the hen is brought about by the action of bundles of smooth muscles which cause the follicle wall to split (Fig. 12). The blood vessels in the follicle wall are so arranged that little or no haemorrhage occurs. If a blood vessel does become ruptured a blood-spot appears in the egg.

Once ovulation has taken place other membranes are added by secretions of the genital tract. In birds, the first one is the outer layer of the vitelline membrane (Plate I), a complex fibrous structure (Bellairs et al., 1963) which extends out at two points as balancers, the chalazae. The second secretion is the egg-white (albumen) which contains a number of proteins in an aqueous medium, some of these being bactericidal. Formerly, it was thought that the vitelline membrane was collagenous or keratinous but a recent analysis has shown it to be formed of a hitherto unknown structural protein (Bellairs et al., 1963). Fertilization probably occurs soon after the ovum enters the oviduct and before the egg-white is secreted on to the egg. The egg-white is particularly interesting since it contains immunologically different proteins in different species of birds (McCabe and Deutsch, 1952; Baker, 1968). This discovery is being utilised to study phylogenetic relationships among birds. During embryonic development the egg-white appears to subserve two functions. In the early stages it supplies water to the embryo and to the amniotic fluid, whilst after 12 days of incubation it provides an additional protein for both the amniotic fluid and for the embryo (Carinci and Manzoli-Guidotti, 1968). Egg-white is present in the eggs of some reptiles though it differs in consistency (Fisk and Tribe, 1949) but is thought to be either reduced or absent in most, if not all, lizards and snakes (Giacomini, 1891; Giersberg, 1922).

(ii) SHELL MEMBRANES AND SHELL

As the bird's egg passes along the oviduct it submits to a number of processes, like a man-made article passing along an assembly belt in a factory (Fig. 13). After the egg-white has been secreted around it, the egg passes further down the oviduct and becomes encased in two fibrous membranes, the shell membranes (Fig. 11). These are in close contact with one another except at the future blunt end of the egg where a fluid separates them. When the hen's egg is laid this fluid evaporates, leaving the air space between the two shell membranes (Fig. 11). This pocket of air is used by the chick embryo in the final stages of incubation, just

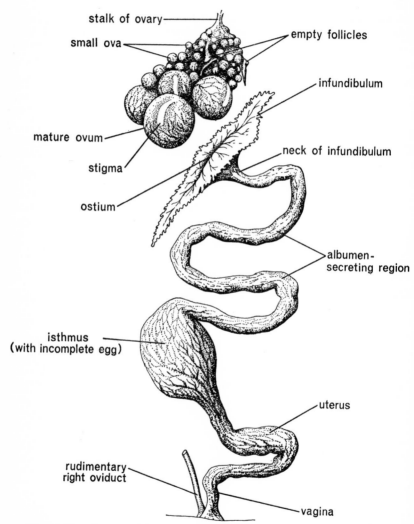

Figure 13. *Ovary and left oviduct of domestic fowl. The right oviduct is vestigial. (After Romanoff and Romanoff, 1949.)*

before hatching. It is said that many reptiles do not possess an air space (Fisk and Tribe, 1949).

The egg now passes into the uterus and the shell is laid down directly on to the outer shell membrane. Indeed, fibres of the outer shell membrane become encased by the calcite crystals of the shell (Bellairs and Boyde, 1969) (Plate VI).

In birds and most reptiles the shell is calcareous. In some reptiles, such as tortoises, crocodiles and most geckoes, the eggshell is hard and brittle like that of birds. In most egg-laying reptiles, however, (e.g. sea turtles, and oviparous snakes) the shell has a leathery texture; but according to Simkiss (1967) it still contains a considerable quantity of calcium. In the domestic fowl there is a continual rhythm of calcium withdrawal from the long bones of the laying hen at the time of shell formation followed by a recovery period of calcium deposition in these bones. Thus, withdrawal of calcium from, and deposition in, the long bones alternate as successive eggs become calcified (Simkiss, 1967). The fine structure of the egg shells of birds shows some variation from one species to another (C. Tyler, 1965). Little is known about the hard, calcareous eggshells of reptiles, though Young (1950) suggested that the shell of the tortoise, *Testudo graeca*, differed from that of the hen's egg in that it possessed no pores; he correlated this fact with his findings that

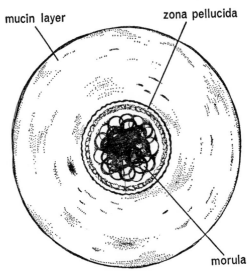

Figure 14.   *Cleavage. Rabbit embryo at the morula stage. The embryo was recovered from the oviduct at three days after fertilization. The rabbit is unusual among mammals in possessing a thick mucin layer.*

water loss through the intact shell of this tortoise was much less than through the hen's egg shell.

The calcareous shells are composed mainly of calcite crystals but in addition they possess a matrix of protein. It is not surprising, therefore, that if the calcium is reduced, the shells are leathery rather than brittle, e.g. those of the English grass-snake, *Natrix natrix*.

Not all reptiles are oviparous, however; many lizards and snakes bear their young alive and possess a primitive type of placenta. To some extent this corresponds with the epithelio-chorial placenta of mammals (p. 157, 174). Viviparous reptiles do not possess calcified or thick leathery shells on their eggs.

The egg membranes of mammals have similarly become modified or lost as a result of viviparity (p. 127 and Fig. 14).

B. SPERMATOGENESIS

(a) Structure of sperm

Our knowledge of sperm structure was rudimentary until the advent of the electron microscope. Although superficially there is a great variation in outward form or in proportion of the parts, the components are much the same in all species. Thus, the headpiece is swollen by the male pronucleus and is covered at its anterior tip by the acrosomal cap; this structure (Fig. 15) contains a lipoglycoprotein, called acrosomin, as well as certain enzymes. An additional structure lies between the nucleus and acrosomal cap in some species, such as the rabbit, and this has been given the name 'perforatorium' (Austin and Bishop, 1958), or 'apical body' (Bedford, 1964) or 'sub-acrosomal space' (Fawcett, 1965), but it is not present in all species and there is as yet no evidence as to its rôle, if any, in fertilization (Fawcett, 1965). Behind the nucleus lies the centrosome and there stretches back from this a filamentous structure consisting of two medial fibres composed of contractile protein, with nine similar ones arranged around them. This structure is the flagellum and it extends throughout the entire length of the middle piece and flagellum of the sperm. In the middle piece it is thought to be surrounded by a spiral coil of mitochondria in many species (e.g. in the bull; Saake and Almquist, 1964). Wimsatt *et al.* (1966) have suggested however that in one species at least (*Myotis lucifugis*, a bat) this configuration is an illusion, since their results indicated that there was an orderly alternation of two kinds of mitochondria which could be distinguished easily from each other

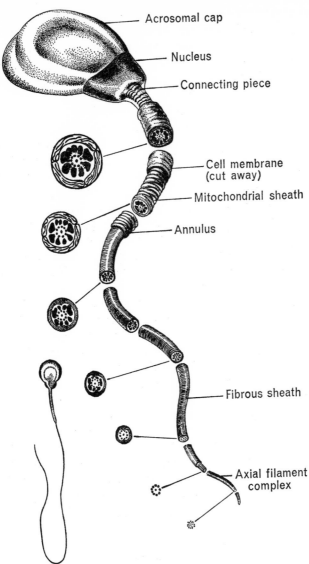

Acrosomal cap

Nucleus

Connecting piece

Cell membrane
(cut away)

Mitochondrial sheath

Annulus

Fibrous sheath

Axial filament
complex

Figure 15. *Spermatozoan. Diagram based on electron micrographs to illustrate the structure of a guinea pig spermatozoan.* (*From Fawcett, 1965.*)

because they stained differently. The distal centriole marks the posterior end of the middle piece of the sperm. The entire sperm is enclosed by a cell membrane which, when examined by electron microscopy, possesses the normal cell membrane structure. The number of flagellae differs throughout the animal kingdom (Austin, 1965). In the normal sperm of vertebrates, however, only a single flagellum is seen (see Nagano, 1962), though striking variations in the proportions of the various parts of the sperm occur. Many of these are illustrated in the atlas of Hadek (1969).

(b) Capacitation

In some animals, the development of the sperm is not completed before they are shed from the testis. In an electron microscopical study, Fawcett and Hollenberg (1963) found that distinct changes took place in the shape and internal structure of the acrosome of the guinea pig during the passage of the spermatozoa through the epididymis. This is particularly interesting since the ability to move by flagellation also appears to be acquired in the epididymis, the sperm actually being propelled by the pressure of the fluid secretions of the seminiferous tubules until they reach this region (Barack, 1968). Fawcett and Hollenberg concluded that their observations supported the view of Young (1931) that sperm development is a continuous process starting in the testis and continuing in the epididymis. It will be interesting to see if evidence will be produced to show that the epididymal secretions affect this process. It is already known that sperm taken directly from the testis and injected by artificial insemination are not capable of fertilizing guinea pig eggs (Young, 1931); even those removed from the epididymis in the rabbit will not fertilize eggs (Orgebin-Crist, 1968). It seems, therefore, that sperm must undergo some further physiological changes before they can fertilize ova. These changes, which are given the term 'capacitation', take place in the female genital tract. Mounib and Chang (1964) have suggested that changes in glycolysis play a large rôle in capacitation. The cytological changes involved in capacitation of rabbit sperm have been described by Bernstein (1966), the most important being a loosening of the sperm membrane and a progressive erosion of acrosomal material (see also Bedford, 1967).

C. SUMMARY

The events that occur during the earliest stages of embryonic development are conditioned by those which have taken place during oögenesis. In birds and reptiles, the yolk which will nourish the embryo is laid down in the oöcyte. Egg membranes and the shell are secreted around the ovum as it passes down the oviduct.

Protein synthesis probably takes place in the earliest stages of oögenesis in all amniotes, but is greatly suppressed before ovulation. These proteins are probably used in the formation of cytoplasmic structures. Nucleic acid synthesis includes the doubling of the DNA in preparation for meiosis. In amphibians, most of the RNA produced during oögenesis is ribosomal, and this may also be true in amniotes though we have less direct knowledge. In the next chapter we shall see that the oöcyte also prepares for its future by synthesising masked messenger RNA.

The development of the spermatozoa is not completed in some animals before they are shed from the testis, various changes in the acrosome taking place in the epidymis.

CHAPTER 3

THE BEGINNING OF EMBRYONIC DEVELOPMENT: FERTILIZATION AND ITS RESULTS

UNFORTUNATELY, most of our knowledge of the events that take place at fertilization, is derived from studies on invertebrates, especially on echinoderms. It is less easy to investigate the process in amniotes for these all have internal fertilization. Such knowledge as we do possess about amniote fertilization is largely derived from studying the process in tissue culture. As Austin (1961) has pointed out, fertilization proper includes a real fusion of two nuclei, followed by an orderly sequence of events. Hence, only if the entry of the sperm is followed by cleavage and gastrulation, can we be certain that we are observing something that corresponds with the normal situation and not just a parthenogenetic stimulation (see p. 62). In considering fertilization aminniotes, therefore, we must distinguish between what is directly known and what is deduced from studies on other groups of animals.

A. EVENTS AT FERTILIZATION

(a) How does the sperm travel toward the egg?

In all amniotes the sperm have considerable distances to travel even after being deposited in the female genital tract, for they must fertilize the egg in the most cranial part of the oviduct before the final egg coverings are put on it. Indeed, it has recently been suggested that the human ovum is fertilized in the peritoneal cavity (Horne and Thibault, 1962). Although there is some evidence that in certain invertebrates the sperm may be attracted to the egg by chemotaxis this does not appear to be the situation in mammals. The basic mechanism of sperm propulsion seems rather to be by contractions of the walls of the uterus with possibly a small rôle played by the cilia (Bishop, 1961). The evidence is based on several approaches. First, the sperm pass the distance from the point of insemination to the ovum at a faster rate than their own mobility. Bull sperm, for instance, move at 100μ per second and would consequently

c

take about one and a half hours to move to the ovum if they went un-aided. In fact, they take about two and a half minutes. It has been reported that even dead sperm pass up the genital tract of the cow in a few minutes, though some authorities doubt whether dead sperm ever pass from the vagina to the uterus (Noyes, 1968). Fowl sperm (Bobr *et al.*, 1964a, b) also pass quickly up the oviduct. The rabbit is apparently unusual in that the process takes several hours.

Second, uterine contractions have been recorded regularly in many mammals, and the activity varies with the phases of the oestrous cycle.

In many animals only a small proportion of sperm actually complete the passage to the egg. For instance (Bishop, 1961; p. 732), only one sperm in about 50,000 in the rat reaches the site of fertilization. This is because a considerable number are lost in different parts of the tract. The cervical mucosa traps large quantities, a fact that has been utilised in certain forms of contraceptive pill which enhance the secretion of mucous. Curiously, however, bull sperm appear to be stimulated to increased activity by cervical mucus (Tampion and Gibbons, 1963). Fur-ther up the female genital tract, the fluid pressures within the lumen may provide an obstacle to the passage of sperm (discussed by Bishop, 1961). Formerly, it was considered that large numbers of sperm were necessary so that a sufficient quantity of hyaluronidase could be produced in the vicinity of the egg, but this proposition now seems improbable. Further aspects of the physiology of sperm in the female genital tract are con-sidered by Restall (1967).

(b) Sperm storage in the oviduct

In some animals the sperm are not used for fertilization immediately after being deposited in the female genital tract, but are stored in a specialised region of the oviduct until they are needed. Domestic hens are able to lay fertilized eggs for up to two weeks after a single insemina-tion (Takeda, 1966) and in certain other amniotes even longer intervals have been reported. Periods of seven months have been recorded in certain bats, whilst the most remarkable examples of all are in reptiles where some snakes and turtles are believed to be capable of produc-ing fertile eggs even after several years in isolation (Fox, 1956). Haines (1940) reports that a female colubrid snake, *Leptodeira annulata poly-sticta*, laid a number of embryonated eggs after it had been in isolation for five years. The possibility that the development which did occur might be parthenogenetic was dismissed at the time by Haines, since there was then no established example of parthenogenesis in higher vertebrates. But, as we shall see (p. 64), it now appears that partheno-genesis can occur in some reptiles, so that this species, and perhaps others, might repay a thorough re-investigation.

Not all reports of sperm storage can, however, be attributed to parthenogenesis, and, inevitably, there has been great interest in the problem of how the sperm remain viable after even a few weeks. In the domestic fowl it is not even clear where the sperm are stored in the female tract, but this seems to be due to the fact that some authors have based their conclusions not just on the results of copulation, but also on the effects of artificial insemination of various types, including the injec-tion of sperm directly into the peritoneal cavity. Most modern authors seem to agree that some sperm storage occurs in glandular sacs at the junction of the uterus and vagina of the hen (Fig. 13) (Bobr *et al.*, 1964a, b; Lorenz, 1966; Takeda, 1966; Schindler *et al.*, 1968) but that sperm may also be found in the infundibular region. There is evidence that these sperm play an active rôle in reaching their storage site, for dead sperma-tozoa become trapped in the oviducal secretions (Schindler *et al.*, 1967), but it is not known if this motility is suppressed during storage. Schindler *et al.* suggest that the glandular secretions in both the sites may provide energy to maintain the sperm. A similar nutritive activity has been suggested for the glandular regions in which sperm become stored in the chamaeleon (Saint Girons, 1962) and in certain bats (Wimsatt *et al.*, 1966).

This concept of the nutritive action of the glands is of especial interest. Wimsatt *et al.* (1966) who examined intra-uterine sperm of the bat, *Myotis lucifugis*, by electron microscopy, were unable to suggest any other mechanisms or any special adaptations which could explain the longevity of the sperm.

Seminal receptacles have been described in a number of snakes and lizards (Cuellar, 1966; Saint Girons, 1962) but apparently not in turtles. In iguanids, the seminal receptacles lie mainly in the anterior segment of the 'vagina' and the mucosa is arranged in longitudinal folds (Cuellar, 1966). In chamaeleons, also, they lie at the posterior part of the oviduct just before it opens into the cloaca (Saint Girons, 1962) and the sperma-tozoa are present there in enormous numbers. Saint Girons correlated the sperm storage with the fact that there is in chamaeleons a long period of time between insemination and ovulation, whilst in those lizards where ovulation takes place soon after insemination there is no necessity for sperm storage and, indeed, none occurs.

(c) How does the sperm contact the egg?

The sperm generally has to penetrate through membranes before it can enter the egg. This is usually by means of a lytic agent, though some vertebrate eggs, such as those of many teleosts, possess a micropyle. The lysin is probably contained within the acrosome, though it is possibly in the perforatorium (Austin and Bishop, 1958).

The cells of the mammalian cumulus oöphorus (discarded follicle cells) are held together by a mucopolysaccharide rich in hyaluronic acid. It is interesting, therefore, that the sperm are rich in hyaluronidase. If egg cells are treated with hyaluronidase, the cumulus disappears although the zona remains. This can be removed with proteolytic enzymes such as trypsin. Thus, it appears that two different lytic agents are necessary to get through the zona. There is evidence that the zona pellucida is a muco-protein with ester-bound sulphate groups (Bishop and Tyler, 1956) and in this it resembles the jelly coats of amphibians.

One of the most stimulating theories about egg sperm interaction is that of the fertilizin-antifertilizin reaction initially proposed by Lillie, but subsequently elaborated by Tyler (1965). Most of the experiments have been carried out on sea urchin eggs. Full accounts are given by Monroy (1965) and by Metz and Monroy (1967) and the concept will not be con-sidered in detail here. Essentially, the facts are that the jelly coats of certain invertebrate eggs secrete an acidic glycoprotein, fertilizin, and this can react chemically with a different acidic protein, antifertilizin, that is found in the head of the sperm. The reaction is a relatively specific one in that it can only occur between closely related species, and for this reason attempts have been made to determine whether this is an antigen/ antibody reaction (Metz, 1967).

There is not much evidence as to how completely the fertilizin-antifertilizin concept can be applied to vertebrates, or to amniotes in par-ticular. In mammals it has been possible to show that there are antigens in the surface of the sperm and that these are the same as those in the seminal fluid, and it seems possible that a similar type of molecular binding takes place between egg and sperm. Tyler (1963) suggests that fertilizin, or something similar, is an antigenic constituent of the egg cell membrane and that this is disrupted when it reacts with the antibodies or antifertilizin in the sperm. Tyler emphasises that there is no evidence that these antibodies interfere with any part of the activation process.

Much interest has centred on the problem of how the sperm enters the egg. Two theoretical possibilities exist. It could be phagocytosed by the egg, or it could actively penetrate through the egg cell membrane

(Fig. 16). It appears that the penetration of the sperm into the egg is an active process. Colwin and Colwin (1967) working on a number of inver-

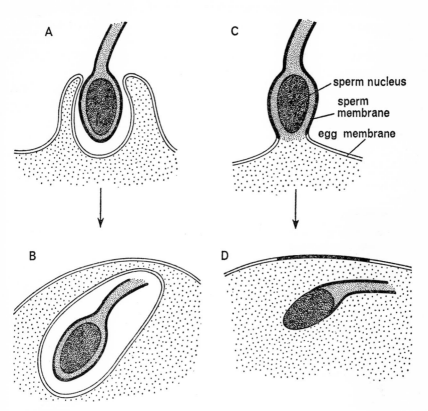

Figure 16. *Fertilization. Diagram illustrating two possible ways in which the spermatozoan may enter the egg. The egg membrane is shown in white, the sperm membrane is shown in black. If the sperm is phagocytosed (as in A) the cell membrane of the fertilized egg will be composed entirely of egg membrane (as in B). Moreover, the sperm will be surrounded by a sac derived from the egg membrane. If the sperm enters the egg as in C the egg and sperm membranes fuse, so that part of the sperm membrane comes to lie at the surface of the egg (as in D). After entry in this way, the sperm head is partially denuded of membrane and lies freely in the cytoplasm. (After Szollosi and Ris, 1961.)*

tebrates have shown that the acrosome dehisces, the acrosome granule disappears, and a filament is put out. Lysins, in particular hyaluronidase, are associated with the acrosome filament. A similar acrosome filament is found in vertebrate sperm, e.g. of the guinea pig (Fawcett and Hollenberg, 1963) and is extruded on capacitation. The Colwins found that the cell membrane of the egg of the worm, *Hydroides*, extends upwards in a cone to meet the tip of the acrosome filament. At that point the vitelline membrane bursts and fusion of the egg and sperm membranes occurs. The contents of the sperm now pass into the egg and the authors emphasise that this fusion permits the sperm nucleus to enter unencumbered by sperm cell membrane. This is important in the sea urchin since, if sperm are injected into an egg with a micropipette, then neither egg activation nor the characteristic fertilization changes of the sperm nuclei take place (Hiromoto, 1962).

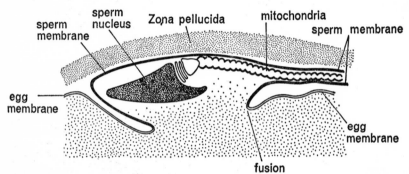

Figure 17. *Fertilization in the rat. The head and the middle piece of the sperm are seen in sagittal section. Note that the membranes of the sperm (black) and of the egg (white) have become continuous with one another. (After Szollosi and Ris, 1961.)*

It is also significant that part of the cell membrane of the sperm has now become included in the cell membrane of the zygote. A similar fusion of the cell membrane of egg and sperm has also been reported for the rat (Szollosi and Ris, 1961) (Fig. 17) and the golden hamster (Barros and Franklin, 1968). This fact—that the sperm affects the egg cell membrane—may turn out to be of importance in helping to establish the future orientation of the zygote, though the usual idea is that it is the centrosome of the sperm which controls the future orientation of the fertilized egg by determining the initial orientation of the first spindle. It is not known how the nucleus and the other components of the sperm are moved into the egg, nor is the mechanism of membrane fusion understood. There is, however, evidence that in some species (e.g. rat) the sperm tail and middle piece move into the egg (as well as the head) though they ultimately degenerate (Szollosi, 1965; Mazanek, 1965).

(d) Prevention of polyspermy

With the entry of the sperm into the egg and the fusion of the two pronuclei, the diploid number of chromosomes has been restored. If more
than one sperm pronucleus were to fuse with the egg pronucleus (polyspermy), too many chromosomes would be present. Most animals therefore possess methods for avoiding this situation. The most elaborate
mechanism is the production of a special membrane, the fertilization
membrane, which prevents the entry of supernumerary sperm. In echinoderms it is formed by the bursting of large granules in the cortex of the
egg, though there is some disagreement as to the exact morphology of the
process (for example, Afzelius, 1956; Wolpert and Mercer, 1961).
Similar cortical granules are present in some mammalian ova, such as the
hamster (Austin, 1956) and the rabbit (Hadek, 1963b), guinea pig,
mouse, coypu and pig (Szollosi, 1962). In rat and hamster eggs they are
formed in close association with several small Golgi complexes (Szollosi,
1967) and disappear when fertilization takes place (Fig. 18), but it is not

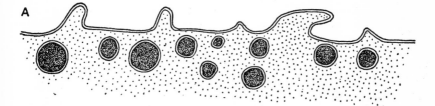

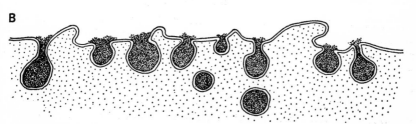

Figure 18. *Cortical granules in the rat. A. Before fertilization cortical
granules are present beneath the egg cell membrane. Each granule is surrounded by a membrane. B. After the entry of the sperm, each cortical
granule moves to the surface of the egg and bursts. The membrane that was
around each cortical granule fuses with the membrane of the egg. The
contents of the granule spill over the surface of the egg and probably prevent
entry of additional sperm. (After Szollosi, 1967.)*

clear whether they fulfil the same functions as those in echinoderms. Histochemical studies have indicated that the cortical granules are rich in acid phosphatase and it has been suggested therefore that they might act as lysosomes, destroying all but the nucleus of the sperm (Hadek, 1969). The question is particularly relevant, since the eggs of many mammals do not appear to have cortical granules so that it is evident that in some species, at least, other mechanisms for excluding super-numerary sperm must be used. In mammals the sperm must tunnel through the zona pellucida (Moricard, 1960), and in some species (e.g. mouse and rat) super-numerary sperm are apparently trapped in the inner layer of the zona (Dickmann, 1964, 1967; Hadek, 1969; also p. 139). There is some evidence that this is due to a specific reaction of the zona pellucida (Austin, 1961; Noyes, 1968; also p. 139). In other mammals, however (e.g. rabbit), the sperm appear to penetrate the zona pellucida but are somehow prevented from entering the egg (Hadek, 1969). It seems possible therefore that at least two methods are used among the mammals.

In large-yolked eggs such as those of the common fowl, many sperm enter the ovum even though only one is successful in fertilizing it. The others, the 'accessory sperm nuclei' are said to cluster around the periphery of the blastoderm and to undergo a rudimentary mitosis before ultimately degenerating. The accessory sperm have been studied by light microscopy by a number of workers (review by Romanoff, 1960) but so far we possess little information about the physiological processes that prevent these nuclei from approaching more closely to the female pronucleus and fusing with it.

(e) Activation of the egg

Soon after the sperm nucleus has entered the egg, the two nuclei fuse. Before they fuse, they are known as the two 'pronuclei'. Little is understood of the mechanism, although an important rôle seems to be played by the centrosome (discussion by Monroy, 1965). In most mammals fusion takes place soon after the sperm enters, since the second meiotic division has already been completed. In others (e.g. the dog and fox) the ovum has not completed its first meiotic division at the time of fertilization; the sperm nucleus, therefore, waits until the second polar body has been thrown off before it fuses with the female pronucleus (Austin, 1961). Once the two pronuclei have fused, the diploid number of chromosomes has been restored. In the rabbit, ovulation takes place about 10 hours after mating, and the fusion of the pronuclei occurs after about another 14 hours (Gregory, 1930).

This is, however, not the only change occurring in the egg, for immediately on contact with the sperm the process known as 'activation' begins. The events that comprise this process are cortical and cytoplasmic changes initiated by the sperm as soon as it touches the egg cell membrane, and can take place even if the sperm fails to enter the egg. Whatever the mechanism involved it is apparently not a specific one for it can be simulated in a variety of ways that lead to parthenogenesis (see p. 62). The cortical changes have been discussed (p. 57). The cytoplasmic changes will now be considered.

(i) PROTEIN SYNTHESIS

The most important event of activation, at least in echinoderm eggs, is that there is a sudden rise in protein synthesis. We have seen that during the early stages of oögenesis a certain amount of protein synthesis takes place; by contrast, during the later stages protein synthesis becomes greatly reduced and the oöcyte almost loses its ability to take up radioactive protein precursors. The small amount of protein synthesis that takes place in the unfertilized egg is thought to be utilised for maintenance.

Immediately after fertilization, however, precursors are readily taken up by the egg. It is now generally accepted that the protein synthesising mechanism of the unfertilized egg is blocked and that activation is the process by which this blockage is removed. The blockage itself is due to two things. First, there is a lack in the late oöcyte of usable mRNA. It is not within the scope of this book to discuss the evidence at length, especially as there are a number of excellent reviews dealing with the subject (for example, Metz and Monroy, 1967; Tyler, 1967; Gross, 1967; Davidson, 1968).

At first it was thought that mRNA was absent in the unfertilized egg but there is now considerable evidence that it is present in abundance, though masked, or stored (i.e. masked messenger RNA). Treatment with trypsin will remove this inhibition, which implies that the mRNA is combined with a protein. Spirin (1966), working with teleost embryos, has called this type of masked messenger complex an 'informosome'. It seems possible that similar informosomes are present in the oöcyte and that the mRNA is released from the complex at fertilization; in other words, the mRNA becomes unmasked at fertilization.

A second mechanism by which protein synthesis is blocked before fertilization may be that the ribosomes lack the ability to organise the amino acids into protein until the egg has been activated. We have seen

(p. 33, 34) that large amounts of rRNA are synthesised during oögenesis, though this process apparently stops once maturation begins, just small amounts of some heterogeneous RNA continuing to be formed in *Xenopus* (Brown and Littna, 1964). In the unfertilized eggs of sea urchins the ribosomes are present almost exclusively as single units, whereas after fertilization increasing numbers become arranged as polysomes (Monroy and Tyler, 1967). Ribosomes from unfertilized eggs seem to be unable to incorporate amino acids. It is possible therefore that during late oögenesis, the protein-synthesising activity may be shut off not only by the apposition of a protein to the mRNA but perhaps also to the ribosome–mRNA complex (Brachet, 1967). Thus, at the end of oögenesis, translation may be inhibited, and activation may involve release from both blocks. Furthermore a block to DNA replication must also be involved in the oöcyte, and this must be released if mitosis is to take place, for DNA replication is one of the main biochemical events of cleavage (Gurdon, 1968b).

The synthesis of mRNA is very low after fertilization, at least in echinoderms and amphibians. The protein synthesis that takes place in these groups during cleavage is entirely under the control of maternally-

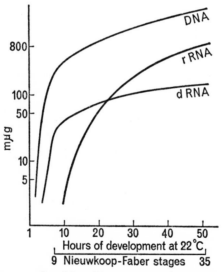

Figure 19.   *Amounts of nucleic acid present in early stage of* Xenopus laevis *Gastrulation begins at stage 9. rRNA is not synthesised until gastrulation, but DNA and some dRNA (i.e. DNA-like RNA, which is probably mRNA) are synthesised during cleavage. Most of the DNA-like RNA is, however, of maternal origin. (After Denis, 1968, and Davidson 1968.)*

derived mRNA which has now been unmasked. It has been suggested that small amounts of new mRNA are, however, built up now and stored for activation and use at a later period (Tyler, 1967). It is even possible that small amounts of paternally-derived mRNA are also produced now and stored in a masked form.

By contrast, no more ribosomal RNA is made until the nucleoli appear in the fertilized egg, which in most species studied is not until gastrulation (Fig. 19). In the normal egg, both male and female pronuclei have fused (p. 58) so that we can assume that information stored in the nucleoli during gastrulation is derived from the full genome. This correlation of the onset of new rRNA with the appearance of nucleoli during gastrulation has been elegantly demonstrated in *Xenopus*. A mutant form of the animal which fails to form nucleoli at gastrulation lacks the ability to develop new rRNA.

I have already pointed out that most of our knowledge of fertilization is based on researches into echinoderm eggs supplemented by a smaller body of work on amphibians. At present we have little knowledge of the position in reptiles and birds, though it is known that in the chick embryo new rRNA begins to be laid down at the mid-cleavage stage (Wylie, 1970). Mouse embryos, however, are able to synthesise new RNA soon after fertilization (Mintz, 1964a; Ellem and Gwatkin, 1968; also p. 134).

(*ii*) OXYGEN CONSUMPTION

Our knowledge of the time of onset of protein synthesis has only been gained in the last few years, but for decades embryologists have been aware of other striking events following on fertilization, the principal one being a visible turmoil in the cytoplasm and an increase in oxygen consumption. The latter was first described by Warburg (1908) using sea urchin eggs, but other workers have not always accepted his conclusions (discussion by Lord Rothschild, 1956). In particular, there is some doubt as to whether the increase occurs in fertilized mammalian eggs; Fridhandler (1968) states that respiratory activity does not increase until the blastula stage. Although it is now generally accepted that such an increase does indeed take place in many species, it is not considered to be so fundamental a change as was formerly suspected but rather to be one of the physiological accompaniments of protein synthesis. The evidence (discussed by Monroy, 1965) suggests that the biochemistry of the oxidative processes involved is no different in the unfertilized and fertilized egg; the change is a quantitative rather than a qualitative one.

(*iii*) THE GERMINAL VESICLE

We have seen that the activation process is initiated normally by contact with the sperm or artificially by various treatments, but the chemistry of activation is not understood. Until recently it appeared that an essential prerequisite was the breakdown of the germinal vesicle in first meiotic division (Grant, 1965). Thus, the presence of some nuclear contents seemed to be essential before the cytoplasm could react. For example, Delage (1901) showed that non-nucleated echinoderm oöcytes could not be activated by fertilization, but that this problem could be overcome by allowing the germinal vesicle to burst. It would now appear that the result of this bursting is to release mRNA into the cytoplasm. Recent experiments, however, show that in *Rana pipiens*, the bursting of the germinal vesicle is not necessary for activation, but that it is required if enucleated eggs are to cleave (Smith and Ecker, 1969). (See also Schuetz (1969) for discussion on germinal vesicle.)

The breakdown of the germinal vesicle probably also has an effect on the chromosomes. Gurdon and Woodland (1968) have pointed out that this process is comparable with mitosis in a somatic cell in that in both situations there is a swelling of the nucleus, followed by a stage when the nuclear membrane disappears and the chromosones become exposed to the cytoplasm. During the S-phase (p. 258), the chromosomes are thought to be acted upon by DNA polymerase, an enzyme existing in the cytoplasm which is necessary for DNA replication. It seems probable that similar interactions play a rôle in stimulating DNA replication during activation. Gurdon and Woodland suggest that this or similar cytoplasmic agents may also play a rôle in reprogramming chromosomes at mitosis. The importance of this idea is that it helps to explain the way in which an embryo is continually adjusting and controlling its own development (p. 10).

B. PARTHENOGENESIS

Parthenogenesis may be broadly defined as the activation of the egg followed by some development without the aid of a sperm. The activation may be brought about experimentally in a variety of ways in amphibian eggs, such as by pricking, or by treatment with acid or alkali, with changes in temperature or with ultraviolet light. In addition, parthenogenesis takes place in some animals, such as certain generations of aphides, as a natural event.

The length of time that a parthenogenetically activated egg may develop is very variable, many dying in the early embryonic stages and relatively few surviving to hatching or beyond. There are several reasons for this failure to survive, perhaps the most important being that there is an unmasking of recessive genes, many of which are deleterious. The problem is discussed by Beatty (1967) who compares the situation with that in inbreeding. Inbred animals have a higher incidence of homozygosity for recessive genes than outbred animals and consequently exhibit a higher incidence of conditions due to deleterious recessive genes. Similarly, parthenogenesis permits a higher incidence of homozygosity (see below) and as a result there is a high death rate from deleterious genes.

Parthenogenesis also affects the sex of the animals concerned. In mammals, the female is the homogametic sex, producing eggs all containing the X chromosome; the male is the heterogametic sex, half the sperm containing a Y chromosome and half an X. The sex of the normal, fertilized egg is thus determined by the sperm. If parthenogenesis were to take place in mammals, the sex of the parthenogenetic individual would always be female, since it would possess no Y-chromosome; the only sex chromosomes would be X, whether it contained the haploid number of chromosomes or whether it managed to double them to form the diploid number.

It is curious, from a phylogenetic point of view, that the male is the heterogametic sex in mammals, since in birds and in those reptiles with which we are familiar, the female is heterogametic. The situation does not, however, appear to be a simple reverse of the mammalian one, for the sex of the individual developing parthenogenetically from such a mother depends on a variety of factors, such as whether or not 'crossing-over' takes place. 'Crossing-over' is a process which often occurs during normal meiotic division and is a device by which genes are exchanged between adjacent chromosomes or even within the same chromosome (Fig. 9).

Usually there is a doubling of the chromosomes to form the diploid number. This may be by retention of the first or second polar body, or by a suppression of cytoplasmic division after replication of the chromosomes at first cleavage (for a consideration of the effect of these events on the sex of the parthenogone, see Beatty, 1967).

Among the amniotes, the best established example of spontaneously occurring parthenogenesis is in turkeys. By careful selection, Olsen (1960, 1965) has obtained a stock of inbred *Beltsville Small Whites* with an incidence of nearly 40% parthenogenesis. Although many of these embryos died *in ovo* a significant percentage survived and were reared to adulthood. At least one male bird was able to inseminate and fertilize virgin female turkeys. All the parthenogones that survived to hatching were male and they appeared to be diploid. They probably doubled their chromosomes by retention of the second polar body (Olsen, 1962).

Spontaneous parthenogenesis has also been described in two wild populations of lizards, *Lacerta saxicola*, in the Caucasus (Darevsky and Kulikowa, 1961, 1964) and in some species of *Cnemidophorus*, the North American whiptail lizard (Cuellar, 1968; Lowe and Wright, 1966). Unlike the turkeys the parthenogenetic lizards were almost entirely male; some, at least, of the females of *Lacerta saxicola* were found to be diploid, but both diploids and triploids were discovered among the whiptail lizards.

In mammals there have been reports of naturally occurring parthenogenesis, though these usually involve only a few cleavage divisions. For instance, Chang (1950) reported some parthenogenetical development of eggs of virgin ferrets. Parthenogenesis can also be induced in some mammals by various treatments, especially cold shock (Braden and Austin, 1954, on mouse eggs; Thibault, 1949 on sheep eggs). The report by Pincus (1939) that diploid, artificially produced, parthenogenetic, female mice were born alive at term has been subjected to some criticism, and later workers have not as yet been able to obtain similar results. The most encouraging results however are those of Tarkowski *et al.* (1970) and Graham (1970). They have, respectively, treated mouse eggs with mild electric shock, or with hyaluronidase, and have obtained development up to the blastocyst stage.

C. CLEAVAGE

Although sexual reproduction solves the problem of how to mix together genes from two individuals, it raises others. This is because it results in the new individual being formed from a single cell. The amount of specialisation that a single cell can undergo is limited, so that one of the tasks of the newly-fertilized egg is to multiply until a mass of cells is formed. This stage of rapid multiplication is the one we call 'cleavage'.

The onset of cleavage coincides with a burst of protein synthesis in most species that have been studied (p. 59) and this synthesis is essential for cleavage to continue. If agents which block protein synthesis are applied to echinoderm eggs, cleavage immediately stops (Denis, 1964). We have seen (p. 59) that in many species this protein synthesis is controlled by maternally-derived mRNA; not surprisingly, therefore, this maternally-derived mRNA in the unfertilized or cleaving eggs of amphibians has no detectable nucleotide sequences in common with the mRNA of later stages, which is manufactured by the embryo itself (Denis, 1968). Despite the addition of the paternal genes, no protein synthesis that is directed by the embryo's own genes occurs during cleavage in most amphibian species and very little new mRNA is produced by the embryo until gastrulation. The cleavage cells lack nucleoli in *Xenopus* and do not regain them until gastrulation when the production of large amounts of RNA is resumed.

The situation appears to be different in the mouse which starts to produce its own protein soon after fertilization and this coincides with the appearance of nucleoli (Mintz, 1964a). The rabbit embryo, on the other hand, manufactures relatively little new protein during cleavage (Manes and Daniel, 1969; also p. 134). There is some indication that the sperm may affect the intermitotic time of some cleavage divisions in mice (Whitten and Dagg, 1961). This has been shown by interbreeding two different strains which differ in their mitotic rates for the first and second cleavage divisions. Although the hybrid eggs always followed the maternal pattern for first cleavage, they did not necessarily do so subsequently. In view of their uniqueness in showing so early a paternal effect, these experiments might repay further investigation.

In amphibian embryos a curious paternal effect is known in certain hybrids. If the sperm of *Rana catesbeiana* is used to fertilize the eggs of *Rana pipiens* the embryos fail to gastrulate (Wilt, 1966). Although these lethal hybrids have been shown to synthesise some hetero-disperse RNA (Kohne, 1965), the rôle of the paternal genes at this stage is not understood.

It has been discovered recently that during cleavage in various species of sea urchins and amphibians, a small amount of new RNA is indeed formed, which is both m and sRNA but which does not appear to be utilised immediately (see reviews by Davidson, 1968; Denis, 1968; Tyler, 1967). Thus, just as mRNA that is formed in oögenesis may be masked and not used during cleavage, so the mRNA that is formed during cleavage may be masked and not used until gastrulation. Thus, the paternally-derived genes are perhaps not completely inactive during cleavage as was formerly thought nor indeed are those provided by the oöcyte, but their products are not utilised by the embryo straight away.

This new idea is especially interesting for, if it is correct, then the necessary mRNAs for many of the processes of embryonic determination may be produced during early development and then be masked until they are required at the appropriate stage and place. Thus, the production of mRNAs may be the key event in the onset of determination (Tyler, 1967).

Similarly, the rRNA synthesised by *Xenopus* during gastrulation after the appearance of the nucleoli (Brown and Littna, 1964) is apparently not needed until the much later swimming stage, since an anucleolate mutant which does not synthesise rRNA during this period will normally develop up to the larva stage. These authors therefore concluded: 'The time of gene expression need not synchronise with the needs of the organism for the gene product'.

D. SUMMARY

Fertilization in amniotes is internal, and so our knowledge about it is limited. The evidence suggests that the sperm are passed up the female genital tract largely as a result of muscular contractions of the oviduct or uterus. Sperm storage occurs in certain species, but little is understood of the physiological processes involved. Some of the mechanisms employed in the penetration of the egg by the sperm or in the prevention of polyspermy resemble the better known ones of certain invertebrates.

One of the main results of fertilization is that the egg becomes activated, and in many species this leads to an increase in protein synthesis. In those amniotes that have been studied, it appears that the embryo begins to form rRNA during cleavage and does not (unlike *Xenopus*) delay the process until gastrulation.

There is now firm evidence that parthenogenesis occurs spontaneously in certain wild populations of lizards and in a highly inbred strain of turkeys, and that the parthenogones may survive to adulthood and be capable of reproduction. Some degree of parthenogenesis has also been obtained experimentally in certain mammals.

CHAPTER 4

ORDER AND ADJUSTMENT IN THE EARLY STAGES OF BIRD DEVELOPMENT

DESPITE the fact that the chick embryo has been studied since the time of Aristotle (Needham, 1959) very little is known about the earliest stages of its development. This is mainly because the embryo has already passed through cleavage and embarked on gastrulation by the time the egg is laid. But it is also because the bird's egg contains a large amount of yolk which makes it difficult to handle.

It is not surprising therefore that most of our knowledge of the events of cleavage and early gastrulation is based on the study of fixed and sectioned material, and that even the biochemical information for this period is limited.

Despite this, the pre-primitive streak stage promises to become one of the most interesting, largely because our traditional ideas on the mechanism of endoderm formation have recently undergone a dramatic change. As a result, gastrulation in the chick is now seen to have more in common with that of amphibians than was formerly realised.

Another reason is that since mouse and *Xenopus* embryos differ in the way in which they control protein synthesis during cleavage (p. 64, 65), it is important to know what mechanisms are involved not only in other mammals and amphibians, but also in a third class of vertebrates, the birds.

The period with which we are concerned in this chapter ranges from the start of cleavage until the primitive streak has formed. It begins immediately after fertilization, continues before laying, and, in the chick, ends after about 18 hours of incubation. By the time the egg is laid, the area pellucida and area opaca are already established and the future orientation of the embryo is apparent.

During this time the endoderm forms and the primitive streak starts to develop. From the 'functional' point of view the main characteristic of these stages is the extensive power of regulation displayed by the blastoderm. Although considerable regulation is possible at the fully-grown primitive streak stage and even later, it is never again so striking as in this early stage.

A. DESCRIPTIVE MORPHOLOGY OF CLEAVAGE

Let us recall that toward the end of oögenesis, the cytoplasm moves to one side of the oöcyte, the future dorsal pole of the egg. At the end of oögenesis the cell membrane of the hen's egg becomes broken up into a number of vesicles lying beneath the vitelline membrane (Chapter 3). It seems reasonable to suppose that these will give rise to the cell membranes of the cleavage cells, though the mechanism by which they do so is as yet obscure.

The traditional concept of the events of cleavage in birds is based on the morphological studies of Patterson (1909) in the pigeon's egg. The type of cell division involved is a highly unusual one and is almost unique in the animal kingdom. Nevertheless, Patterson's conclusions have been substantiated in a thorough, modern investigation (Bekhtina, 1960).

The first cleavage division takes place at about three to five hours after ovulation. As the daughter chromosomes separate during the anaphase stage of mitosis, the yolk granules disperse so that a clear area of cytoplasm lies between them, and it is into this area that the first cleavage furrow extends (Fig. 20). This furrow is brought about by the downward extension of cell membrane. Two blastomeres are thus formed but they each remain open to the yolk on their ventral sides. Even at the 10-celled stage, all the blastomeres are still open ventrally, but by the time 100 cells have formed only the most marginal blastomeres are still in open communication with the yolk (Bekhtina, 1960). Between the stages of two and eight blastomeres the mitoses seem to occur synchronously though they later cease to be in step with one another. It seems possible that the fact that these cells are still 'open' enables them to collaborate with one another so that they divide in unison. Bekhtina, who stained his sections with Feulgen's stain, concluded that there was an increase in the DNA of the chromosomes during successive divisions but that nucleoli did not appear until the stage when the blastomeres had become too numerous to count. We may assume that the visible appearance of nucleoli is an indication that ribosomal RNA is probably being produced (Chapter 3).

During these early stages there is no definite edge to the blastoderm, but there is at its periphery a region called the periblast which is thought to be part yolk, part cytoplasm. This region contains a number of nuclei, the 'free' or periblastic nuclei, and since these are not contained in proper cells the region is said to be syncytial. It is supposed that if an 'open' cell at the periphery undergoes mitosis, one of the daughter nuclei migrates out to form a periblastic nucleus. Subsequently, each

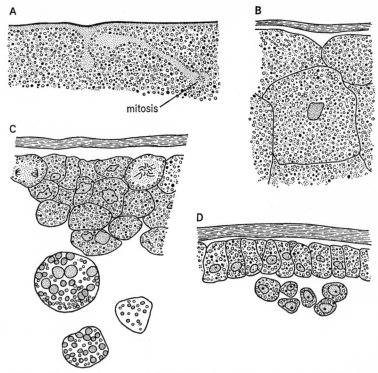

A

B

mitosis

C

D

Figure 20. *Development of the chick embryo before the egg is laid: All the cells are more yolky than in the later stages. A. Early cleavage: A mitotic figure of the first cleavage division lies at the end of the 'cleavage track', a relatively clear area of cytoplasm formed as the nucleus migrated through it. It is into this area that the first cleavage furrow will extend. B. Mid-cleavage: About 100 blastomeres are present. At this stage the egg is enclosed in shell membranes. C. Late cleavage: The cells are smaller but now contain nucleoli. A thin shell has been deposited on the shell membranes (not shown). D. Early gastrula: By this stage the egg is enclosed in a fully formed shell (not shown). (After Bekhtina, 1960.)*

periblastic nucleus, together with some periblastic yolk and cytoplasm, becomes surrounded by a cell membrane. Thus, a central, cellular region grows at the expense of the syncytial zone. In this way the number of cells increases and the blastoderm spreads a little way over the yolk. The nuclei of the syncytial zone are frequently large and lobulated and possess many nucleoli (Bekhtina, 1960).

At the end of cleavage the blastoderm is a round disc about five or six cells thick in the centre but tapering out to one or two cells deep at the periphery.

Gradually, however, the arrangement of the cells changes so that two regions become recognisable in whole mounts examined by transmitted light. In the centre is the thinner *area pellucida*, and at the periphery is the thicker *area opaca* (or germ wall) (Fig. 21). Small yolk droplets are present within the cells of the area pellucida but larger ones are contained in the cells of the area opaca. At a later stage, the primitive streak

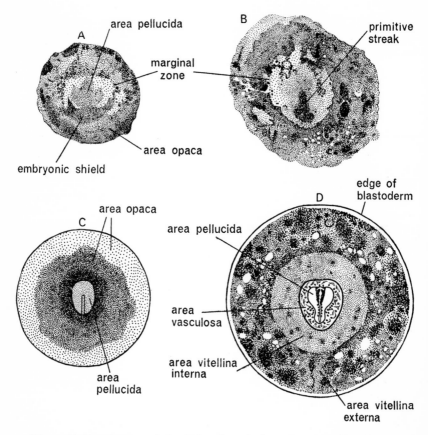

Figure 21. *Development of the chick embryo in the early stages of incubation. A. At the time of laying. B. Primitive streak stage as it often appears after dissection of the embryo from the yolk. C. Primitive streak stage after the adherent yolk granules have been washed from it. Stages B and C are reached after about 18–20 hours of incubation. D. Early embryonic axis with paired somites. The area opaca has now been divided into the area vasculosa (in which blood starts to circulate) and the area vitellina. The area vitellina externa possesses larger, more yolky endodermal cells than the area vitellina interna. This stage is reached after about 36 hours of incubation.*

forms in the area pellucida, and subsequently gives rise to the embryo; the yolk sac develops from the area opaca.

The arrangement of the cells into area opaca and area pellucida indicates that cleavage is over and gastrulation is about to begin.

At first, the area pellucida is not arranged in clearly defined layers and is only about two or three cells deep. At its future posterior end is a small opaque crescent known as the embryonic shield (Fig. 21) and this may be as many as about five cells deep. The superficial cells of the blastoderm gradually become arranged into an epithelium of irregularly-shaped cells both in the area pellucida and in the area opaca. In the embryonic shield the cells are arranged in a less orderly manner. No endodermal layer is present in the blastoderm but a number of loose cells lie beneath the epithelium just anterior to the embryonic shield.

The earliest stages of development take place whilst the egg is still passing down the genital tract of the hen. The exact stage reached by the time the egg is laid varies with the species. Duck embryos are usually only at the beginning of gastrulation when the egg is laid whereas hens' eggs have generally become two-layered. There is also some individual variation, some hens laying their eggs at an earlier stage than others (Romanoff and Romanoff, 1949). Also, external factors, such as length of daylight at different times of year, or a frightening experience, can affect the hormonal control of the laying cycle and result in the eggs being retained for longer or shorter periods. When it has been laid, the temperature of the egg drops and the metabolic rate slows down. When incubation begins the temperature of the egg is raised and development proceeds at a faster rate. At the time of laying, about 60,000 cells are present in the chick blastoderm (Spratt, 1963).

B. INTEGRATION OF DEVELOPMENT IN THE EARLY STAGES

(a) Orientation

Within a short time of fertilization the amniote egg possesses not only a dorsal and a ventral side but also an anterior and a posterior end. The dorso-ventral axis is established in the oöcyte before fertilization takes place for, as we have seen (p. 28), the germinal vesicle migrates to one side of the oöcyte. A flat disc of cells forms from the fertilized ovum and lies upon the yolk. The side of the disc touching the yolk becomes the ventral side of the embryo and the side furthest from the yolk becomes the dorsal side. The first visible sign of a cranio-caudal orientation, however, is the appearance of the embryonic shield, which marks the future caudal (tail) end of the embryo. With the establishment of these two major axes, the third (viz. right–left axis) automatically becomes apparent.

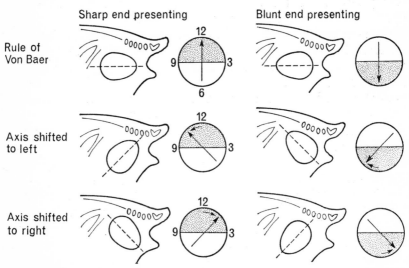

Figure 22. *Orientation of the chick embryo. Diagram to illustrate how the position taken up by the egg in the oviduct of the hen shortly before laying, determines the orientation of the embryo. The arrow indicates the orientation of the embryonic axis. The head of the arrow corresponds with the head of the embryo. (From Clavert, 1963.)*

It is convenient to have some indications of where the embryonic shield will arise and for this purpose it is usual to apply the rule of von Baer (1828). This states that if the egg is held with its sharp end toward the right of the investigator, then the future long axis of the embryo will lie at right angles to that of the egg and the head of the embryo will point away from him. Unfortunately, this rule is probably reasonably accurate for only about 60–70% of the eggs of domestic fowls, ducks and quails (Lutz, 1965); other embryos deviate to left or right by varying degrees. However, if the eggs from a single female quail are considered they are usually found to be consistent in their orientation (Fargeix, 1963). Thus, von Baer's rule is highly accurate for some female birds and inaccurate for others.

Von Baer believed that the orientation of the future axis was laid down in the ovary. Experiments by Clavert and Vintemberger (reviewed by Clavert, 1963), however, demonstrate that there is a relationship between the position taken up by the egg in the uterus shortly before laying and the future orientation of the embryo (Fig. 22). Thus, if the egg is laid sharp end first, the embryo is orientated according to the rule of von Baer. Conversely, if the egg is laid blunt end first, the orientation is reversed. These investigators based their conclusions not only on radiographic evidence to establish the position of the egg in the oviduct but they also studied the effects of experimentally altering its position in the oviduct. They found that in this way they could alter the orientation of the embryo. Finally, in a series of careful experiments they showed that if uterine eggs were placed in a cylinder *in vitro* the orientation of the embryo could be altered by tilting and rotating the cylinder. The way in which this process is brought about is not clear but it seems reasonable to assume that gravity plays a rôle. Following these experiments, Rogulska and Komar (1969) demonstrated that the orientation of the embryo is largely dependent on the shape of the shell, and that the alignment is more likely to correspond to von Baer's rule if the egg shell is elongated than if it is almost rounded. They suggest that this is because the more elongated the shell, the fewer rotations it is likely to undergo in the hen's uterus.

Although the future orientation is thus partially established by the time the egg is laid, it is not completely fixed, as certain authors have implied, for, as we shall see below, it can be overcome and the cells can re-integrate to form an embryo with a new orientation. More important is the fact that the cells themselves are in a highly labile state and, even if their orientation remains unchanged, they are capable of extensive regulation. This lability of the young embryo was recognised by some of the earlier workers. For instance, Morita (1936) and Twiesselman (1938) showed that if an unincubated blastoderm was cut in two as it lay *in situ* in the egg, two embryos could form. Similarly, if it was cut partially in two, conjoined twins could develop. Thus, the young avian embryo was shown to have similar powers of regulation to the amphibian embryo.

Lutz (1949) and his collaborators systematically investigated the extent of this regulative capacity at this stage of development. Primarily they used the duck embryo since it is at an earlier stage of development than the chick when the egg is laid (Fig. 23). When the blastoderm was cut into as many as four parts, each part was found to be capable of regulating to form an embryo. Similarly, if most of the blastoderm was destroyed even a small, lateral portion was capable of regulating. The orientation of these embryos was especially interesting. It appeared

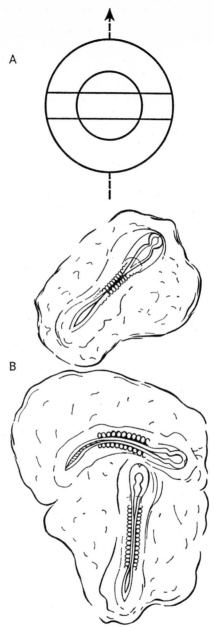

Figure 23. *Regulation and orientation. A. A young duck blastoderm was cut into three pieces. The original orientation is indicated by the arrow which points anteriorly. B. Each of the three pieces has regulated to form an embryonic axis. Two are orientated in the same direction as the original embryo. (After Lutz, 1949.)*

already to be determined in the embryos developing at the posterior side of the area pellucida, for they always had the same orientation as the original blastoderm. But the embryos developing at the anterior side were frequently reversed in orientation.

A possible explanation is that this was because the endoderm determined the orientation of the embryo. Thus, presumably, if a complete layer of endoderm was present at the time of operation, the embryonic axis that developed tended to have the same orientation as the original area pellucida. Conversely, if the endoderm was only partially formed, orientation of the anterior axis was haphazard. Although Lutz and his colleagues did not investigate the intermediate stages in the formation of these axes it is apparent that a primitive streak developed in each isolated piece of blastoderm, and that the antero-posterior orientation of this streak determined the orientation of the embryo. The orientation of each primitive streak was therefore determined by the underlying endoderm.

In a further series of experiments (Lutz, 1962) a full-length primitive streak was grafted into an unincubated blastoderm. If the graft was orientated in the same way as the host then a single embryo developed, but if the graft was placed in reverse orientation to the host two axes formed. In the latter case the grafted streak disappeared and two new ones appeared, one on either side of it. Thus, it appeared that in this experiment the graft orientation was overcome by that of the host. This conclusion is of especial interest in view of the rôle that is played by the area pellucida in controlling axis orientation in later stages of development (p. 114).

These experiments of Lutz and his co-workers drew attention to the importance of the posterior end of the blastoderm in integrating the development of the embryo. In this they confirmed the conclusions of previous investigators who had isolated fragments of blastoderm upon chorioallantoic membrane. Some of the early workers also believed that regulation was possible only in fragments of blastoderm containing this region (e.g. Butler, 1933) but the work of Lutz has now demonstrated that, in the duck at least, all regions of the area pellucida of the unincubated blastoderm are capable of regulation.

(b) *The controlling rôle of the posterior end of the early embryo*

Most of the experiments discussed so far have been carried out directly on the embryo as it lies on the yolk *in situ*. The following experiments however were performed *in vitro* (for discussion of experimental techniques used in studying the chick embryo, see p. 108; Fig. 31).

Several authors have demonstrated that endodermal cells migrate out from the ventral part of the posterior germ wall (Spratt and Haas, 1961a; Vakaet, 1967; Dubois, 1968). The moving cells are arranged in a fan-like pattern; hence, some authors speak of a fan-like movement.

Spratt and Haas have carried out a number of ingenious experiments aimed at analysing how these movements arise, and as a result have postulated the presence in the unincubated blastoderm of a 'dominant growth centre', or 'lower cell surface movement centre'. Both terms appear to apply to the same region. They locate this centre in the

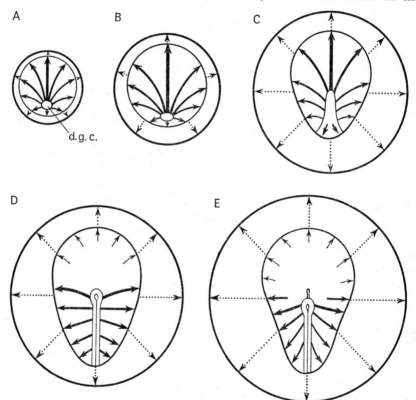

Figure 24.   *Diagram to show the morphogenetic cell movements in the lower (endodermal) side of the chick embryo as envisaged by Spratt and Haas (1960a). These authors report that the cells migrate from a dominant growth centre (d.g.c.). Thick, black arrows show fan-shaped, endodermal cell movements; dotted arrows and thin, black arrows show general expansion of the area opaca and the area pellucida respectively. A. Unincubated blastoderm. B, C, D and E. Three stages in the formation of the primitive streak and early head process.*

posterior part of the area pellucida. These authors have also introduced the term, 'junctional' or 'marginal zone' to describe the peripheral cells of the area pellucida. Thus, in their terminology, the dominant growth centre lies at the posterior part of the marginal zone (Figs. 21A and 24).

These authors propose that there is a high mitotic rate in the 'dominant growth centre'. The evidence is based mainly on mitotic counts (Spratt, 1966). They suggest that, because of the high mitotic rate, the number of cells increases in this region and so they become more closely packed together. Indeed, they state that in the marginal zone there is a gradual drop in the population density of the cells in the regions further and further away from the dominant growth centre. Their evidence is based on gross observations of whole mounts, but a more detailed investigation seems desirable since other authors do not appear to have observed the phenomenon. Spratt and Haas suggest that because many cells are present and closely packed together in the dominant growth centre there is some migration of cells from this region (i.e. the fan-like movement, mentioned above).

This migration normally takes pace only from the region of the dominant growth centre and not from any other part of the marginal zone. Spratt and Haas deduce, therefore, that the other regions are dominated by the growth centre; hence the introduction of the term *dominant* growth centre.

This conclusion is based on a number of experiments. For instance, Spratt and Haas (1961a) rotated almost the entire area pellucida, including the marginal zone, and in all those explants that formed an embryo its head to tail polarity was reversed (Fig. 25). They therefore concluded that the posteriorly situated dominant growth centre had dominated and controlled the position of the future embryo, even though it had itself been transplanted to the position normally occupied by the anterior part of the marginal zone.

The way in which this posterior region 'dominates' the rest of the marginal zone is not easy to understand. According to Spratt and Haas (1960c) the dominance is a direct result of having a higher growth rate and cell population density. They suggest that the cells deriving from the dominant zone are thus able to spread rapidly beneath the entire area pellucida before the cells have begun to migrate out from other regions of the marginal zone which have a smaller cell population density. Having spread out in this way, the cells of the dominant region are able by some chemical means to suppress any migration from other regions. If, however, (Spratt and Haas, 1961a and b) the 'dominant growth

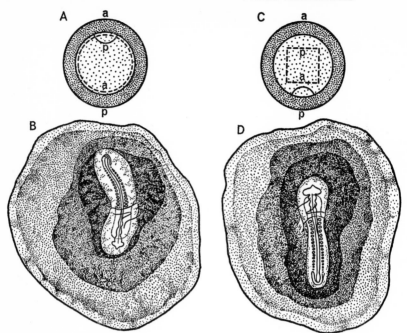

Figure 25. *Evidence for the 'dominant growth centre' of Spratt and Haas (1960). The region assumed to be the dominant growth centre is indicated by a semi-circular line in A and C. A and B. Almost the entire area pellucida, including the marginal zone, is cut from the area opaca, rotated, and replaced; thus the original posterior end (p) has exchanged places with the original anterior end (a). The embryo develops with its polarity reversed. C and D. If a smaller region is cut from the area pellucida and reversed, the original posterior end remains in situ. The embryo develops with unchanged polarity. (After Spratt and Haas, 1961a.)*

centre' is damaged or absent, then the region of highest cell population density becomes the dominant region.

As a result of the experiments of Spratt and Haas, there can now be little doubt that the posterior end of the area pellucida plays an important rôle at this stage of development, but it seems likely that their theories may undergo some modification during the next few years. It is, for instance, possible that the dominance of the so-called growth centre may be due not so much to a high mitotic rate and density of population, as to the co-ordination of cells throughout the entire area pellucida. Some evidence for this idea comes from the 'combination' experiment of Spratt and Haas (1961b). In these experiments three unincubated blastoderms were grafted together in such a way that the three 'dominant growth centres' were in the 'topographical centre of the system' (Fig. 26) and yet, apparently, in none of the twenty-four combinations did a single embryonic axis arise from one of the original growth centres.

It seems clear that there is still much to learn about the rôle of the posterior end of the blastoderm in controlling the development of the early embryo. The destination of the cells that spread out in a fan-like movement is discussed in section 3(c).

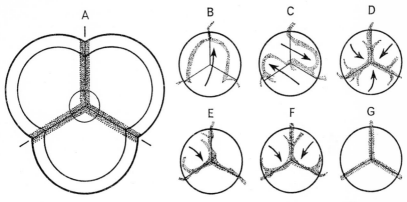

Figure 26. *Experiments with unincubated chick embryos. A. Three unincubated chick embryos were grafted together so that the three original 'dominant growth centres' were in the middle (shown by circle). Carmine particles (stipple) were used to mark the boundaries between adjacent blastoderms. B, C, D, E, F and G. These show the main types of result. Each arrow represents an embryonic axis. The head of the arrow is at the head end of the embryo. (After Spratt and Haas, 1961b.)*

(c) Origin and formation of the endoderm

(i) THE PROBLEM

One of the most lively and persistent controversies has been concerned with the process of endoderm formation in birds. The process is difficult to study because it has often been completed by the time the egg is laid. Furthermore, we must concern ourselves not only with the origin of the endoderm of the area pellucida but also with that of the area opaca. Until recently most of the investigations were carried out upon fixed and serially sectioned material in which it is not easy to decide whether the gaps between the cells are artefacts of preparation or not. Investigators have often been driven to make deductions from the shape of a fixed and sectioned cell as to whether it is migrating this way or that. It is hardly surprising therefore that so many different theories have been put forward.

A number of workers have, however, tackled the problem experimentally, though once again the conclusions are far from unanimous. To some extent this may be due to the fact that different techniques have been used. Some workers have employed vital dyes (e.g. Hunt, 1937; Peter, 1938) whereas others have utilised carbon or carmine particles (Spratt and Haas, 1960a) or radioactive markers (e.g. Modak, 1965); some experiments have been carried out *in ovo*, whereas others have been performed *in vitro*. It is sometimes argued that the technical differences between marking experiments performed *in ovo* and *in vitro* are unimportant since an embryo can form under either set of conditions. This argument is not completely convincing in view of the well-known regulative ability of the early blastoderm. If we are concerned to find out what happens normally, it is distracting to be presented with what may be a regulation effect. It may well be that tissue culture conditions inhibit some of the normal movements and promote regulative ones. Though of interest, this may lead to a faulty conception of the normal processes of endoderm formation. The possibility that this may occur is demonstrated by the fact that Vakaet (1962) who carried out identical experiments on two sets of blastoderms growing according to the New (1955) and to the Spratt (1947) techniques, obtained different results from the two experiments. (Culture techniques used for maintaining chick embryos removed from the egg are discussed on p. 90.)

It is unfortunate that these doubts must exist about the efficacy of tissue culture methods for interpreting these movements, for, as we shall see, some of the most important information on endoderm formation has recently been obtained from experiments carried out *in vitro*.

For many years the theory of delamination was most widely accepted. This supposed that the endoderm forms by separation of a sheet of the lowermost cells of the epiblast (Kionka, 1894; Peter, 1938; Wetzel, 1929). Generally, separation was supposed to start at the posterior end and gradually progress forwards. In a rather similar theory, that of *polyinvagination*, the individual cells were thought to migrate ventrally and only subsequently join together to form a coherent layer (Merbach, 1935; Pasteels, 1937b). According to Merbach, the individual cells were derived from naturally occurring folds in the blastoderm, but other workers who have seen similar folds have interpreted them as artefacts of fixation (e.g. Jacobson, 1938a). The theory of polyinvagination was supported by Pasteels (1937b) though he subsequently (1945) revised his opinion in favour of delamination.

Some experimental support for the polyinvagination theory was obtained by Hunt (1937), who stained cells in the upper layer of the blastoderm and subsequently kept both pieces of area pellucida under continuous observation in Tyrode's solution. According to this author there was a migration of cells from the upper to the lower layer. But other authors have objected to these results on the basis that the stain (i.e. Nile blue sulphate) has a tendency to diffuse.

Opposed to these two theories of an origin of endoderm from the lower surface of the area pellucida was the concept of *an origin from the germ wall*. The term 'germ wall' at this stage of development is generally used as being synonymous with area opaca. There have been several variations on this theory.

A few authors have supposed that the endoderm originates from the periphery of the blastoderm. This concept was put forward on morphological grounds by Disse (1878) and revived by Lutz (1955) on the basis of an experimental investigation.

Patterson (1909) reported that a temporary blastopore arose at the posterior edge of the blastoderm wall and that this later closed by 'concrescence'. This theory was comparable with a similar theory of 'concrescence' in fishes (e.g. His, 1876) and perhaps for reasons of homology received some support at the time. Later authors have been almost unanimous in rejecting it on the basis that 'blastopores' in this region have seldom been seen except in damaged embryos.

Some authors have considered that the region at which the posterior germ wall abuts the area pellucida may be the source of the endoderm. This theory supposes that there is a forward migration of cells from that region, so that a progressively enlarging sheet of tissue expands from posterior to anterior beneath the area pellucida (Rauber, 1876; Köllicher, 1879; Butler, 1935). Jacobson (1938a) believed that an invagination of cells took place in this region through a blastopore, but other workers in the field have not described such a structure.

(ii) RECENT EXPERIMENTAL WORK

Recent workers have aimed at testing these theories by experimental means. Spratt and Haas (1960a, b) and Vakaet (1962) have concluded on rather different bases that the posterior germ wall does not contribute to the *embryonic* endoderm. In both sets of experiments the blastoderm was grown *in vitro* with its endodermal side uppermost and marked with particles either of carbon or carmine, or with vital dyes. The main results obtained were similar though differing in details. That is, a movement of lower layer cells begins in the region of the embryonic shield and fans out towards the area opaca–area pellucida border. (These are the fan-like movements mentioned in Section B(b) above.) According to Spratt and Haas (1965) these cells give rise to the mesoderm and endoderm of the area vasculosa. Vakaet (1962) on the other hand has found that these cells, which he calls 'endophyll cells', having almost reached the anterior border of the area pellucida, are met by other cells that migrate centripetally into the area pellucida from the anterior germ wall; the region in which they meet establishes the border of the definitive anterior germ wall. Until recently it had been thought that the primordial germ cells arose at a later stage in the anterior germ wall (p. 230). Recently, however, Dubois (1967) has located the primordial germ cells in the posterior germ wall of the unincubated blastoderm. He has demonstrated that they are then carried forward passively in the fanlike movement toward the anterior germ wall. Their further history is discussed in Chapter 11.

It now appears, therefore, that the forward migration of cells from the posterior germ wall mainly forms extra-embryonic endoderm and does not lead to the laying down of the endoderm of the area pellucida. On the contrary, the latter is probably formed by the migration of cells from the primitive streak. The evidence is of three main kinds. Firstly, when cells from the middle layer of the early primitive streak are marked with carbon particles they are subsequently found in the embryonic endoderm (Spratt and Haas, 1965); secondly, if the embryonic endoderm is extirpated immediately after its formation a new sheet of endoderm will form. Vakaet and Mareel (1964) and Vakaet (1967), who have followed the events involved in this process by means of cine-film, have concluded that the regenerating endoderm originates from the lower layers of the primitive streak. A similar conclusion was reached by Modak (1963, 1965, 1966) who labelled the regenerating endoderm with carbon and tritiated thymidine. Modak (1965) concluded, however, that there was also a small but essential contribution to the regenerating tissue from the cells of the posterior germ wall.

Despite the ingenuity of these experiments on endoderm regeneration and the apparent care with which they have been carried out, it may be pertinent to enquire whether they are strictly relevant to our problem, for it is questionable whether the process of endodermal regeneration is necessarily identical with the normal mechanism of formation of endoderm in the undisturbed blastoderm.

The third type of evidence is derived from the experiment of Rosenquist (1966), although these were apparently not performed until the primitive streak was well established (stages 3 to 5 of Hamburger and

Hamilton, 1951; see appendix). In this experiment, grafts of mesoderm and endoderm labelled with tritiated thymidine were inserted into homologous regions of non-labelled embryos. Subsequently, the fate of the

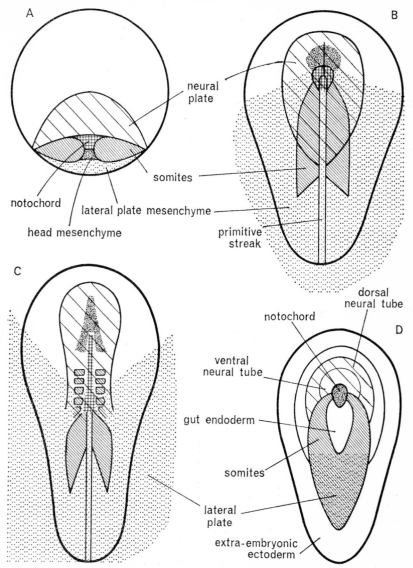

Figure 27. *Presumptive areas of the chick embryo. A. The short primitive streak stage. B and D. Head process stages. C. The stage of three to four pairs of somites. (A, B and C after Pasteels, 1937b; D after Rosenquist, 1966.)*

grafts was determined by means of serial sections and autoradiography. Rosenquist was able to demonstrate that even at these stages the endoderm radiates from the base of the primitive streak. However, Rosenquist's views on endoderm formation are unusual, for he concluded that a pre-endoderm region existed in the epiblast near the anterior end of the primitive streak and that this was progressively invaginated to form first of all extra-embryonic endoderm and, subsequently, foregut endoderm (Fig. 27). Thus he deduced that the endoderm lying medially beneath the streak at stage 3* (streak practically full length) was destined to migrate bilaterally to form extra-embryonic endoderm. By contrast, the tissue that migrated from beneath the streak at stage 5 (head process stage) was destined to form foregut endoderm. Similar experiments involving homologous exchange of radioactively labelled grafts were carried out by Nicolet (1965a, 1967) who also concluded that the gut endoderm formed from the base of Hensen's node, the anterior end of the primitive streak.

These modern views on the origin of the embryonic endoderm (area pellucida endoderm) all tend to suggest that it arises from the base of the primitive streak. Mesoderm also begins to migrate out from the primitive streak at about the same time as the endoderm and, indeed, continues to do so whilst the primitive streak is forming. Thus, the primitive streak is depleted of a considerable number of cells and it is generally assumed that these cells are continuously replaced by others. It is, of course, possible that there is an actual diminution in the number of cells in the primitive streak during this process but this does not seem to have been investigated directly, probably because of the difficulty in deciding which part of the short length primitive streak may be compared with which part of the full length primitive streak. If, however, we accept the probability that there is a constant renewal of cells in the primitive streak, there seem to be only two ways in which this could happen. Either there must be a high rate of mitosis or else cells must enter by immigration from the epiblast.

(d) Mitotic rate of the primitive streak

The possibility that the primitive streak might be a region of high mitotic rate, proliferating cells which give rise to mesoderm, has attracted many workers at different times. But the evidence is confused and it is difficult to make a firm statement at the present time. The mitotic rate (i.e. (number of dividing cells/number of non-dividing cells) $\times$ 100%) has been estimated for the primitive streak and compared with

* See appendix.

other regions of the area pellucida. Some investigators (e.g. Derrick, 1937; Bellairs, 1957) found no evidence for a high mitotic rate in the primitive streak, whereas others (Chen, 1932; Spratt, 1966) obtained the opposite result.

(e) Morphogenetic movements during primitive streak formation

Many ingenious hypotheses have been put forward to account for the way in which the primitive streak develops (reviewed by Romanoff, 1960). The earlier ones were based entirely on the study of fixed material, an approach that can give no direct evidence about the cell movements that occur. The first direct evidence came from the pioneer experiments of Gräper (1929) who marked patches of the blastoderm in ovo with vital dyes and followed the fate of these patches with cine-photography.

Experiments of this type have been repeated with various modifications by others. Broadly, these workers may be classified into two groups, i.e. those who have carried out their marking experiments in ovo (e.g. Wetzel, 1929; Pasteels, 1937b; Malan, 1953) and those who have done them in vitro (e.g. Spratt, 1946; Spratt and Haas, 1965). The results have not always been the same and, once again, the discrepancies have some-times been attributed to differences in technical approach.

The first point of almost general agreement, however, is that morpho-genetic movements occur within a few hours of the onset of incubation, probably as soon as the temperature of the cells rises near the normal level of 38·5°C. Spratt and Haas (1965) alone failed to detect movements in the epiblast at this stage, but, since they explanted the embryo with this surface against the agar of the culture medium, it seems possible that the cell movements were mechanically inhibited. For this reason the conclusions of Spratt and Haas (1965) will not be considered here; instead, the earlier work of Spratt (1946) will be discussed.

The second point of general agreement is that there is a movement of cells toward the posterior median end of the blastoderm. This is of importance since, as we shall see, it is related to the layout of the pre-sumptive areas at this stage. According to Spratt (1946) almost the entire posterior half of the area pellucida at the pre-primitive streak stage is destined to enter the primitive streak and become mesoderm. Pasteels (1937b) (Fig. 28) and Malan (1953), however, illustrate a much smaller region as converging upon and entering the primitive streak.

It is not clear whether there is any mass movement of cells at the anterior end to balance that at the posterior. The maps of Wetzel (1929) and Spratt (1946) exhibit a medio-lateral movement of cells at the

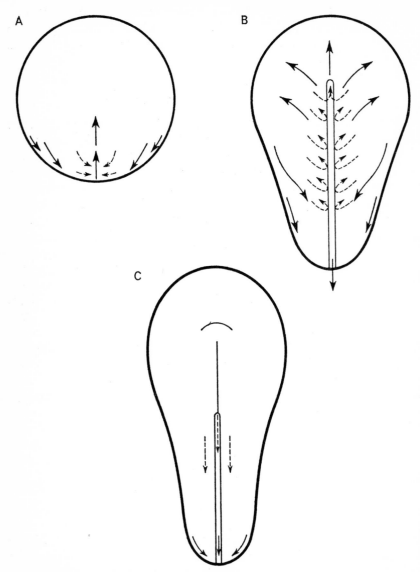

Figure 28. *Morphogenetic movements in the young chick embryo. A. At the short primitive streak stage. B. At the full length head process stage. C. At the head fold stage. Dotted arrows: invagination movements, and movements of the mesoderm cells. Black arrows: movements of cells other than invagination. (After Pasteels, 1937b.)*

anterior end of the area pellucida, but none of the experiments of Pasteels (1937b) nor those of Malan (1953) offer any information on this point and none appear to have been designed to investigate it.

All workers who have carried out marking experiments appear to agree that some forward movement takes place in the mid-line as the primitive streak forms; Spratt (1946) believed it to be considerably less than Pasteels (1937b) had supposed.

(f) Co-ordination of tissues

One of the main themes of this book is that the tissues co-operate with each other during development so that an orderly embryo forms. It appears that the earliest endoderm cells (which give rise to the anterior germ wall, p. 82) actively migrate beneath the lower surface of the epiblast cells. By extirpating a strip of area pellucida and replacing it in an inverted position, Spratt and Haas (1962) were able to demonstrate that the cells moved most successfully if they were in contact with the underside of the upper layer, and that their performance was less satisfactory when presented with other surfaces. We know little about the factors controlling the direction taken by these migrating cells. It seems unlikely that there is a pre-formed antero-posterior pattern in the lower surface of the upper layer cells, for otherwise it would be difficult to obtain regulation of the embryo after experimental interference. It is more probable that just as cells growing in tissue culture may spread out to fill the available space so do these migrating lower layer cells of the embryo. Unable to move posteriorly because of the bulk of the germ wall, they spread out beneath the area pellucida in a fan-shaped pattern.

We have seen (p. 82) that both the area pellucida endoderm and the mesoderm arise from the primitive streak. The formation of the endoderm rapidly outstrips that of the mesoderm (Malan, 1953) and an endodermal layer is complete at a stage when the mesoderm is still only partly invaginated.

This premature formation of the endoderm has an important consequence for development in that it determines the orientation of the primitive streak. This can be shown by removing the endoderm from a blastoderm at the early primitive streak stage and then replacing it at right angles to its original position. The primitive streak now completes its formation but in so doing becomes bent at its anterior end toward the original anterior end of the endoderm (Waddington, 1932; Vakaet, 1967). Thus, the endodermal layer plays a rôle in directing the formation of the primitive streak and of the cell movements in the upper layer (see also experiments of Lutz, p. 75).

The mesoderm itself is formed as a result of the epiblast cells converging on the midline and invaginating (Fig. 29). As they invaginate and transform into mesoderm these cells become elongated and flask-shaped; pinocytotic vesicles appear in them that are so small as to be visible only by electron microscopy (Balinsky and Walther, 1961; Ruggeri, 1966). Similar changes take place in the invaginating cells of amphibian gastrulae, so it is possible that they are related to the migratory movements.

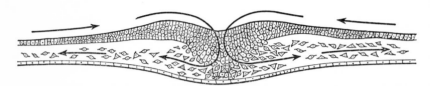

Figure 29. *Transverse section of chick embryo to show invagination through the primitive streak; the arrows indicate the path taken by the invaginating mesoderm. (After Bellairs, 1960.)*

It is not known whether some aspect of the gastrulation movements in themselves is essential for converting the ectoderm cells into mesoderm in the chick. One could, for instance, imagine that the effect of stretching of the cell surface might affect the structure of the cytoplasm and thus lead to a switch in protein synthesis. Waddington and Taylor (1937), who implanted pieces of ectoderm from the periphery of the area pellucida directly into the primitive streak, concluded that the movements were necessary because they involved the breakdown of the epithelial structure.

It is noteworthy that during the time when they are invaginating the cells are continually in contact with one another (Balinsky and Walther, 1961). Special modifications of the cell membranes are visible where the cells touch one another (Balinsky and Walther, 1961; Trelstad *et al.*, 1967). The possible significance of these points of contact as regions where communication may take place between the moving cells has already been discussed (Chapter 1).

The possibility that some sort of extra-cellular layer, such as a surface coat, is present to control the integration of cell movements has sometimes been postulated. It has also been suggested by some authors that intercellular bridges may exist between certain mesodermal and ectodermal cells. But neither of these suggestions has been borne out by electron microscopical studies (e.g. Bellairs, 1958; Trelstad *et al.*, 1967).

(g) The presumptive areas

With the formation of the primitive streak the cells of the area pellucida have become arranged in three main layers, ectoderm, mesoderm and endoderm, and the presumptive areas are now laid out in positions where they can start to develop into their respective tissues. Thus, marking experiments aimed at plotting the cell movements have also given information about the layout of the presumptive areas at different times.

Once again although there is agreement about the relative positions occupied by the presumptive areas, there is no unanimity about the details of the maps. In particular there has been a controversy about the position of the presumptive areas at the primitive streak stage. On the one hand, Pasteels (1937b) and Malan (1953) consider that all the presumptive axial areas lie far back in the area pellucida and are brought anteriorly by the (disputed) forward migration in the midline (see Fig. 30 of Malan). Thus, the presumptive notochord is placed far back in the area pellucida at this stage. On the other hand, Spratt (1946) believed that almost the entire posterior half of the area pellucida invaginated through the primitive streak and a presumptive map based on his results was published by Rudnick (1948); in this, the presumptive notochord

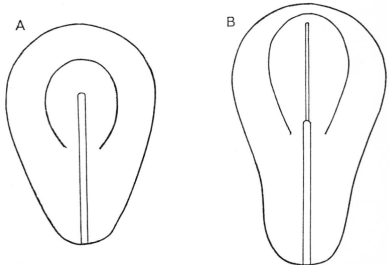

Figure 30.   *Presumptive foregut area. (Shown as horse-shoe shaped region.) A. At the primitive streak stage. B. At the head process stage. Area opaca not shown. (After Bellairs, 1953.)*

occupies almost the same position as it does at the full length primitive streak stage.

There has been less controversy about the presumptive maps of the later stages of the chick blastoderm, and those prepared by Pasteels (1937b) are illustrated in Fig. 27. These maps are concerned only with the ectoderm and mesoderm layers; the layout of the presumptive foregut is illustrated in Fig. 30.

The fact that so many attempts have been made to map the presumptive areas is a measure of their importance to the embryologist, for it is necessary to understand something of the normal development of an organism if we are to interpret the abnormalities that result from our experiments. The significance of the presumptive areas is that they enable a fairly accurate forecast (or presumption) to be made about the subsequent fate of each region of the blastoderm. It should be noted, however, that the presumptive fate is most likely to be achieved if the blastoderm remains intact. If pieces are cut from the blastoderm and grown in isolation from one another they may not develop according to their presumptive fate. For this reason reliable presumptive maps cannot be drawn up from the results of isolation experiments.

(h) Isolation experiments

An interesting line of experimentation in the 20th century has been the study of how small pieces of tissue behave when they are isolated from their normal surroundings. The tissues can be maintained in complete isolation from other types of cells by growing them in tissue culture, or they can be transplanted to the chorioallantoic membrane of another egg. A full discussion of these and other techniques is given by New (1966) and by Wilt and Wessels (1968).

The *in vitro* method most commonly used for explantation is the watch-glass method of Waddington (1932) which has subsequently been modified by others (New, 1955; Spratt, 1947; Gallera and Nicolet, 1961; Gallera and Nicolet, 1963). Whole blastoderms, or fragments of them, can be grown by these techniques for periods of up to about 48 hours (Fig. 31). The technique of explanting on to the chorioallantoic membrane can only be used for relatively small pieces of tissue but these can be grown for periods of up to about ten days.

The effect of isolating pieces of blastoderm has been extensively studied, especially by using the technique of chorioallantoic grafting. The results are of special interest when they are compared with the results of marking experiments, for pieces of blastoderm tend to give rise

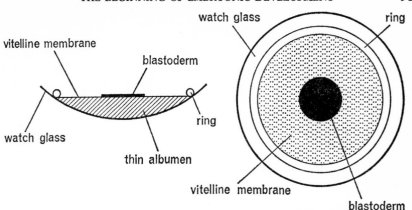

Figure 31. *The New culture technique for young chick embryos. The vitelline membrane, with the blastoderm still attached to it, is pulled taught around a glass ring and placed in a watch glass. Left. Watch glass seen from side. Right. Watch glass seen from above. Albumen is placed beneath the vitelline membrane. The whole watch glass is then enclosed in a petri dish in a moist atmosphere. (From New, 1955.)*

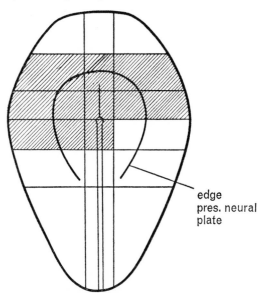

Figure 32. *Isolation of tissues on the chorioallantoic membrane. The chick embryo was cut into pieces, each of which was grown separately as a graft on the chorioallantoic membrane of another egg. The shaded areas are those from which neural tissues were obtained. Superimposed on the drawing is an indication of the region of the presumptive neural plate shown in Fig. 27. (After Rawles, 1936.)*

to a wider variety of tissues when they are isolated than when they remain *in situ*. For instance, although the presumptive neural plate is limited to the anterior part of the primitive streak and the regions on either side of it (Fig. 27), Rawles (1936) obtained neural tissue from isolates taken from a much larger region of the area pellucida (Fig. 32). Thus, although the results of the isolation experiments cannot give us accurate information about presumptive areas, they demonstrate that the embryo is capable of extensive regulation at this stage.

C. SUMMARY

This chapter covers the period of development in the chick embryo from the onset of cleavage until the stage when the primitive streak is fully formed. During this time the embryo is capable of extensive regulation. Recent experimental work on the formation of the endoderm shows that it develops in two stages. Firstly, there is a forward migration of cells from the posterior germ wall which gives rise to the endoderm of the area opaca as well as to the primordial germ cells. Secondly, the endoderm of the area pellucida develops from the lowermost cells of the primitive streak.

Even at these early stages, the cells co-ordinate their activities. In particular, the mesoderm and the endoderm appear to affect each other's development.

Information on the nucleic acids of the early chick embryo is given in Chapter 6.

LAYING DOWN OF THE ORGANS IN BIRDS

WE have already seen that one of the main factors affecting embryonic development is the interactions that take place between cells. In this and the following chapter, in which the laying down of the organs will be considered, two important mechanisms for interaction and co-ordination of development will be discussed. These are (Chapter 5) by the mass migrations of cells accompanying node regression and (Chapter 6) by the processes involved in embryonic inductions.

A. REGRESSION OF HENSEN'S NODE

The earliest experiments on bird embryos were performed *in ovo* and consisted essentially in damaging or interfering with the developing embryo in some way. For instance, Peebles (1898) inserted small sable hairs into the chick blastoderm and then subsequently noted how the hairs had shifted into new positions when the egg had been re-incubated. She concluded that 'as the embryo elongates, the position of the primitive streak changes by a pushing back'. Although the true significance of these results was not realised for many years, this was the first demonstration of node regression.

Node regression begins shortly after the primitive streak is fully developed and consists essentially of the events shown in Fig. 33. At the full length primitive streak stage the presumptive notochord is situated in the mesoderm in and around the node (Fig. 27). Some of this material migrates anteriorly beneath the ectoderm and endoderm to form the beginning of the head process but the remainder moves posteriorly. This posterior migration (or regression) is not limited to the node but occurs also in the tissues on either side of the primitive streak; the regression is, however, maximal in the midline, i.e. down the primitive streak.

The significance of node regression is not just that it leads to formation of notochord but that it is closely followed by the laying down of the initial organs of the embryonic axis (Fig. 34). Indeed, it seems possible that regression is necessary for axis formation to take place and that the movements themselves act as a controller of development at this stage (p. 103).

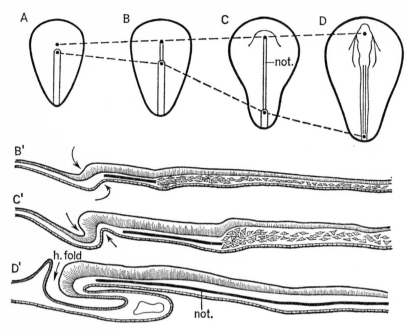

Figure 33. *Regression of Hensen's node A, B, C and D. Diagrams of experiments to show regression of Hensen's node and the formation of the head. (After Spratt, 1947.) A. Cells at the anterior end of the primitive streak and just anterior to it were marked with small lumps of carbon particles, shown here as black dots. During the following stages Hensen's node was seen to move backward and the marked cells anterior to it were seen to move forward. The increasing distance between the two lumps of carbon is indicated by broken lines. (After Spratt, 1947.) B^1, C^1, D^1. Longitudinal sections through three of the stages shown above, illustrating the formation of the head fold and foregut. Arrows indicate the folding movements leading to head fold and foregut formation. As the node regresses, the notochord is laid down. (After Bellairs, 1960.)*

(a) Evidence for regression

It has occasionally been suggested that regression of the node is not a real, but an apparent phenomenon (Jacobson, 1938b). Other experimental evidence, however, favours the theory of genuine migration of node tissue. Several workers who have marked different regions with vital dyes (e.g. Gräper, 1929; Wetzel, 1936; Pasteels, 1937a) or with powdered carbon (Spratt, 1955, 1957b) have all concluded that regres-

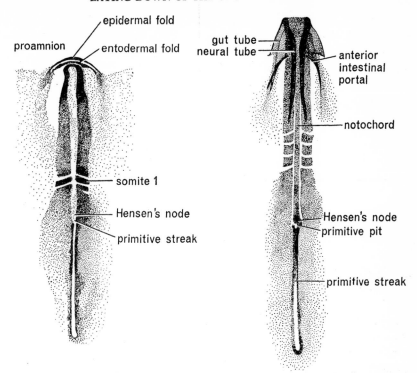

Figure 34. *Surface views of chick embryos at two stages during node regression. (After Nelson, 1953.)*

sion occurs. More recently, the problem has been re-investigated by Nicolet (1965, 1970) and Rosenquist (1966) using autoradiographic techniques, and further evidence in favour of regression has been obtained.

(b) Effects of regression

There is evidence that regression is not confined to the node and can, indeed take place in the area pellucida in the absence of the node (p. 115). Furthermore, a considerable amount of axial differentiation can occur in blastoderms from which the node has been removed (Wetzel, 1936; Abercrombie and Bellairs, 1954; Grabowski, 1956). Typically, in these nodeless embryos the trunk region is marred by a split; this split is formed by the backward extension of the wound left in the blastoderm when the node was extirpated (Fig. 3). On either side of the split a partial axis is present which occasionally contains fragments of notochord.

Anterior to the split, the head may be deformed. Behind the split region the trunk is well developed and may possess some notochord. This notochord probably develops only if the operation fails to remove all of the node. It is, however, possible that some regulation of notochordal tissue has occurred in these embryos.

B. REGULATION OF HENSEN'S NODE

It has been demonstrated many times that if, in the experiment described above, the hole left in the blastoderm heals up, a normal embryo can form, and it is generally concluded that the entire node has regulated (Waddington, 1932, 1952; Grabowski, 1956; Spratt, 1955). A regulation of this type occurs only in embryos operated upon at the primitive streak stage and does not appear to be possible when the operation is carried out at the head process or somite stages. The ability of the node to regulate was also demonstrated by the experiments of Abercrombie (1950) (p. 114). A fragmented node is also able to regulate well if the fragments are replaced in the hole in the blastoderm (Shoger, 1960) or even if they are just placed close together on the surface of an albumen-agar clot (Spratt, 1957a). A pre-requisite for this regulation, however, appears to be that the node should have been cut up with a knife rather than have been disaggregated with EDTA (Shoger, 1960). The effects of disaggregation on nodal tissue have also been studied by Zwilling (1960b, 1963; see below).

Under certain circumstances a node will form by regulation in an unusual situation. For instance, Abercrombie and Bellairs (1954), who replaced the node by a graft of presumptive lateral plate mesoderm taken from the posterior end of the primitive streak, found that a double-headed embryo was frequently formed (Fig. 35). Many of these embryos possessed notochord in one or both of the anterior twinned axis as well as in the posterior single axis. These authors argued that since in the normal embryo notochord is only formed from nodal tissue, the presence of notochord in the experimental embryo implied that a new functional node had formed by regulation on either side of the implanted graft. The alternative explanation, that the two notochords were derived entirely from fragments of node that had not been extirpated, seems improbable; the total volume of notochordal material was just as high in an embryo that was doubled in this way as in a normal single one. Furthermore, the incidence of twinning was reduced if the graft was taken from different regions (Fig. 35), though the reason is not understood.

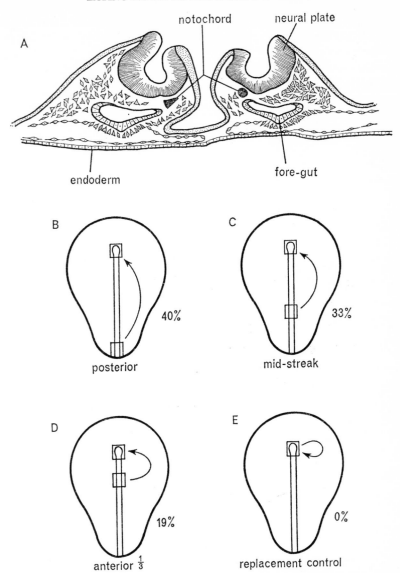

Figure 35. *Experimental production of double-headed chick embryo. A. transverse section across the anterior end of the embryo showing two embryonic axes, each with its own notochord. (From Abercrombie and Bellairs, 1954.) B, C, D and E. Scheme of operations. The incidence of twinning (shown as percentage of total experiments) was reduced as grafts were taken from regions lying progressively nearer and nearer the node. (From Bellairs, 1969.)*

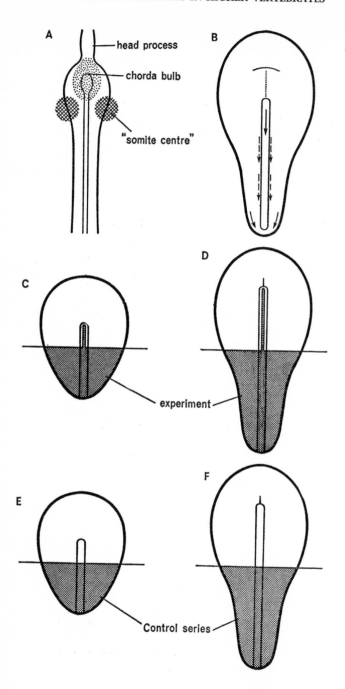

C. THE FUNCTIONAL NODE

These experiments emphasise a difficulty. We can see the node as a morphologically distinct region and we know that this contains the entire presumptive notochord material. But the node is also important in playing a controlling rôle in development, even though as we shall see (p. 114) this control is subject to that exerted by the area pellucida as a whole. Our problem is now to try to understand something of the extent of this control. In other words, how far does the 'functional node' extend beyond the visible one?

This is a problem that has been especially investigated by Spratt (1955, 1957a, b) who has introduced the concept of chorda and somite* 'centres'. He considers that these are part of a general 'organiser centre' located around the anterior end of the primitive streak. He described the chorda centre as corresponding with a morphologically visible structure, the chorda bulb (Fig. 36). 'It is a single, median chorda centre which lies beneath the primitive pit and extends anteriorly to the edge of the node where it becomes continuous with differentiated notochord.' Further, 'It is about twice the width of the notochord proper and often has a rather sharp posterior boundary usually directly below the posterior edge of the pit'. He concluded that the chorda centre played an important rôle in notochord formation (see below).

By contrast, the somite centres are not morphologically distinct (Fig. 36) and their presence is simply deduced by Spratt from his experiments. He concluded that the somite centres extended 0·2–0·3 mm on either side of the primitive groove and that they regressed along with the node and induced somites from presumptive somite mesoderm.

* Spratt's 'somite centre' is discussed on p. 102.

Figure 36. *Formation of somites in the chick embryo. A. Concept of Spratt. This author suggests that a general 'organiser' is located at the anterior end of the primitive streak, comprising two somite centres and a chorda bulb. B. Regression movements during the head process stage of development C, D, E and F. Experiments to test the 'somite centre' hypothesis. The stippled region is explanted, the white region discarded. If that part of the primitive streak that lies between the "somite centres" is retained (as in C and D) regression movements and stretching of the posterior region of the area pellucida continue and somites develop. If the anterior part of the primitive streak is not retained (as in E and F) no stretching and regression occur and no somites develop. (From Bellairs, 1963b.)*

D. FORMATION OF THE NOTOCHORD

The evidence that the notochord forms as the node regresses is nowadays undisputed. Most authorities are of the opinion, as a result of marking experiments, that the notochord (including head process) arises from the node and from the region of the primitive streak immediately behind it (Pasteels, 1937b; Wetzel, 1936; Nicolet, 1965a; Rosenquist, 1966). The exception is Spratt (1955, 1957a, b) who considers that other cells also contribute to the notochord. He regards the so-called chorda bulb as a group of specialised cells which as they regress rapidly down the streak overtake, pick up, and transform (i.e. induce) additional mesodermal cells to form notochord. Leaving these transformed cells behind, the chorda bulb passes on, transforming other cells lying further posteriorly. His evidence is of two types. Firstly, when he stained any part of the anterior two thirds of the primitive streak with Nile blue sulphate the marks were subsequently found in the notochord (1955). Similarly (1957b), when carbon marks were jabbed into the mesoderm of the anterior half of the primitive streak of a head fold embryo they sometimes became incorporated into the notochord.

Secondly, in another series of experiments (Spratt, 1957b) he marked the chorda bulb with carbon particles, and the region some way behind the node with carmine (Fig. 37). Subsequently, most of the carbon was found still in the chorda bulb whereas some of the carmine was in the notochord. He concluded that as the chorda bulb regressed the carmine-marked cells were picked up and transformed into notochord. Spratt

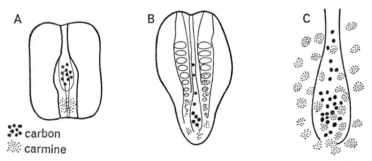

Figure 37. *Formation of the notochord in the chick embryo, according to Spratt (1957b). A. Hensen's node was marked with carbon particles and the region some way behind it was marked with carmine particles. B and C. enlargement of node. As the node regressed some of the carbon particles became enclosed in the notochord. (After Spratt, 1957b.)*

suggested that these cells were all organised into notochord by the chorda bulb, and he based his idea on the fact that when he transected the blastoderm he obtained notochord only in those isolates that contained at least part of the chorda bulb (1955). In view of the great elongation undergone by the developing notochord, he concluded that the chorda bulb was a centre of proliferative activity.

The main difficulty in accepting Spratt's hypothesis is that no other author who has placed marks on the mesoderm in the middle of the primitive streak has subsequently found them in the notochord. For instance, Pasteels (1937b) found the marks later in the somites. Similarly, in a more recent analysis by autoradiography, Rosenquist (1966) and Nicolet (1970) both found that marked cells posterior to the node (i.e. within the anterior $\frac{1}{3}-\frac{1}{2}$ of the primitive streak) were subsequently incorporated into the somites. Conversely, it was not possible to obtain marked cells along the length of the notochord other than by applying the marks to the node (Nicolet, 1965a; Rosenquist, 1966).

It will be apparent that these recent experiments throw considerable doubt on the chorda centre hypothesis, for if there is normally no contribution to the notochord from the primitive streak, it is not surprising that notochord fails to develop in pieces of blastoderm that lack nodal tissue. It would seem advantageous therefore to abandon the idea of the chorda centre. In doing so, however, we are still faced with an important problem, namely that of explaining how the node, which consists of a small number of cells, can give rise to the much larger volume of the notochord. Most, though not all, investigators who have carried out mitotic counts (p. 259) have found no evidence for an increased proliferation rate. Jurand (1962), who carried out a careful study of the chick notochord, concluded that changes in the closeness of cell to cell contact were one of the main factors in causing the chorda-mesoderm to become fashioned into notochord. It would be interesting to know if these changes lead to an increase in the antero-posterior length of individual cells. If they do not, it is difficult to understand where the extra material comes from.

E. FORMATION OF THE SOMITES

Several theories have been put forward to account for the way in which the somites form from the undifferentiated presumptive somite mesoderm. Three of these will be considered. Each ascribes an essential, inductive rôle to a special region of the embryo.

(a) The rôle of somite centres (Spratt, 1955, 1957a, b; Fig. 36)

Spratt's evidence was based on various types of experiment. In the main experiment however, he transected the area pellucida and found that if the cut was about 0·3 mm posterior to the primitive pit, somites failed to form in the posterior part. He concluded that somites failed to form because the posterior presumptive somite material was not in contact with the somite centres. He emphasised the importance of the regression of the somite centres in this process. In examining the results of the many experiments carried out by Spratt, however, the reader may feel that although the general process of regression throughout the area pellucida is an essential for somite formation, the regions known as the somite centres are not. Two experiments have therefore been designed specifically to test this possibility (Bellairs, 1963b). The first is illustrated in Fig. 36 C and D. In this the blastoderm was transected across the presumptive somite region but a very narrow strip of primitive streak, about 0·1 mm wide, was allowed to remain attached to the posterior portion. This strip, which was made as narrow as possible in order to eliminate the 'somite centres', was retained to promote regression movements down the primitive streak. The majority of these specimens formed somites in the posterior piece, whereas in all the control experiments (Fig. 36 E and F) where no part of the anterior streak was included, no regression took place and no somites formed.

The second experiment consisted of removing an area from one side of the anterior primitive streak and area pellucida. This extirpated piece was designed to be considerably larger than Spratt's somite centre. Typically, a split embryo formed which possessed somites even on the operated side. As a result of these experiments, Bellairs (1963b) concluded that the so-called 'somite centre' was not essential for somite formation but that regression movements were necessary. It was suggested that the concept of the somite centre be abandoned.

(b) The rôle of Hensen's node and notochord

It has at times been suggested that the node itself (e.g. Nicolet, 1967) or the notochord directly induce somites from the presumptive mesoderm. There is little support for this idea nowadays since many authors have reported the presence of somites in notochordless embryos (e.g. Waddington, 1932; Waterman, 1936; Wolff, 1936; Abercrombie and Bellairs, 1954; Spratt, 1955, 1957b; Grabowski, 1956; Fraser, 1960).

Moreover, attempts to induce somites from dissociated mesenchyme cells by the insertion of notochord grafts produced at best a non-specific clumping of cells (Fraser, 1960).

(c) The rôle of the neural tissue

The possibility that neural tissue plays a rôle in somite induction has been suggested by several authors (e.g. Fraser, 1960). The evidence is based mainly on the fact that the two tissues are usually closely associated. Bellairs (1963b) however described seven specimens which, while totally devoid of neural tissue, still possessed somites.

(d) The rôle of regression movements

The final idea is that regression movements are essential for somite formation. This concept was put forward by several authors as a corollary to one or other of the theories outlined above (e.g. Spratt, 1955; Fraser, 1960). The experiments of Bellairs (1963b) (described on p. 102), however, suggested that regression movements were the main influence in promoting segmentation of the somites. It seems likely that the changes in cell shape which occur during the regression movements lead to important alterations in the surface properties of the cells. These, in turn, may affect the internal structure and metabolism of the cells. We do not know whether this theory may be applied to all vertebrates. It may be noted, however, that although there is no direct counterpart of node regression in amphibians, there is, nevertheless, a general elongation and stretching of the embryo at the time of somite formation.

F. SUMMARY

This chapter illustrates one of the ways in which cells co-ordinate their activities in the laying down of the organs. This is by an active migration down the primitive streak (regression movements) of most of the cells of Hensen's node to form notochord. This migration also occurs in the cells on either side of the primitive streak and is controlled by the area pellucida as a whole. Regression movements probably play an important rôle in stimulating the differentiation of the somites. Regression movements are a characteristic of amniote embryos and there is no direct counterpart of them in Amphibia.

CHAPTER 6

EMBRYONIC INDUCTIONS

PERHAPS the most famous embryological experiment ever performed was carried out, not on amniotes, but on amphibians. Spemann (1918) grafted a piece of the dorsal lip of an amphibian blastopore into the flank of another similar embryo at the same stage of development and obtained there a secondary embryonic axis (Fig. 38). Subsequently, Spemann and Mangold (1924) modified the experiment so that they were able to decide which of the tissues of this axis were formed from the host and which from the graft. This was done by taking the graft and host from two separate amphibian species which differed from one another in their intra-cellular pigmentation; it was then possible to see which were host tissues and which were graft. It was found that most of the neural tissue was formed from the host and that, in general, there was a collaboration of host and graft tissues to form a unified axis.

It became apparent that the graft had influenced the host tissues around it to form an embryo. Grafts taken from other parts of the donor embryo could not achieve this result, so that Spemann was able to conclude that the dorsal lip of the blastopore possessed special properties which enabled it to 'organise' the tissues around it to form an axis.

The effect of these classical experiments was to stimulate a great deal of embryological investigation. The story of the excitement engendered

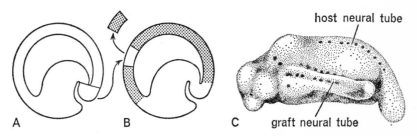

Figure 38. *The classical 'organiser' experiment. A and B. A graft from the dorsal lip of a lightly pigmented amphibian gastrula* (Triturus cristatus) *was implanted into the flank of a heavily pigmented amphibian gastrula* (Triturus vulgaris) *(shown in section). C. A secondary embryonic axis formed. (After Spemann and Mangold, 1924.) In sections (not shown), it was possible to identify the tissues that had formed from the graft because they were heavily pigmented.*

by this early work and the disappointments that followed are well known. Much of the early work was reviewed by Spemann (1938, reprinted 1965). The work done after 1938 has been analysed in some detail by Saxén and Toivonen (1962). A rather briefer account will be found in the book of Weber (1965). Our present task is to consider the relevance of the amphibian work to our understanding of similar processes in amniotes. Our first question must be to ask whether amniotes resemble amphibians in having 'organisers'.

A. TERMINOLOGY

To some extent, our answer to this question depends on how we define the term 'organiser' as well as on the way we use the related term 'induction'. Spemann (1938) defined the term 'organiser' as follows: 'I will use, at present, the term only in a descriptive sense for the whole region of the presumptive chorda-mesoderm which, . . . after transplantation . . . is capable of inducing a secondary axis'. Others, however, have employed it in a wider sense as a means of describing a supposed chemical substance passing from the graft to the reacting tissue. The term has been used also as being synonymous with the term 'inductor'.

An 'inductor' may be defined as a tissue (or indeed any other agent) that brings about an induction; for instance, 'the optic cup calls forth the formation of the lens which, without its influence, would not come into existence. It *induces* the lens in the epidermis; the lens arises through *induction*' (Spemann, 1938). In other words, whereas the word 'organiser' is usually used to refer to the tissue which initiates the development of the embryonic axis under 'normal' conditions in the embryo, the term 'inducer' is applied to any tissue, or agent, that can initiate development in another tissue or organ or even entire embryonic axis.

It might be argued that since a considerable volume of work has been carried out on the amphibian organiser and induction systems since Spemann's definitions were published, more recent definitions would be appropriate. However, in view of the confusion that has surrounded these terms most recent authorities have apparently been reluctant to re-define them, and for this reason I shall use them as originally used by Spemann. However, it is relevant to note that the term 'primary induction' has frequently been used to mean the activity of the 'organiser' region in amphibians. The term 'secondary induction' is generally used to refer to inductions that take place later in time than the 'primary induction' and tend to be more specific (e.g. the induction of the lens by the eye-cup). The terms 'acting' and 'reacting' tissues refer, respectively, to the inductor and to the tissue on which it acts.

B. EVIDENCE FOR EMBRYONIC INDUCTIONS

The existence of a specific embryonic induction system is generally demonstrated by means of one or more of a series of experiments as follows:

(a) If the inductor of an organ is extirpated or destroyed then no development of that organ will occur (provided, of course, that the operation has been carried out before the inductor has exerted its effect). For example, the eye-cup is the normal inductor of the lens. If the eye-cup is destroyed the lens fails to form (Spemann, 1938).

(b) If the reacting tissue is transplanted away from the inductor or cultured in isolation it will not be induced and will not develop along its normal path. The ectoderm from which the lens forms is a reacting tissue. If it is isolated from the eye-cup it fails to form a lens.

(c) If the inductor is transplanted to another part of the embryo it brings about an induction in tissue that would not normally behave in this way. If the eye-cup is grafted beneath flank ectoderm it may induce a lens in that tissue.

(d) If the inductor is combined *in vitro* with tissue with which it would not normally react, an induction takes place.

In general it is necessary to have contact or fairly close apposition between the inducing and reacting tissues, though exceptions to this rule do exist. It is, for instance, often possible to obtain an induction in a tissue that has been separated from its inductor by a Millipore filter (Fig. 39). A more dramatic exception, however, is that furnished by Niu and Twitty (p. 119).

C. EMBRYONIC INDUCTION IN BIRDS

One of the results of Spemann's demonstration of primary induction was that it became important to establish whether this type of system was restricted to amphibians or was universal throughout the animal

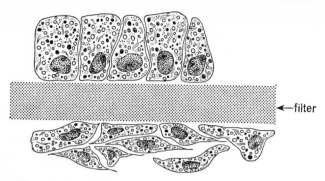

←—filter

Figure 39. *Testing for the presence of a chemical inductor in an embryonic induction system. The acting and reacting tissues are separated by a Millipore filter.*

kingdom. Although it quickly became apparent that similar sorts of inter-tissue influence existed in certain invertebrate embryos, it was not until 1932 that definite proof of embryonic induction was obtained for any amniote embryo. The reasons for this delay were technical ones brought about by the difficulty of implanting grafts into the large, yolky eggs of birds or reptiles, and by the impossibility of operating upon young mammalian embryos *in utero*. The first of these obstacles was overcome when Waddington (1932) demonstrated that chick blasto-derms could be dissected away from the yolk and grown *in vitro*, using the watch-glass technique of Fell and Robison (1929). The technique of New (1955) (p. 90) is a modification of Waddington's technique. Sub-sequently, it was found that mammalian blastocysts could also be grown in this manner (Waddington and Waterman, 1933). Under these conditions it now became possible to carry out a wide range of experiments. Waddington (1932 and later: summarised by Waddington, 1952) thus showed that in general the same sort of processes were at work in birds as in amphibians. In particular, he demonstrated that if the anterior end of the primitive streak (Hensen's node) was extirpated, stripped free of any adherent endoderm and implanted beneath the ectoderm at the side of the area pellucida of another embryo it could induce there a secondary embryonic axis* (Fig. 40). The ectoderm in this region does not give rise to neural tube in the normal embryo, and so it was concluded that the graft had the power to act as an embryonic inductor and was to some

* The term 'embryonic axis' is usually applied to the median structures of the embryo; viz. neural tube, somites, gut and notochord. An induced axis may however sometimes be deficient in one or more of these tissues.

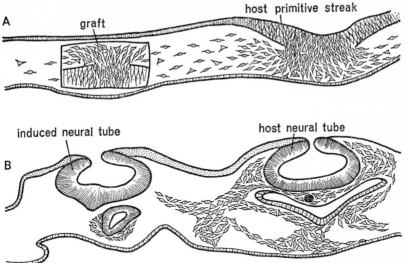

Figure 40. *Embryonic induction in the chick embryo. A. A graft from the anterior end of the primitive streak of a donor embryo is inserted beneath the ectoderm of the host. B. After further incubation for about 24 hours, the graft has induced a secondary axis.*

extent, therefore, comparable with the amphibian primary organiser. In both birds and amphibians the induction appeared to be brought about by the mesoderm of the graft rather than by the presumptive neural plate ectoderm. A similar process of neural induction was also shown to occur in a mammalian embryo (p. 152).

It soon became apparent, however, that the ability to induce a secondary axis was not restricted to Hensen's node but was present in at least the anterior third of the primitive streak (Waddington and Schmidt, 1933). Similarly, Bautzmann (1926) had shown that the organizing ability of the amphibian embryo was not narrowly limited to the dorsal lip of the blastopore but extended for some distance around it. It was found in the chick, furthermore, that an embryonic axis could also be induced using grafts of the head process or of the sinus rhomboidalis (i.e. the most posterior part of the neural plate), although the inductive capacity of these regions was not so great as that of Hensen's node.

It was considered by most workers at this time that the anterior part of the primitive streak and, indeed, possibly its entire length, corresponded with the amphibian organiser. In addition, it was found that in birds, as in amphibians, simple neural placodes could be induced by a variety of substances, living, dead and inert (Waddington, 1938a;

Abercrombie, 1939). More recently, Pasternak and McCallion (1962) have shown that neural inductions can be obtained in chick by implants of liver and kidney; Viswanath *et al.* (1968) demonstrated that alcohol-killed Hensen's node is able to induce various tissues from competent ectoderm. In these responses, also, the chick embryo parallels amphibian embryos.

Perhaps the most well-established finding about the node is that its powers of bringing about an induction of neural tissue are greater than that of any other part of the primitive streak, or, indeed, of any other region of the entire blastoderm. Indeed, so striking is the superiority of the anterior end of the primitive streak and of Hensen's node in particular in promoting inductions, that it was believed for a considerable time that regions of the area pellucida other than the anterior end of the primitive streak, head process and sinus rhomboidalis were incapable of doing so. But in recent years there have been demonstrations by several investigators that other regions also have inductive ability.

For instance, neural inductions have been obtained by grafts taken from behind the node, or from regions immediately lateral or anterior to it (Mulherkar, 1958; Gallera, 1966); these regions were probably presumptive somite and presumptive head mesenchyme, according to the map of Pasteels (1937b). Similarly, the middle of the primitive streak (Gallera, 1964; Gallera and Nicolet, 1969; Vakaet, 1964) and the head process mesoderm of embryos possessing not more than four pairs of somites (Gallera, 1965) are also capable of bringing about neural inductions when transplanted to the area opaca of a primitive streak stage host. Grafts taken from the middle of the primitive streak, however, do not induce neural tissue directly; instead, they first induce an additional primitive streak in the host and this then gives rise to an embryonic axis (Gallera, 1968). If some embryonic endoderm is formed by this new primitive streak it may play a rôle in inducing neural tissue (Gallera and Nicolet, 1969).

Most attempts to obtain inductions by grafts taken from the posterior third of the primitive streak have been less successful (Waddington, 1932, 1952) although one case was recorded by Waddington and Schmidt (1933). This difficulty is perhaps not surprising when it is remembered that this region consists largely of presumptive lateral plate mesoderm overlain by ectoderm (Fig. 27) and is not normally capable of differentiating when isolated (Spratt, 1952; Butros, 1960) (but see p. 122).

The induction (or organisation) of an entire embryonic axis (including neural tube, notochord, somites, lateral plate and gut) is clearly a more complex process than the induction of a neural placode, particularly as the axis is generally formed partially by graft and partially by host tissues. Our next task is to consider how this complexity is brought about.

D. REGIONALISATION AND ORIENTATION

(a) Head and tail organisers?

One of the most striking facts about an embryonic axis is that it normally has an orderly arrangement and is not merely a jumbled mass of organs and tissues. Much of the earlier work on amphibian embryos was therefore concerned with the problem of whether or not there are different organisers for head, trunk and tail. The most anterior part of the archenteric roof of amphibian embryos which normally induces the head has therefore sometimes been called the head organiser while the more posterior regions which normally induce trunk and tail have been termed trunk and tail organisers respectively. The problem is, however, a complex one, for the result of any induction is affected by the location of the reacting tissue as well as by that of the inducer. Thus, a graft of the so-called amphibian 'tail organiser', when inserted into the head region of the host embryo, may induce there a secondary head (Hall, 1937).

When we turn to consider the regionalisation of the bird embryonic axis a similar situation is apparent. That is, the result of any induction appears to be a compromise between the actions of the inducer and of the host, and there is no real evidence for the existence of regional organisers. Initially, it was thought possible that, by analogy with the supposed situation in amphibians, the anterior part of the primitive streak might act as a head organiser whereas the posterior part might act as a trunk organiser. There is, however, no evidence that a graft taken from the posterior end of the primitive streak can induce a trunk or tail. Indeed, if it is placed at the level of the host's head, it can very occasionally induce a head (Waddington and Schmidt, 1933). The evidence for an interaction of the host and graft tissues will now be considered.

(b) The position of the node

One of the difficulties encountered in assessing the results of these experiments is that whilst the amphibian embryo which has received a graft will survive, if conditions are suitable, up to and even beyond the stage of hatching, the chick embryo will only live for about 24–36 hours *in vitro*. It is not easy to assess the degree of regionalisation of induced structures in the chick during this limited period. For this reason recent investigators have introduced new methods for enabling the induced structures to undergo more prolonged development. Hara (1961) has combined ectoderm and mesoderm from different levels of the chick blastoderm and grown them as sandwiches within the coelomic cavity of older embryos. Others (Grabowski, 1957; Tsung *et al.*, 1965) have developed techniques for grafting tissues into blastoderms lying *in situ* on the yolk. It is of interest that all these authors appear to have achieved similar results, though their interpretations vary somewhat. Grabowski (1957) and Tsung *et al.* (1965) both took grafts of Hensen's node from full length primitive streak stage blastoderms and inserted them beneath the ectoderm of hosts of the same age. Both sets of authors found that forebrain, mid-brain and hind-brain were induced, and concluded that Hensen's node is initially a head organiser, but that as it retreats it becomes successively a trunk and then a tail organiser. Hara (1961) concluded that the regional determination of the neural tissue was dependent largely on the origin of the chorda mesoderm. Thus, for instance, the compact mesoderm lying just anterior to the node at the full length primitive streak stage induced forebrain and midbrain structures, whereas the posterior head process mesoderm at a slightly later stage induced hindbrain and spinal cord.

There is, however, little contradiction with the findings of Grabowski (1957) and of Tsung *et al.* (1965), for the chorda mesoderm of the head process is, like that of the trunk, derived entirely from the node (Pasteels, 1937b). So, we can conclude that differences in regional character of the induced neural tissue are to a large extent dependent on the position of the node. As we shall see below, this conclusion is related to another, namely that the morphogenetic movements within the area pellucida are a key factor in differentiation of the chick embryo.

(c) Influences on the reacting tissue

When we now consider the influences exerted by the host upon the development of both the graft and of the induced axis, we must return to some of the earlier literature. One of the most important facts to be taken into account is that the anterior end of the primitive streak possesses a considerable amount of autonomy. Thus, if the node and the anterior half of the streak are isolated, either *in vitro* or on the chorioallantoic membrane, the node will regress down the streak and an embryonic axis will differentiate. Furthermore, the orientation of this axis will correspond with the original orientation of the isolate. It will be apparent that in the grafting experiments any influence exerted by the host upon the graft must be one which affects and probably suppresses the inherent developmental tendencies of the graft itself. This becomes especially apparent when we consider the effect of the host tissues upon the orientation of the secondary axis.

The host blastoderm is of course not morphologically uniform throughout, and so the influence it exerts varies in its different regions. Thus, if a graft from the anterior part of the primitive streak is placed at the lateral border of the area pellucida the graft maintains its own

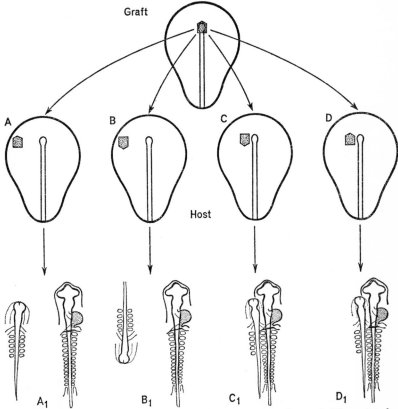

Figure 41. *Effect of different regions of the host on the graft. The pointed end of the graft is the original anterior end. A and B. If the graft is placed near the edge of the area pellucida the induced embryonic axis (shown in A_1 and B_1) has the same orientation as originally possessed by the graft. C and D. If the graft is placed close to the primitive streak of the host, the orientation of the induced axis (C_1 and D_1) conforms to that of the host. (Based on the results of Waddington and Schmidt, 1933.)*

orientation and imposes it upon that of the secondary axis. This occurs whether or not the graft is placed with its antero-posterior orientation to correspond with that of its host, or whether it is placed in reverse orientation to its host (Fig. 41) (Waddington and Schmidt, 1933; Abercrombie and Waddington, 1937). If, however, the graft is placed close to the host primitive streak, the secondary axis that develops has the same orientation as the host, quite irrespective of the original orientation of the graft. Thus, in these embryos, the initial antero-posterior orientation of the graft can be overcome and, indeed, may be reversed by the host. Waddington and Schmidt (1933) concluded that this effect of the host was brought about by the activity of its 'organisation centre', i.e. the anterior end of the primitive streak and possibly the regions immediately adjacent to it.

(d) Subordination to the entire area pellucida

It subsequently became apparent, however, that the primitive streak itself is subordinated to the area pellucida as a whole. Abercrombie (1950) extirpated lengths of primitive streak and grafted them back into the blastoderm in reversed orientation so that the original anterior end, the node, now lay toward the posterior end of the area pellucida (Fig. 42). The reversed piece included the entire dorso-ventral thickness of the primitive streak, but care was taken to maintain this dorso-ventral orientation. It might have been expected that where successful healing followed the operation, the embryo would develop with its head directed towards the posterior end of the area pellucida. Contrary to expectations, Abercrombie found that in a significant number of specimens the orienta-

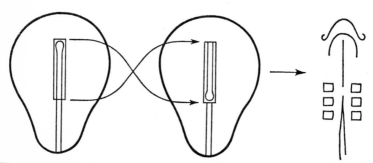

Figure 42. *Subordination of the primitive streak to the entire area pellucida. The anterior 2/3rds of the primitive streak is reversed and grafted back into the blastoderm. In many of the specimens the orientation of the embryo is unaffected by the experiment. (After Abercrombie, 1950.)*

tion of the embryo corresponded with that of the area pellucida as a whole. Thus, the fate and development of the node, as well as that of the anterior part of the primitive streak, was strongly influenced by the surrounding tissues. This influence of the area pellucida upon the orientation of the graft is, however, restricted to its action on the antero-posterior organisation. It is unable to affect the dorso-ventral axis, for Abercrombie found that if the primitive streak was reinserted with its ventral side uppermost its original orientation persisted.

(e) Regression movements and embryonic induction

The influence of the host upon the graft is not limited to its action upon the antero-posterior organisation. There is generally in addition a regional correspondence between the organs of the host axis and those of the secondary axis (Waddington and Schmidt, 1933; Abercrombie and Waddington, 1937; Grabowski, 1956, 1957). Thus, both heads tend to be on the same level in the blastoderm and the somites correspond level for level in the two axes. A further effect of the host is that where the two axes are in close apposition, the tissues may become 'fused' and co-ordinated together (Abercrombie and Waddington, 1937). It is even possible for the host to convert a graft of ectoderm into mesoderm (Waddington and Taylor, 1937). Finally, the host also affects the shape of the graft which tends to become elongated posteriorly (Abercrombie, 1937; Abercrombie and Waddington, 1937; Abercrombie and Bellairs, 1954). Abercrombie (1937) argued that this shape change was due to the active migration backwards of the graft cells and that there had been an induction of this movement in the graft by the host. The implications of this idea are far-reaching and underline for us the major difference in the organiser systems of birds and of amphibians.

We have already seen (p. 93) that during the normal development of the chick embryo, Hensen's node retreats down the primitive streak, paying out notochord as it goes. This movement is to some extent autonomous, for if the primitive streak is removed from the area pellucida and grown in isolation in tissue culture, the node can undergo regression in the normal way.

This autonomous movement of the node, however, can be overcome by its surroundings. For instance, if the node is grafted into the posterior end of the primitive streak of another embryo, the graft is frequently resorbed and fails to differentiate. We can surmise that in these grafts regression has probably failed to occur, since not even notochord has been laid down (Abercrombie and Bellairs, 1954).

E

When we now study the maps of morphogenetic movement, we see that the regression of Hensen's node is merely the high point in a process of cell migration that affects the whole area pellucida (Fig. 28). The node itself is not essential to the movement, despite its degree of autonomy, for if it is extirpated (and not allowed to reform) regression still occurs and a relatively notochordless embryo generally results (Wetzel, 1936; Abercrombie and Bellairs, 1954; Grabowski, 1956). Despite its subordination to the rest of the area pellucida, however, the node is distinguished by the fact that it is the only part of the blastoderm capable of regressing autonomously.

In considering the significance of this regression movement in the process of embryonic induction, it is important to bear in mind that the regression of the node in the chick has no exact parallel in amphibians. In the latter, the presumptive notochordal mesoderm becomes laid out *in situ* beneath the presumptive neural plate during the process of invagination and archenteron formation. In the chick and presumably in all amniotes, gastrulation has undergone modification because of the large amount of yolk that is such a characteristic of most amniote eggs. The laying down of the notochord in birds has thus become separated from the main events of gastrulation. It is difficult to understand the significance of this wave of regression to normal embryonic development, but, as we have seen, the evidence strongly suggests that it plays an important rôle (p. 103).

E. THE TIME SCALE OF INDUCTION

The ability of a particular tissue to react to an inductor is called *competence*. A tissue is competent only during a specific period of time. Indeed, it is fortunate for the embryo that this is so, since uncontrolled inductions might result in the development of abnormalities.

The study of time as a factor in induction has been applied more to amphibian embryos than to chicks, and one of the more significant aspects of it is related to the ability of the host tissues to respond. Woodside (1937) investigated this problem in chick embryos by taking grafts from long primitive streak stage donors and transplanting them to hosts of varying ages, the whole process being carried out *in vitro*. He found that the younger the host at the time of transplantation the greater the response, and he concluded that the host ectoderm is able to react right from the short, broad primitive streak stage, until the time when the neural folds are forming.

But such a conclusion should be treated with caution, since there is no evidence to show that the younger hosts actually responded for a longer period than the older ones, or whether they reacted for the same length of time and merely began their response earlier. It is even possible that the difference in results indicates that the younger hosts were more easily able to overcome the trauma of the experiment.

The most instructive investigation in this field has been carried out recently by Gallera (1964, 1965, 1966) His experiment consisted of explanting orientated grafts of a standard size taken from the anterior primitive streak. Each host normally received one graft implanted into its area pellucida and another into its area opaca. Gallera found, contrary to Woodside, that there was practically no difference in response with increasing age of the host, at least until the development of the head process.

The same author (Gallera, 1966) investigated the effect of implanting grafts taken from donors of different ages, into hosts which were at the long primitive streak stage (Fig. 43). The grafts were of notochord and somites taken from embryos at the head fold stage, and of notochord (of the head process) and associated mesenchyme taken from embryos at a younger stage (Gallera, 1966). He found that the head process induced

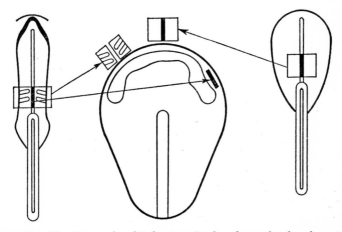

Figure 43. *The time scale of induction. Grafts of notochord and somites were taken from two donors of different ages and were implanted into the same host. The grafts of notochord and of somites from the older donor (left) were less efficient as inducers than were those of the notochord and associated mesoderm from the younger donor (right). Note: this diagram illustrates the experiments that were performed, but does not show the results. (After Gallera, 1966.)*

neural tissue in only about half his experiments, whereas he had pre-viously found that Hensen's node produced inductions in about 95% of cases (Gallera, 1964). Moreover, the inductive power of notochord and somites from the older embryos was even lower than that of the head process. Thus, he concluded that once the notochord (and somites) have invaginated and come to lie in their definitive position they lose a great deal of their inductive ability. In this respect birds resemble amphibians, for the inductive ability of the amphibian archenteric roof also falls with time.

Finally, Gallera (1965) has enquired into the length of time that it is necessary for graft and host to be in contact. He removed the graft at a known time after it had been implanted and subsequently studied the host for inductions. It had already been shown in amphibians that there was a great variation between species in the time required for neural induction to take place. Using the axolotl, *Amblystoma*, neural induction occurred in the host if the graft was allowed to remain in contact with it for only half an hour (Johnen, 1964), whereas in *Triturus* sixteen hours of contact were needed (Gallera, 1959). In the chick, Gallera (1965) found that at least six hours of contact was necessary between inductor and ectoblast in order to obtain a minimal response. Longer contact was required for the differentiation of cerebral structures.

F. THE IMPORTANCE OF CLOSE CONTACT

These findings have especial relevance when we come to consider the importance of contact in induction systems. The situation in amphibians is reviewed by Saxén and Toivonen who point out that, historically, embryologists have had to discriminate between two opposing theories, *viz.* that induction is brought about by diffusion of a chemical or, alternatively, that induction is due to reactions at the cell surface.

The two principal approaches to this problem have been as follows. (1) Studying the effects of interposing some semi-porous material between the two tissues. In amphibians, Saxén (1961) studied the effect of interposing Millipore filters between graft and host and obtained neurulation in some cases. Similarly, Gallera *et al.* (1968) obtained neural inductions when Millipore filters were inserted between the grafted node and the host ectoderm. When these specimens were examined by electron microscopy, microvilli from the graft were found to have penetrated for short distances into the filter but never to have made contact with the host. These authors, therefore, concluded that induction was due to the diffusion of some substance through the filter.

(2) Testing the capacity of the supposed chemical to diffuse through culture medium. For instance, Niu and Twitty (1953), while culturing urodele ectoderm and normal inductor in the same vessel, kept them separated and obtained a certain amount of differentiation. A little differentiation also occurred if the ectoderm was cultured alone, but in a medium that had previously been conditioned with the inductor.

It is perhaps necessary to consider the meaning of 'close' contact. With the advent of electron microscopy and other new techniques it has become apparent that the cell surface is a more complex structure than was formerly supposed. Whether or not we should restrict the term cell membrane to the structure visible in electron micrographs and known as the 'unit membrane' (Robertson, 1963), or whether we should also include intercellular materials that adhere to it, is a problem discussed by many authors (for instance, Curtis, 1967).

The closeness of contact achieved between inducing and reacting tissues may, therefore, be partly dependent on the thickness and properties of the intercellular matrix. There is evidence that when a Millipore filter is inserted between cell layers in a chick embryo, material is exuded into its pores (England, 1969). There is some indication from electron microscopical studies that in the chick, the intercellular matrix may become slightly reduced in thickness during induction of the neural plate by the underlying head process (Bellairs, 1959b).

It has occasionally been suggested in the literature that intercellular bridges develop between inducing and reacting tissues, but no evidence could be found for this by any of the investigators who have examined the tissues by electron microscopy (e.g. Bellairs, 1959b).

G. THE CHEMICAL NATURE OF INDUCTIONS

One of the most tantalising problems of embryology in the 20th century has been that concerned with attempts to identify a chemical inductor. Most of the work was carried out on the so-called primary organiser of amphibians during the 1920s and '30s. Excellent résumés can be found in the texts mentioned on p. 106. Recent work on the chemical basis of induction has been reviewed by Toivonen (1967), Tiedemann (1967a, b) and Yamada (1967). Our present task is to consider what bearing their conclusions have upon the situation in amniotes. Practically nothing is known of the chemical aspects of inductions in reptiles and mammals and our discussion will thus be restricted to the little we know about the processes in birds.

Formerly, it was assumed that a single chemical was involved in induction but this now appears to be unlikely, for different results are brought about by different types of chemical inductor. For instance, killed tissues, whether adult or embryonic, tend to induce an 'archencephalic' (forebrain) type of neural tissue which is not regionally distinct, whereas living tissues bring about specific types of induction depending on the inductor. Guinea pig liver induces archencephalic structures in *Triturus*, whereas guinea pig kidney induces deuterencephalic (mid- and hind-brain) and spinocaudal ones. Guinea pig bone marrow, by contrast, induces only trunk-mesodermal and endodermal structures (see table by Yamada, 1967). Saxén and Toivonen (1962) state (their p. 150): 'There is today strong evidence for the fact that we have to look for several inductive agents with different specific actions, and that these agents are obviously large molecular protein-like factors'. In particular, they conclude that ribonucleoproteins are important in neutralisation.

Other substances that may play a part in induction are the normal metabolites of the cell. For instance, Wilde (1955) has produced evidence to suggest that phenylalanine is essential for the initiation of neural crest development and it is possible that some as yet unidentified metabolites may pass from the inducing to reacting tissue during embryonic induction.

Another current idea is that some component of yolk may act as an inducing agent, at least in Amphibia (Flickinger, 1961; Brachet, 1967; Yamada, 1967). This concept is based largely on the finding of nucleic acids in yolk (p. 41). Although it may also apply to those amniotes possessing yolky eggs, this idea cannot be extended to the yolkless eggs of eutherian mammals.

It is with concepts of this type in mind that various investigators have studied the effects on amniote embryos of nucleic acids and their inhibitors, and of antimetabolites of various types, but it is not always clear whether the results should be considered as evidence of embryonic induction or of a more generalised physiological reaction to changes in the environment. It is of course true that the visible changes in morphology that result from embryonic inductions are brought about mainly by changes in the patterns of protein synthesis in the cells. These are dealt with elsewhere (p. 190). Nevertheless, changes in protein synthesis cannot always be regarded as firm evidence of embryonic inductions. For instance, Amos and Kearns (1962) were able to 'persuade' embryonic chick fibroblasts growing in culture to manufacture a new specific protein when treated with bacterial RNA.

(a) Effects of nucleic acids on the posterior end of the area pellucida

(i) ASSESSING THE RESULTS

In considering the series of investigations described below on the effects of nucleic acids on development and induction of the blastoderm, it is as well to bear in mind that the explanation of the results may turn out to be far more generalised and less specifically associated with the concepts of molecular biology than we may suppose at present. Recent experiments by Sherbet (1963) and by Sherbet and Mulherkar (1963, 1965) remind us that follicle stimulating hormone is capable of calling forth inductive capacity in chick blastoderms. We must remember, as Abercrombie (1937) demonstrated, that the chick resembles the amphibian embryo in responding to a wide variety of inductive agents.

(ii) ON THE REACTING TISSUE

Butros (1963) has studied the effect of various ribonucleic acids on the posterior part of the area pellucida. As we have seen, this region, if taken from sufficiently far behind the node, is incapable of differentiating *in vitro* any axial structures of its own, although with prolonged growth on the chorioallantoic membrane it is possible to obtain muscle, gut and skin derivatives (Rawles, 1936). Axial differentiation can, however, be elicited if anterior fragments of primitive streak are grafted into it (Butros, 1962; Bellairs, 1963b). Butros (1963) cultured pieces of blastoderm obtained from the posterior end of head process stage embryos *in vitro* on a protein-deficient medium to which RNA of various types was added. After two days *in vitro* the fragments were transferred to the chorioallantoic membrane for 8–9 days. The main effect of the various RNAs was to cause early keratinisation, though the reason for this is not clear. Perhaps the most striking feature was that Butros could not obtain neural tissue, notochord or somites in these experiments.

By contrast, when he carried out similar experiments (Butros, 1960) in which the posterior part of the blastoderm was treated with DNA of various origins (calf thymus, rooster testis and 5-day chick embryo) he obtained neural, notochord-like, somite and lateral plate tissues. From these results Butros concluded that neural tissue was developed because of the direct action of DNA (but see Hillman and Niu, below).

A converse approach has been adopted by Gallera (1970) who has incubated chick blastoderms in a physiological saline solution containing actinomycin D (p. 186). He found that although the development of the host axis was delayed, the reacting tissue was still able to respond (by producing a neural plate) if the Hensen's node from a normal, untreated donor embryo was grafted into it. He concluded that neural competence is not under the direct control of nuclear DNA but is dependent on the changes that have already taken place in the cytoplasm.

(iii) ON INDUCTORS

The experiments of Butros (above) were concerned with effects of the nucleic acids on the differentiation of post-nodal fragments themselves. Others have been concerned with the effects that nucleic acids and other substances have on the inducing ability of post-nodal regions. As we have already seen, grafts from the hind end of the primitive streak are not capable of bringing about inductions. If, however, the graft is pre-treated with RNA it may become capable of inducing axial tissues (Sanyal and Niu, 1966). These authors reported that the type of tissue induced depended on the source of the RNA. Thus, grafts treated with kidney and heart RNA induced tubular and vesicular structures respectively, whilst those treated with RNA from chick brain could induce neural tissue. A similar result was obtained by Hillman and Niu (1963). It would be interesting to see if these results can be substantiated by other investigators since they have been subjected to a certain amount of criticism. In contrast to the experiments of Butros, Hillman and Niu found that DNA had an adverse effect in development (see also p. 185).

(b) The –SH containing compounds

In amphibian induction the –SH containing ribonucleoproteins formerly appeared to be important (Brachet, 1950). The implications of –SH compounds in chick induction have been studied by various workers. Lakshmi and Mulherkar (1963) treated chick blastoderms with chloroacetophenone—a specific and irreversible inhibitor of –SH groups—and produced malformations especially in the brain region. They concluded from a paper chromatographic analysis that the agent had a specific effect on the amino acids, glutathione and cysteine. This was confirmed when it was found that the effects of choroacetophenone could be overcome by treating the embryos with glutathione (Mulherkar et al., 1967) or with cysteine (Mulherkar et al., 1965). When Hensen's node grafts were treated with chloroacetophenone their ability to induce neural tissue in normal hosts was reduced. Similarly, when hosts were treated with this reagent the ability to respond to normal grafts of Hensen's node was reduced (Lakshmi and Sherbet, 1964). Conversely, it was found that if post-nodal pieces of the chick embryo, which do not normally differentiate axial structures when cultured *in vitro*, were treated with glutathione or with cysteine, they differentiated neural tissue, somites, notochord and nephric tubules (Chauhan and Rao, 1970). This stimulation could however be inhibited by simultaneous treatment with actinomycin D, an inhibitor of RNA synthesis (p. 186).

I. THEORIES OF INDUCTION

Little though we know of the chemical nature of inducing agents, we are even more ignorant of their mode of action. Saxén and Toivonen have critically reviewed the main theories in relation to primary induction in amphibians, and their concluding ideas may be summarised briefly as follows:

(1) At least two active principles or types of reaction can be distinguished, i.e. neuralisation and mesodermalisation.
(2) It is possible that small-molecule neuralising agents diffuse from the archenteron roof more rapidly than mesodermalising agents of larger molecular size. Thus, differences in distribution of the 'organiser' material might facilitate the formation of two gradients, which by their interactions could lead to regional differences in inductive effect and response throughout the entire organism.

An alternative concept is that of Nieuwkoop (1952, 1967), who suggested that in neural induction there is first an *activation* of the ectoderm, which is then followed by a *transformation*.

Theories of this kind are highly stimulating but serve merely to emphasise our ignorance of the processes—even in amphibians where we are surrounded by a mass of data. We are much more ignorant about the validity of these theories for amniotes. Deuchar (1969) has tested on the chick embryo two factors (Tiedemann's inductors, 583/2 and 583/12) which have a mesoderm-inducing activity on amphibian embryos. The effects produced in the chick were however small compared with those in amphibians.

J. NEW ATTITUDES TO THE PROBLEM OF EMBRYONIC INDUCTION

Recent ideas on the mechanisms controlling the induction processes are derived not so much from the study of primary inductions as from that of the induction of tissues developing at a later stage (secondary inductions). As we shall see, the first indication of differentiation in any tissue is the fact that specialised molecules have been synthesised (p. 190) and this takes place before any histological differentiation is visible. Thus, the first effect of an induction will also be a chemical one.

In some tissues, such as muscle and cartilage, it has been possible to maintain a population of cells in culture that have all been derived from the same initial cell. Such a population is called a *clone*, and its main advantage to the experimenter is that all the cells possess the same genetic make-up. The study of differentiation in clones of cells has led some investigators to re-allocate the emphasis laid on the two components of an inducing system, the reacting tissue and the inducer.

(a) The reacting tissue

It has long been accepted that only certain tissues are capable of responding to a particular inductor. We say that these tissues are *competent*. Recent work is aimed at trying to find out the meaning of competence in biochemical terms (for example, discussion by Holtzer, 1968).

The first probable meaning is that the cell already possesses the appropriate RNAs to enable it to form the proteins specific to that tissue. Thus, to some extent a competent cell is one that has already embarked on differentiation; it has already restricted the number of possible ways in which it can become histologically differentiated, although it still retains an element of choice. For instance, the mesoderm of the chick at the primitive streak stage is competent to form somites or lateral plate, but it has lost the ability to form neural plate.

Thus, it seems that cells have already undergone some differentiation even before any induction takes place. For this reason, it has been suggested that 'there is no such thing as an undifferentiated, uninstructed cell' (Holtzer, 1968), and that the inducers are, therefore, acting on cells which have already received some genetic instruction. So far we possess little direct information to help us understand the changes that have already occurred in competent cells. But it will be recalled that masked mRNA was prepared during oögenesis, ready for use during cleavage (p. 59). Similarly, masked mRNA is perhaps also synthesised at other stages of development in preparation for subsequent events (p. 65). It is possible therefore, that induction is the process that either directly or indirectly unmasks the mRNA. Whatever the mechanism, however, the ultimate effect is that the activity of certain genes is either apparent for the first time or else becomes greatly enhanced as a result of induction.

We have seen that certain tissues are only able to react for a limited period to the stimulus of an inductor. For example, neural plate may be induced in the lateral region of the early area pellucida (p. 108). If the experiment is not carried out, however, until the stage when several pairs of somites have formed, the ectoderm is no longer able to respond, i.e. it has lost its competence. Its potentialities have undergone some restriction, and this is presumably a reflection of further biochemical changes within the cell.

(b) The inductor

The relation between inducing substances and such agents as hormones (p. 272) and growth factors (p. 271) has recently been discussed by Holtzer (1968). He suggests that many of these exogenous substances that affect cells may simply enable cells that have been 'pre-programmed' to proceed with various synthetic processes. Thus, the stimulus is perhaps less specific than we normally suppose. Holtzer has suggested that many of the inductive substances now appear to act in the same way. Certainly, that would appear to be the situation with the experiments in which neural induction can be brought about in certain amphibian tissues by such non-specific agents as sodium or potassium ions (Barth, 1941).

Many of the organ inductors (secondary inductors), however, do appear to have a specific action on the tissues they induce. For instance, kidney (metanephric) mesenchyme will only differentiate if induced by epithelium; but the type of tissue that is formed is determined by the inducer and not by the mesenchyme itself (discussion by Wolff, 1968). Thus it seems unlikely that we shall ever completely abandon the concept of specific inducing molecules. Nevertheless, the time has now come when interest is shifting to a consideration of how the cells become pre-programmed so that they are competent to respond in appropriate circumstances.

K. SUMMARY

This chapter illustrates another way in which cells of amniote embryos co-ordinate their activities in the laying down of the organs by embryonic induction. In this they resemble amphibians.

Many experiments have shown that primary induction of a similar kind takes place in both classes. There is no evidence that intercellular bridges exist at this period but close contact is usually necessary.

Our knowledge of the chemistry of embryonic inductions is briefly considered. It seems likely that in future greater emphasis will be laid on the changes that take place in the reacting tissue *before* induction. In other words, emphasis is shifting to an interest in what makes a cell *competent* to respond to a particular inductor.

CHAPTER 7

SPECIAL ASPECTS OF THE EARLY STAGES OF DEVELOPMENT OF MAMMALS

THE early stages of mammals and reptiles have much in common with those of birds. For instance, they all develop a primitive streak and probably undergo the same kind of morphogenetic movements; at a later stage they all form an amnion.

It is, however, perhaps too easy to stress the similarities and to forget that the three classes have diverged so much that important differences in development have also arisen between them. This chapter is therefore an attempt to emphasise special aspects of the early development of mammals, and the following one summarises some of the peculiarities of reptile embryos.

It is often implied that one of the main differences between eutherian (i.e. higher) mammals and other amniotes lies in the viviparous habit of mammals, but, as we shall see (p. 172), viviparity is also found extensively among the reptiles. Mammalian viviparity does however differ from that of other forms in two important ways. Firstly, mammals are unique in that they suckle their young after birth; secondly, the eutherian mammals alone among the higher vertebrates possess eggs that are practically yolkless. There can be little doubt that the reptilian ancestors of modern mammals laid large, yolky eggs similar to the so-called 'cleidoic' eggs (p. 37) of modern birds and reptiles. As the mammals evolved, however, and developed their own elaborate system of viviparity, their embryos began to receive food through the placenta and, perhaps as a result, the amount of yolk in the egg became reduced. The shell and shell-membranes were no longer necessary to protect the embryo since these functions were largely taken over by the body of the mother, and they have disappeared in most modern mammals.

The reduction in yolk of the mammalian egg has led to other effects on early development. For instance, as we have seen in Chapter 3, mammals avoid super-fertilization by a different method from that of the large-yolked birds and reptiles. Furthermore, the cleavage of mammals differs from that of the large-yolked forms. Thus, in mammals the entire ovum becomes divided up by cell divisions to give a ball of cells (total cleavage), whereas in birds and reptiles, only a small protoplasmic region becomes divided in this way (discoidal cleavage). The yolk-sac of many species (e.g. man) has become a transitory and vestigial structure and in others (e.g. rabbit) it has been adapted to a new function as a placental organ.

Despite these modifications the mammalian embryos still retain features that are characteristic of all amniotes and many of the processes that they undergo bear witness to the fact that their ancestors possessed large-yolked eggs. In particular, gastrulation is strikingly similar to that of reptiles and birds and quite unlike the process in the moderately yolky amphibian egg.

There are, of course, other features of mammalian development that are characteristic of all vertebrates rather than of amniotes in particular. For instance, the early embryos of certain mammals are capable of extensive regulation (p. 135) and appear to be subject to types of embryonic induction (p. 152) similar to those in every other class of vertebrates. Apart from morphological studies, however, research on mammalian embryos has been confined to relatively few species. Although it seems likely that the developmental processes will be found to be similar throughout all the mammals, we should be careful not to assume that this will inevitably be so, for the mammals show a wide structural and functional variation among themselves and should not be considered as a homogeneous group. In particular, the differences in the anatomical and physiological relationships of the placenta are so pronounced (p. 158) that they are likely to have some effect on the embryo.

Before considering the results of recent experimental work on the embryos of eutherian mammals however, we must turn aside briefly to discuss the monotremes and marsupials.

A. MONOTREMES AND MARSUPIALS

The monotremes are entirely oviparous and lay eggs that by mammalian standards are large and yolky. The egg of *Echidna* is about 4·0–4·5 mm in diameter including the shell (Flynn, 1930), whereas that of *Ornithorhynchus* may be 9·0–13·0 mm across (Wilson and Hill, 1907). The effect of this yolk on the earliest stage of development is much the same as in other large, yolky eggs; cleavage is discoidal, a disc of cells forming on the surface of the yolk. This is followed by formation of a primitive streak. Subsequently the mesoderm on either side of the primitive streak in *Ornithorhynchus* develops in an unusual way and forms a series of 'primitive' segments, named as 'protosomites' by Wilson and Hill. According to these authors, the protosomites gradually thin out and disintegrate. They do not give rise to the somites proper.

The marsupials, although viviparous and having much smaller eggs than the monotremes, nevertheless possess a certain amount of yolk which is present as droplets. It is not clear whether this is the same type of inclusion as the fatty droplets sometimes found in the eggs of eutherian mammals such as the cat. More yolky material is however present in marsupial than in eutherian eggs and in consequence they are larger in diameter. The eggs of *Dasyurus* are about 0·25 mm in diameter, and those of *Didelphys* are about 0·13–0·14 mm (Hill, 1910); by contrast, the eggs of the domestic cat, *Felis domestica*, are only about 0·09 mm (Hill and Tribe, 1924). Marsupial eggs differ from eutherian ones, however, in that they possess external to the *zona pellucida* a layer of albumen and a shell membrane, which are laid down in the Fallopian tube and which Hill (1910) has considered to be homologous to the egg coverings of monotreme eggs. Embryologically then, marsupials appear to be intermediate in the amount of yolk they possess between the primitive egg-laying type of mammal, exemplified by present-day monotremes, and the more advanced viviparous type that is found in modern eutherians. The three groups cannot, of course, be placed in order of direct evolutionary sequence and it is thought that the monotremes arose independently from the reptiles, and that the marsupials and eutherians originated from a common ancestral stock of primitive mammals in the early Cretaceous (Romer, 1966).

Unfortunately, our knowledge of the early embryology of the monotremes and marsupials is largely confined to morphological studies. One of the reasons is that some of these animals, such as *Ornithorhynchus* and certain of the marsupials, are rare and consequently are most properly protected by the Australian government. However, this cannot be the sole reason for the small quantity of experimental work, for many marsupials are present in abundance in Australia and South America, and some, like the opossum in North America are useful laboratory animals, ideally suited for research into the late stages of development by virtue of their accessibility at the pouch stage (Burns, 1942). A more telling reason is perhaps that embryology has not attracted many biologists in Australia and the South American countries. A considerable amount of information can, however, be culled from the classical detailed descriptions of Hill and his associates.

B. EUTHERIAN MAMMALS

The earliest stages of development of most species of eutherian mammals are very similar to one another in their morphology and it seems likely that the same sorts of developmental processes take place. The main differences lie in the mode of implantation and in the structure, and perhaps functioning, of the placentae.

Only one example, the mouse embryo, will be considered here in any detail, for this is perhaps the most widely used mammalian embryo. More is known about its development and genetics than of any other species (Grüneberg, 1952; Snell and Stevens, 1966; Rugh, 1967) and pure strains of mice, including mutants, are maintained in many laboratories. Mice and rats are also widely used in tests for the teratological activity of drugs (p. 202). Furthermore, experimental work on the early stages of mammalian embryos has been largely centred on these animals because of their availability.

It is, of course, possible to carry out certain types of enquiry just as easily with mammalian embryos as with those of any other group, provided that the experiments do not necessitate keeping the embryo alive. For instance, the histochemical analysis of the early stages is as well advanced as in any other class of vertebrates. Electron microscopical studies too, have not been hampered by the viviparous habit of mammals. Those aspects of the work that involve maintaining the embryo under somewhat abnormal environmental conditions are the ones that have lagged behind comparable studies in amphibians and birds.

When it is ovulated, the mouse egg is surrounded by a clear capsule—the zona pellucida. Together, egg and zona measure about 0·1 mm in diameter. The first few cleavages occur as the egg passes down the Fallopian tube. It is possible that these early stages may be affected by the hormones present in the tubal and uterine fluids since in rabbits, at least, the cleavage of ova growing *in vitro* can be inhibited by adding 10 μg/ml of progesterone to the culture medium (Daniel and Levy, 1964). It is difficult to culture mouse embryos *in vitro* between the zygote and the blastocyst stages unless they have been allowed to stay for a short period in the Fallopian tube, or are cultured in ovarian bursal fluid.

The mitotic divisions take place more or less synchronously at first and as there is little or no yolk to impede them, the cells that are produced are about the same size; a cluster of cells, known as a morula, forms. In the mouse, 16 cells have appeared by about 60 hours after fertilization. When about 32 cells are present a fluid-filled cavity, the blastocyst cavity (or blastocoel), arises. The entire mass of cells is now known as the blastocyst and consists essentially of a thin-walled vesicle, the trophoblast, which encloses the fluid, and a small mass of cells at one side, the inner cell mass (Fig. 46).

The trophoblast develops into part of the chorioallantoic placenta, and the inner cell mass forms the embryo, the amnion and the yolk sac. By the time the inner cell mass has formed in the mouse, the blastocyst has reached the uterus, lost its zona pellucida and started to implant. As soon as this process has begun, a highly complex relationship with the uterine wall becomes established, although the details are not the same for all species. However, the result is much the same from the point of view of the embryologist. From now on the practical possibilities for an experimental analysis of organ formation become severely restricted. Many have attempted to circumvent these difficulties in two main ways. The first, and more successful, approach has been to carry out the experiments before the stage of implantation.

The second has been to try to maintain the implanted blastocyst under artificial conditions.

(a) Pre-implantation

Simple experiments were performed on young mammalian embryos as early as 1890 by Heape (for summary of older work, Nicholas, 1947). Many refinements of technique have taken place in recent years (New, 1966).

The usual process involves washing out the oviducts of a pregnant female mammal with suitable salines at perhaps 24 hours after fertilization and collecting the embryos into a dish. By this time the mouse embryo is usually two-celled. These early embryos can then be cultured *in vitro* for about two or three days which allows time for operations to be carried out. The conditions of culture, which are more critical than for chick embryos, are discussed by New (1966). In many experiments the embryo reaches the blastocyst stage at which it would normally implant, and if it is maintained in culture after this period it tends to die or become malformed. It is fortunate, therefore, that techniques have been devised for re-implanting mammalian blastocysts into the uteri of other individuals, which of course must be carefully chosen to be in the correct physiological state (see for example Nicholas, 1947; McLaren and Michie, 1956). Under suitable conditions the embryo may implant and develop and subsequently undergo an apparently normal birth.

Other techniques which have been evolved consist of implanting the blastocyst into various adult organs. For instance, Kirby (1962) has implanted mouse morulae beneath the capsule of the kidney. Although some limited development was obtained it does not seem that these procedures are so promising as that of re-implantation into a uterus, though they have yielded useful information on the mechanisms of implantation (p. 142). Techniques for maintaining the already implanted blastocyst are discussed on p. 142.

It seems clear that with the advent of these new methods we are at last on the threshold of a real understanding of mammalian embryos. At this stage of optimism, it is appropriate to enquire whether the development of these early stages is controlled and co-ordinated in the same way as are those of the early stages of other vertebrates.

(b) Developmental influences in the early stages of mammals

(i) PROTEIN SYNTHESIS AND NUCLEIC ACIDS

During cleavage the rate of protein synthesis remains low and there is little growth in size of the embryo, the cells becoming smaller as they increase in number. Quantitative measurements show that in the rabbit embryo the rate of protein synthesis per cell drops a hundred-fold during cleavage, but increases ten-fold when the blastocyst has formed (Manes and Daniel, 1969). The proteins synthesised by the rabbit blastocyst are, however, not identical electrophoretically to those synthesised during cleavage, and it is not clear whether they are specified by mRNA of purely maternal origin (as in amphibians, p. 60, 64) or are directly controlled by the zygote.

In the mouse embryo maintained *in vitro*, H³-thymidine is readily incorporated (Mintz, 1964a; Ellem and Gwatkin, 1968). The synthesis of DNA, however, ceases in these embryos during blastocyst formation (Ellem and Gwatkin, 1968). The latter authors found that after the blastocyst had attached itself to the artificial substrate, DNA synthesis began again. It seems probable that a similar resumption would take place after attachment to the uterine wall, which is an event associated with increased mitosis.

Samoshkina (1965), however, who injected pregnant mice with H³-thymidine, found that no new DNA was formed during cleavage, but that some was synthesised during the blastocyst stage. The striking difference in the results in embryos treated *in utero* and those treated *in vitro* is probably due to the fact that DNA precursors did not have direct access to the embryo when injected into the mother; they were perhaps not easily available until blastocyst formation, when contact with the uterine wall took place.

It will be recalled that a characteristic feature of the cleavage stages of amphibians is that the nucleoli are absent from the cells and there is no synthesis of new ribosomal RNA (p. 61). The mouse embryo (and probably other mammalian embryos also) is strikingly different in that nucleoli are present throughout cleavage (Dalcq, 1954; Mintz, 1964a; Izquierdo and Vial, 1962), and synthesis of new ribosomal RNA apparently takes place from about the 8-celled stage (Mintz, 1964a; Izquierdo and Roblero, 1965; Ellem and Gwatkin, 1968). Both mouse and rabbit embryos are unable to cleave if treated with actinomycin D, a substance that specifically inhibits the synthesis of ribosomal RNA. Mintz (1964a) has suggested that this early synthesis of ribosomal RNA in mammalian eggs may be correlated with their low level of food reserves and their dependence for nutrition on their surroundings.

It seems possible that the explanation is a more complex one, for certain mammals which have yolky eggs, nevertheless also have nucleoli at the cleavage stages. For instance, structures that appear to be nucleoli are already present in the cleavage stages of the marsupial, *Didelphys aurata* (Hill, 1918, his Fig. 15), and are even visible in the highly yolky cells of the monotreme embryo, *Ornithorhynchus* (Gatenby and Hill, 1924, their Fig. 2). Although we have no information about ribosomal synthesis in these mammals, it seems reasonable to suppose that it takes place in all cells that possess nucleoli. It may be that the formation of nucleoli, together with the concomitant ribosomal RNA synthesis, is programmed for an earlier stage in all amniotes than it is in amphibians. Certainly, nucleoli are visible in the mid-cleavage cells of the chick embryo (Bekhtina, 1960), and this corresponds with the time when synthesis of rRNA begins (Wylie, 1970).

Once the nucleoli have appeared in the rat embryo, visible changes take place in them which can be correlated with morphological changes in the cytoplasm. From the two-celled to the eight-celled stage the ribosomes are arranged in a linear manner; after the eight-celled stage when the nucleoli acquire a different structure, the ribosomes are visible in rosette-like groupings (Mazanec, 1965) or as polyribosomes (Schlafke and Enders, 1963). Other events in the segmenting rat embryo include changes in the structure of the mitochondria, which become longer (Mazanec and Dvorak, 1963; Schlafke and Enders, 1963); such visible changes are probably related to changes in the enzyme content of the mitochondria.

(*ii*) ORIENTATION

One of the characteristics of yolky eggs is that from the time of fertilization an upper and a lower side can be distinguished. This is because the cytoplasm is relatively light in weight and floats to the upper, (animal) pole of the egg whereas the heavier yolk sinks to the lower (vegetal) pole. In the typical cleidoic egg, the quantity of the yolk is so much greater than that of the cytoplasm that, as we have seen (p. 36, 39), it occupies most of the egg and the cytoplasm is merely a thin disc at the surface. In amniotes, the animal/vegetal polarity corresponds with the future dorso-ventral axis. Having lost their yolk, the mammalian embryos have no clear-cut animal/vegetal (dorso-ventral) orientation in the morula stage, and do not acquire any until the inner cell mass becomes distinguishable. All the cells appear to be more or less identical in the morula, and it is difficult to understand how regional differences begin to arise. Why, for instance, does the inner cell mass form at one side?

(*iii*) MOSAIC DEVELOPMENT, OR REGULATION?

One possibility favoured by Dalcq (1955) and his colleague Mulnard (1965a, b) was that a certain asymmetry is present right from the beginning of cleavage and even from the stage of the oöcyte. This hypothesis was based on histochemical surveys which seemed to show definite localisations of materials at successive stages. For instance, they found that RNA basophilia, which was first present in the Balbiani body subsequently spread, as in most other types of oöcyte, to the periphery (Fig. 44), and in the rat it could even be traced throughout cleavage until it became most strongly localised in the ectoderm of the inner cell mass (Dalcq, 1955). However, in a similar investigation in the mouse Potts and Wilson (1967) were unable to confirm these findings. On the contrary, they found a higher level of RNA, in the form of ribosomes, visible by electron microscopy, in the trophoblast than in the inner cell mass. Other histochemical findings of Dalcq (1955) were that in the rat mucopolysaccharides eventually become localized in the trophoblast, and acid phosphatases, which are not present at first, gradually become concentrated in the cells of the inner cell mass. In the rat, the latter reaction appears to be associated especially with large granules within the cells, the so-called metachromatic granules of Dalcq. Despite their acid phosphatase content, these metachromatic granules do not appear to be lysosomes (p. 267 for discussion of lysosomes) but are more likely to be multi-vesicular bodies (Dalcq, 1963; Pasteels, 1966). Multi-vesicular bodies consist of a number of small vesicles enveloped in a larger vesicle, and are common in many tissues examined by electron microscopy; their function is not understood.

On the basis of these histochemical findings, Dalcq and his colleagues suggested that in mammals there was a genuine differentiation of cells

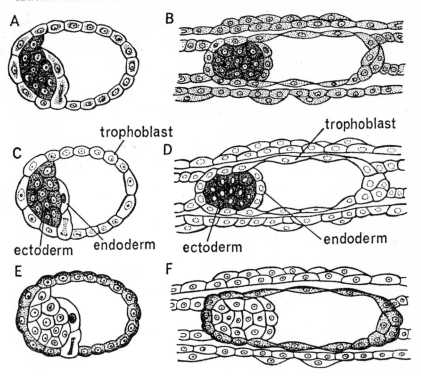

Figure 44. *Histochemical studies on the rat embryo at two stages of the blastocyst. A and B. Ribonucleic acid basophilia. C and D. Alkaline phosphomonoesterase. E and F. Mucopolysaccharides.* (*After Mulnard, 1960.*)

even in the cleavage stage, an event unknown in any other class of vertebrates, although one that is not uncommon in certain invertebrates and protochordates (discussion on 'mosaic' eggs, Chapter 1).

It is important to remember that early localisation of cytoplasmic materials, of the kind demonstrated by Dalcq, is not necessarily correlated with early determination. Indeed, there is overwhelming experimental evidence to show that young mouse and rat embryos are highly regulative. For instance, if one of two blastomeres is killed, a normal embryo may form from the remaining one (Seidel, 1952; Tarkowski, 1959). In a brilliant exploitation of the new techniques described above, Tarkowski (1961) found that if he fused together two mouse eggs before reimplanting them into another mouse mother, they could regulate to form a single embryo (Plate VII). Subsequently it was shown that regulation was possible until at least the 16-cell stage (Mintz, 1965;

Tarkowski, 1965) and that as many as 16 mouse morulae could be united to form an enormous single blastocyst (Mintz, 1965). It is not clear whether these multiple mosaics could give rise to newborn mice of standard size, though when only two eggs are fused, a normal-sized baby mouse is born (Tarkowski, 1961). An interesting aspect of this regulation was that it applied to cells destined to form both embryonic and extra-embryonic tissues. According to Tarkowski (1965) all blastomeres of the 4- and 8-cell stage mouse embryo are capable of forming trophoblast. There can be little doubt that, despite the early localisation of certain substances in the cytoplasm, these embryos are as highly regulative as those of other vertebrates.

How then can we reconcile the histochemical results of Dalcq (p. 134) with these experimental ones? We can perhaps do so if we suppose that a cell will react in a specific way, depending on its position in the mass of developing cells. Thus, as the cells move around, they acquire the histochemical characteristics of the region in which they find themselves. Probably these characteristics should be considered as indications of events in the cell, rather than as demonstrating the segregation of fixed cell types. Tarkowski and Wroblewska (1967) suggest that up to the 8-celled stage, all blastomeres in the mouse are able to form trophoblast and that those cells that form the inner cell mass are simply those that have become enveloped by the trophoblast.

(*iv*) APPLICATIONS OF EXPERIMENTS ON REGULATION

An interesting aspect of these experiments is that if male and female blastocysts are combined the resulting mosaic individual will be part male, part female. So it is not surprising that some of the embryos that develop are hermaphrodites (Tarkowski, 1961; Mintz, 1965), though the number is less than would be expected. Mystowska and Tarkowski (1968) reported that the sex ratio in twelve of these chimaeric mice was nine males: one hermaphrodite: two females. The expected combination was that 50% of the chimaeras would have both XX and XY cells and that these animals would develop into hermaphrodites. The authors concluded that some of those which possessed both male and female sex cells became normal males; thus in some way the activity of the XX cells had been suppressed or subordinated to the male ones. They suggested that this elimination probably occurred when the degree of the mixing of cell types in the gonad was very thorough.

An important subsidiary finding was that the sperm of these males or hermaphrodites was entirely genetically male in character and so was derived from the XY cells, even though the female germ cells were present. They concluded that the female germ cells had contributed to the somatic tissue of the gonad.

In a similar study, Mintz (1968) also found that the proportion of hermaphrodites that developed was much lower than expectation (1·3% instead of 50%) and was able to show that some of the males and the females possessed both XX and XY somatic cells.

Another development of the techniques used in making mosaic (or chimaeric) embryos has been that of combining morulae from different genetic strains. The earliest acting mouse mutant is tailless ($T^{12}t^{12}$) which is lethal in the recessive homozygous state ($t^{12}t^{12}$), these embryos dying at the late morula stage, after exhibiting various cellular anomalies including a failure to increase their cytoplasmic and nucleolar RNA (Mintz, 1964b, c). When these homozygotes were amalgamated as mosaics with normal embryos, the mutant cells seemed to be able to survive for slightly longer periods even though they continued to show nucleolar and cytoplasmic deficiencies (Mintz, 1964a). It has not yet been established how it comes about that they are able to survive in this way, although they seem to live longer if they become located in the inner cell mass of the blastocyst than in the trophoblast. Interestingly enough, there appeared to be no persistent pattern of distribution, mutant and normal cells being present in an apparently haphazard arrangement anywhere in the blastocyst, a fact which supports the evidence for the general lability (i.e. ability to regulate) of the early embryo.

Using other combinations of mutant and normal mouse morulae, Mintz and Palm (1965, 1969) produced mosaic embryos, some of which contained two genetically different kinds of red blood corpuscles. The two genetically different kinds of cells were still present in these individuals when they had become adult, thus showing that a state of immunological tolerance existed. Similarly, using two genetically different mouse strains, each with a different complement of enzymes in the skeletal muscle, Mintz and Baker (1967) have been able to demonstrate that a hybrid enzyme system is present in the mosaics. This finding, apart from its contribution to our understanding of the development of immunological tolerance (also p. 213), provides an elegant and striking confirmation of recent ideas on development of muscle syncytia. For many years there have been two main theories as to how the syncytia are formed, the first one supposing that a repeated division of nuclear material took place without cytoplasmic cleavage, and the second that there was a fusion of separate myotubes. Recently, Königsberg (1965), using clones of muscle cells in tissue culture, has been able to show that the second theory is correct. Mintz and Baker have confirmed this finding, for hybrid enzymes were found in each syncytium, demonstrating that different genetic material had been brought together in the same cytoplasm (Fig. 45).

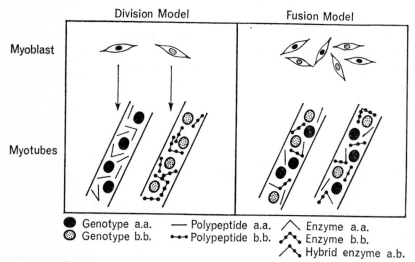

Figure 45. *Formation of muscle syncytia. The diagram shows two hypothetical ways in which muscle syncytia might form: by division of a single initial myoblast (left) or by fusion of several myoblasts (right). These theories were tested by obtaining mouse morulae from two genetically different strains, the adults of which differed from each other in possession of specific enzymes in the skeletal muscle. The skeletal muscle fibres of the hybrid embryos were found to contain hybrid enzymes, thus confirming that the fusion model is correct. (After Mintz and Baker, 1967.)*

(c) The coverings of the early mammalian embryo

When the mammalian ovum is shed from the ovary it is still enclosed by a capsule, the *zona pellucida*, as well as by follicular, (*cumulus oophorus*), cells that surrounded it when it was still in the ovary. It has sometimes been suggested that the cumulus cells, together with the epithelium of the Fallopian tube, play an essential rôle in providing nourishment for the cleavage stages in the mouse.

The zona pellucida has a fairly uniform, non-granular appearance by electron microscopy though the outer part is sometimes tunnelled by spaces left by the penetrating sperm. Histochemical studies suggest that it is composed principally of acid mucopolysaccharides (Stegner and

Wartenberg, 1961): it is rich in sialic acid in many species (Soupart and Noyes, 1964). The zona pellucida is present until the blastocyst stage is reached and the embryo is ready to embed (Fig. 46). It seems that in some animals, at least, it plays a rôle in preventing polyspermy (Chapter 3) but it is not known whether its usefulness ceases at this stage. It has

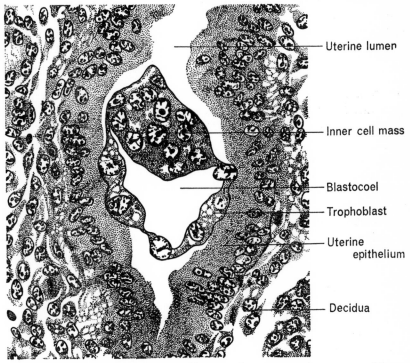

Figure 46.   *Section of a mouse blastocyst lying in a uterine crypt, and just about to embed. (From Snell and Stevens, 1966.)*

also been suggested that it holds together the cells of the early morula at a time when their adhesive properties are low, although eight-celled mouse eggs can develop quite normally *in vitro* in the absence of the zona (Tarkowski, 1961; Mintz, 1962). (But see Modlinski, 1970.) Rabbit embryos, on the other hand, develop better if the zona pellucida remains intact (Cole *et al.*, 1966). Perhaps it is best to regard the zona pellucida as being simply a vestigial egg covering, corresponding not so much with the shell or shell membranes of other amniotes (for these structures possess a totally different form) but being comparable with the vitelline membrane of birds. Certainly, when viewed by electron microscopy, it bears a strong resemblance to the fine granular matrix that surrounds the developing ovum of the hen's egg and in which the vitelline membrane becomes laid down.

Before a blastocyst can implant in the uterine wall it must escape from the zona pellucida. In general, there appear to be two methods of escape; the zona pellucida may be ruptured by the blastocyst, or dissolved by uterine secretions. Either, or both, of these methods may be used depending on the circumstances (McLaren, 1970). The blastocysts of rats and mice growing in tissue culture appear to rupture the zona pellucida and there is evidence from cine-films to show that in the mouse at least, this event is preceded by rhythmical bursts of contraction and expansions of the blastocysts (for instance, Cole and Paul, 1965; Cole, 1967). Unfertilized eggs are unable to contract in this way (McLaren, 1970). It has been suggested that the large amounts of glycogen present in the late cleavage stages of the mouse may provide the energy source for this activity (Thomson and Brinster, 1966). In the rat, the escape may take place from either the embryonic or abembryonic pole (Dickman, 1967).

The escape from the zona pellucida appears also to be under some environmental influence. For instance, the rat uterus is normally under the influence of both oestrogen and progesterone and deviation from the normal hormonal conditions affects shedding (Dickmann, 1969). There is some evidence that the uterus produces a lysin as a result of stimulation by oestrogen (McLaren, 1969), but nothing is known of its chemical composition. The importance of the lysin varies, however. For instance, mice grown in culture routinely escape from the zona pellucida at the appropriate stage of development, whereas rabbit embryos cannot do so (Glenister, 1965). Unfertilized mouse eggs, unable to escape by their own efforts, eventually become freed because the zona pellucida becomes lysed (McLaren, 1970). A variety of techniques have therefore been devised for removing the blastocyst (especially of rabbits) from the zona pellucida (see New, 1966). Perhaps the most useful is that of Mintz (1962) who dissolves away the zona pellucida with pronase. It is of especial interest that the zona pellucida of unfertilized eggs of pseudopregnant hamsters and of mice is also lost, after an appropriate time interval, in much the same way as with that of the normal fertilized egg (see discussions in CIBA Symposium), though it is not clearly established whether this loss is due to an inherent decay of the zona pellucida itself or to a dissolution of it by uterine or hormonal secretions.

The rabbit egg is distinguished by possessing a mucin layer in addition to the zona pellucida (Fig. 14). This layer, which varies in thickness in different individuals (from about 65–130 μ), is an acid mucopolysaccharide (Braden, 1952) and is secreted on to the ovum by the epithelium of the Fallopian tube. It has been suggested that the mucin acts as a reservoir of water which is drawn upon by the blastocyst to help it expand (Gregory, 1930). In an experimental analysis, Greenwald (1959) showed that if oestrogen was injected into the uterus immediately after coitus, a much thinner layer of mucus than normal was laid down and that this led to a higher mortality. Since it was not clear whether the deleterious effect was a direct one of having less mucin or an indirect one due to the hormonally-abnormal uterine environment, the ova were removed from the mother and transplanted to the uterus of another rabbit in the right stage of the oestrus cycle. For convenience, a pseudo-pregnant rabbit was chosen. It was found that the proportion of embryos with reduced mucin that implanted was much less than of the controls (Greenwald, 1962).

(d) Metabolism of the pre-implantation blastocyst

Much of the energy needed by the early mammalian embryo is provided by metabolising carbohydrates, especially glycogen. Glucose is also used, so it is not surprising that it is present in the blastocyst fluid, both before and after implantation (Lutwak-Mann, 1954). Different metabolic pathways are followed at various stages. For example, in the rabbit, the hexose monophosphate pathway is employed in the first three days, the Embden-Meyerhof from the third to the sixth day, and the Krebs cycle subsequent to that time (Fridhandler et al., 1957).

During this period the organism is very vulnerable to abnormal situations and obnoxious substances. The embryo itself is often more sensitive to drugs than is the trophoblast (Adams et al., 1961). Even weak electric light is said to have an adverse effect upon cleaving rabbit embryos (Daniel, 1964).

(e) Implantation

The difficulties encountered in attempts to culture mammalian embryos that have already begun to implant are notorious, and have largely been responsible for the slow progress of mammalian experimental embryology. Many investigators have attempted to repeat the experiments of Waddington (1932) in culturing chick blastoderms on plasma clots. The most successful have been New and Stein (1963, 1964) who grew mouse blastocysts in culture for up to 40 hours, during which time the yolk sac not only develops a functional circulation, but the whole blastocyst excavates for itself a cavity in the clot and sinks into it as if it were sinking into the endometrium of the uterus. Other workers have implanted mouse or rat blastocysts into the anterior chamber of the eye of the adult, but development tends to be slow and morphogenesis poor. Indeed, if the embryo is transplanted to this site before its mesoderm has formed, it appears to be incapable of developing any mesodermal derivatives (Levak-Švajger and Škreb, 1965). Older embryos apparently differentiate better, though according to these authors this is not because they are bigger at transplantation, since even parts of them differentiate well if transplanted separately.

Some workers have attempted to culture blastocysts along with pieces of endometrium, e.g. Glenister (1961, 1963, 1965) was able to promote implantation of rabbit blastocysts into strips of endometrium although the development of the embryo was usually abnormal.

Perhaps the most promising technique so far has been that of New (1967) who has developed a method for culturing rat embryos explanted together with their extra-embryonic membranes into an environment of

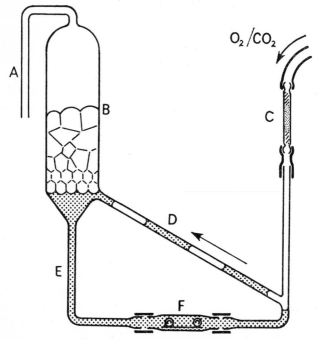

Figure 47.   *Culture of rat embryos* in vitro. *The embryos are in the detachable tube, F. The triangle of glass tubing contains serum. Oxygen and carbon dioxide enter through column C and pass as a stream of bubbles up tube D, the bubbles collapsing in chamber B. The stream of bubbles ensures a continual circulation in the triangle D, E, F. (From New, 1967.)*

circulating homologous serum (Fig. 47). Embryos explanted at somite stages often live for about 40 hours, during which time differentiation takes place and a certain amount of new protein is formed. Even better results were obtained when hyberbaric oxygen was supplied (New and Coppola, 1970).

(*i*) The act of implantation

It is difficult to give a generalised account of the processes that occur at implantation for although our knowledge is derived from several approaches it is as yet inadequate. Also, the relationship between embryo and mother varies in different groups of mammals. For instance, in most carnivores the blastocyst does not actually embed beneath the wall of the uterus but remains in the lumen (central implantation of Bonnet, 1882) whereas in other mammals (e.g. man and hedgehog) it burrows beneath the surface (interstitial implantation); in others such as the mouse, the blastocyst implants in a recess (excentric implantation). Detailed descriptions of three types are given by Boyd and Hamilton (1952). Here we will restrict ourselves mainly to the events that take place in the rodents and the rabbit.

(*ii*) Factors affecting implantation

Much of the work on mammalian embryos at this early stage has been concerned with the many factors affecting implantation. Perhaps the most important is that the trophoblast is a highly invasive tissue. This is evident from the fact that mouse blastocysts will implant into abnormal sites, such as the spleen and the testis (Kirby, 1965); 90% of the eggs transplanted by this author to the testis embedded and developed. In the cryptorchid (undescended) testis which has a higher temperature than the descended one, as well as in the spleen, the trophoblast grew to an immense size with an adverse effect on the embryo proper (Kirby, 1963a, b). Curiously, the adult organ which was most able to resist the invasion of the blastocyst was the uterus. It appears, therefore, that the uterus can only be invaded at certain specific periods of the oestrous cycle. Thus, one of the most interesting aspects of the subject is the problem of how the uterus prevents implantation, a special piquancy being added to the question by the importance of such knowledge in its application to human contraception.

The second major factor affecting implantation therefore is that the uterus shall be in a receptive stage when the blastocyst is ready to implant, and this is under the control of the maternal hormone system. Progesterone is produced by the corpus luteum and is needed for the growth of the endometrium before implantation, as well as for the maintenance of pregnancy after fertilization in many species. Some oestrogen is also necessary but the proportion seems to vary with the species (Adams, 1965, 1967). (There is even some possibility (Shelesnyak, 1960) that oestrogen is produced by the blastocyst as well as by the mother.) Rabbit blastocysts will implant *in vitro* in pieces of uterine wall not under hormonal control (Glenister, 1965; also p. 142); it would therefore appear that the progestational phase of the oestrous cycle is the time when the uterus is released from an inhibition, as well as the one where it is actively prepared for implantation. On the other hand, the secretion of a luteo-tropic hormone seems to be necessary for the maintenance of pregnancy (Gardner and Allen, 1942; Bland and Donovan, 1966).

A third factor affecting implantation is that the blastocyst must make contact with the uterine lining. Several hours before it attaches, a multilayered 'attachment cone' appears in the ab-embryonal pole of the guinea-pig blastocyst. In embryos growing in culture the outer layer of the cone can be seen extending out in protoplasmic projections through the zona pellucida (if that is still present) and into the culture medium (Amoroso, 1959). Protoplasmic projections of this type (villi) have also been seen by the electron microscope at the time of implantation in the armadillo (Enders, 1964) and in the rabbit (Larsen, 1961). Nilsson (1966) has shown that in the mouse and rat they are also present before implantation, but become more irregular at the time of initial contact with the blastocyst. He suggests that this may be due to an increase in adhesiveness; this increase is said to be induced by oestrogen.

Kirby (1966) and Kirby *et al.* (1967), who have discussed six hypothetical mechanisms which would enable the blastocyst to become implanted with the correct orientation, have concluded that only one mechanism fits in with the available evidence; that is, the trophoblast implants with random orientation and that the position is then 'corrected' by the inner cell mass which migrates around the inner surface of the trophoblast. It would be interesting to test this remarkable theory by filming a blastocyst as it implanted under tissue culture conditions.

Unfortunately, it is not possible to present an all-embracing theory to explain the cytology of the events that take place at the point of contact between the embryonic and maternal tissues. This is because the range of animals studied by modern techniques is still small. Enders (1964), who examined the early stages of the armadillo by electron microscopy (Fig. 48), thought that the disruption of the epithelium seen in his sections was due to the activity of the blastocyst. Subsequently, however, he revised his interpretation and concluded that this disruption was an artefact (Enders and Schlafke, 1969), and that the trophoblast

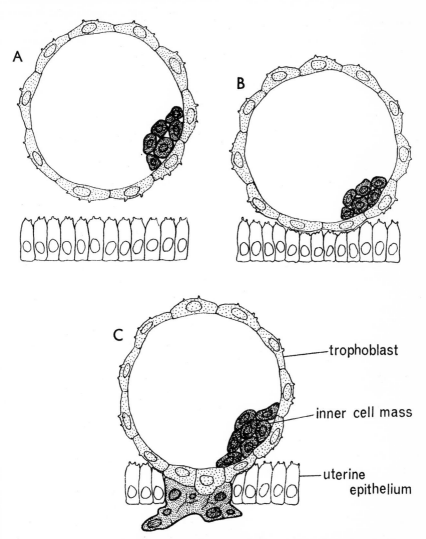

Figure 48. *Implantation. Postulated mechanism of implantation in the nine-banded armadillo,* Dasypus novemcinctus. *A. Blastocyst is free in the uterine cavity. B. Trophoblast cells indent the surface of the cells of the uterine epithelium. C. The trophoblast extends through the epithelium and increases in size. (After Enders, 1964.)*

does not produce substances that cause lysis of the uterine epithelium. On the contrary, in the blastocysts of rabbits (Larsen, 1962; Glenister, 1965; Enders and Schlafke, 1969), of rats (Enders and Schlafke, 1967), of mice (Potts, 1969) and of guinea pig, ferret, armadillo (*Dasypus novemcinctus*) and of a bat (*Myotis lucifugus*) (Enders and Schlafke, 1969) the initial contact appears to lead to a 'fusion' of the tissues. In some of these species the apposition is so close that desmosomes are formed between the trophoblast and the uterine epithelial cells (Enders and Schlafke, 1969). These authors suggest that the attachment is so strong that the epithelial cells are now pulled apart by the pressure of the intruding trophoblast.

The changes that take place in the endometrium at about the time of implantation again vary according to the species. Generally, however, there appears to be an increase in the numbers and size of the cells, which then become known as the decidual cells; many of these are rich in glycogen. There is also an increase in the vascularisation of the region. Zhinkin and Samoshkina (1967) who studied the process in the mouse found that the population of decidual cells in the implantation area is supplemented by the migration of cells from adjacent areas which have a high level of proliferation. These areas are first in the epithelium and then in the endometrium. In the mouse (Zhinkin and Samoshkina, 1967) and in the rat (Galassi, 1968) each phase is accompanied by a burst of DNA synthesis in that region. Other changes that occur in the endometrium during the decidualisation process include an increase in basophilia (L. J. Smith, 1966) and an increase in mitochondria, microsomes, lysosomes and fibrillar structures (Jollie and Bencosme, 1965).

In rats, but not in guinea pigs, these changes can be brought about by introducing glass beads into the uterus (Blandau, 1949). According to Shelesnyak *et al.* (1963), the decidualisation of the uterus is necessary before implantation can occur in the rat, although not in all species. For instance, very little decidualisation occurs in the human (Fridhandler, 1968) or in the rabbit (Glenister, 1965) until after implantation.

It is a curious fact that implantation in the uterus is normally accompanied in the rat and the mouse by the complex decidual reaction, since implantation can occur without it. We have seen that in the rat and the mouse at least it will take place in sites such as the testis. Various workers have therefore attempted to explain the function of the decidual reaction in normal development. One possibility is that the decidua forms an easily sloughed-off layer at birth. Another is that the glycogen cells in the decidua may have a nutritional value. McLaren (1965), however, suggests that the decidua may play a 'protective' rôle in the sense that it may prevent blastocysts from implanting after pregnancy has already been in progress for some time. Some evidence for this has been presented by Wilson (1963) who showed that if tumours were introduced into the pregnant uterus of the mouse they implanted only in the regions which had not undergone a decidual reaction. Further evidence was provided by Kirby (1965), who utilised the fact that a non-pregnant (and, indeed, pseudopregnant) mouse is not in the correct hormonal state to produce a decidual reaction; he found that pieces of trophoblast introduced into such a uterus were highly invasive.

The biochemical changes that take place at implantation are still not clear, though a number of enzyme changes have been studied. Hafez and White (1967) have shown that under the influence of progesterone there is an increase in the activity of both succinic dehydrogenase and of glucose-6-dehydrogenase in the uterus. Since both these enzymes are concerned in oxidation, these authors suggest that the differences in vascularity that take place at this time are connected with them. Similarly, in the mouse embryo there is a change at the time of implantation from a predominance of lactic dehydrogenase-1 to lactic dehydrogenase-5 (Auerbach and Brinster, 1967).

Lutwak-Mann (1963) reported among other biochemical changes an increase in carbonic anhydrase, though she cautiously refrained from drawing any conclusions as to its importance at the time of implantation. Some stimulating though essentially hypothetical schemes have however been devised by others building on her observations. Böving (1968), who has shown that there is a rise in pH in the pregnant uterine wall of rabbits, has suggested that this is due to residues of alkaline carbonate which are deposited when bicarbonate passes from the blastocyst and produces carbonic acid, which in turn gives off carbon dioxide to the maternal system. His theory is that this reaction is brought about by endometrial carbonic anhydrase, and that the action of the enzyme itself is promoted by progesterone. Böving's idea has come under heavy criticism from other embryologists (for example, discussion by Fridhandler, 1968).

(f) The spacing of implantations

The number of blastocysts that implant is initially dependent on the number of oöcytes that mature and are subsequently ovulated and then fertilized. A survey of the normal litter size in various species is given by Asdell (1965). On the whole, those species that possess bicornuate uteri tend to have larger litters and it might be argued that this is because they possess a greater uterine surface area on which implantation can take

F

place. However, in many mammals, including man, it is possible, by suitably adjusting the hormonal balance of the mother, to obtain *super-ovulation*, i.e. to increase the numbers of oöcytes that mature and ovulate at any given time. Loewenstein and Cohen (1964) report that as many as 90 eggs may be obtained in one day from the uterus of a super-ovulated mouse. Super-ovulated embryos are especially interesting in that they throw light on a problem of mammalian development that has tantalised investigators for decades; that is, the question of why embryos in a multiparous pregnancy such as the rat's or the pig's, should implant at more or less regular intervals along the uterine horns, rather than at random or all in the same region. Mossman (1937) suggested that the fertilized ova implanted serially down the uterine horns, the first one to arrive in the uterus implanting immediately, the next one to arrive passing further down the tube, the third one moving further down still before implantation and so on. He suggested that once a blastocyst had implanted, the uterine wall in the vicinity became changed so that it was not possible for the next embryo to implant too close to it. In this way, the embryos became spaced out.

McLaren and Michie (1959) argued that if this theory were correct, then the additional eggs present in a superovulated mouse would be unable to find suitable implantation sites and would be lost. This apparently does not happen, however, for in a super-ovulated animal many more foetuses than usual may be found; in the rabbit, as many as 30 implantations may occur (Hafez, 1963). For this and other reasons, Mossman's theory has now been supplanted by the view that the regular spacing of implanting blastocysts is due to the churning up of the contents of the uterus by waves of uterine contractions. Certainly, if glass beads are placed in the uterus they tend to become spaced out along its length, but the complaint has often been made that these beads are so unlike blastocysts that little credence can be paid to results obtained in this way. An amusing variation of this experiment was carried out by Markee (1944) who injected large numbers of sea urchin eggs into the cranial end of a rabbit uterine horn and found that they became evenly distributed. He attributed this to uterine contractions which he could see taking place through a window he had made in the abdominal wall. Böving (1968) has recently followed these contractions in the rabbit uterus by an elaborate recording device and found that the waves of contractions originated from those parts of the uterus that were distended by blastocysts. The significance of these contractions in human females fitted with intra-uterine contraceptive devices is discussed by Böving (1968) and by various authors in the CIBA Symposium (1966).

Similar contractions occur in the Fallopian tubes, as well as in the uterus and are probably under hormonal control. They appear to play a major rôle in transporting the egg to the uterus (Greenwald, 1968).

Before leaving this subject however it should be noted that although the presence of these uterine contractions is well established, there is still no direct evidence that they are responsible for regulating implantation. Indeed, one might expect that such contractions would lead to a random scattering rather than a regular one. It is of interest that Krehbiel and Plagge (1962) have denied that the distribution is regular in the rat and suggested that it is random. Similarly, bovine embryos which have been super-ovulated tend to be unevenly spaced (Hafez, 1964).

In addition to this rôle that it appears to have in spacing the blasto-cyst, the uterine wall probably also determines the precise site of implantation. Wilson (1963) found that tumours introduced into the uterus of a non-pregnant rat or mouse always implanted on the normal, antimesometrial, side. The way in which the uterus controls this location is not understood.

(g) Immunological relations between the implanting embryo and the uterus

This topic is discussed on pp. 213 et seq.

(h) Delayed implantation

This occurs as a natural event in certain species of mammals (see Hamlett, 1935). Delayed implantation can, however, be brought about in rodents by several means which involve prolonging the time when the uterus is in a receptive state. One method is by mating lactating animals, which takes advantage of the natural hormonal balance; other ways involve interfering with the hormonal state experimentally by hypo-physectomy, ovariectomy, administration of tranquillisers or hormone treatments (Mayer, 1959; New, 1966; Psychoyos, 1966).

It is generally assumed that under conditions of delayed implantation, the blastocyst rests unchanged in a 'state of suspended animation'. It is perhaps not surprising however that an electron microscopical study of the rat blastocyst shows that the cells undergo quite extensive cytological changes (Schlafke and Enders, 1963), though the significance of these changes is not understood. These authors found that no gross morpho-logical differentiation of the embryo proper occurred during this delay. This may be correlated with the fact that RNA synthesis is reduced in the rat embryo during delayed implantation (Prasad et al., 1968). (For a general survey of delayed implantation see the symposium *Delayed Implantation*, 1963, Ed. Enders.) No significant changes appear to take place in the endometrium of the mother during delayed implantation (Warren and Enders, 1964). A type of delayed implantation also occurs in certain marsupials, though it differs from that in eutherian mammals in that the blastocyst is thought to remain completely quiescent (Sharman and Berger, 1969).

(*i*) *Development of the embryo after implantation*

The embryo proper develops only from the inner cell mass. This becomes flattened out to form the embryonic shield (embryonic disc). The endoderm forms from the side of the shield nearest the blastocyst cavity and the primitive streak develops on the other side.

Until recently, many of the experiments on the early stages (discussion by Nicholas, 1947) were impeded by the lack of suitable culture media and salines. Fortunately, a number of satisfactory recipes now exist (New, 1966) and improvements are continually being announced.

Little work has been carried out on the morphogenetic movements of early mammalian embryos, but Daniel and Olson (1966) followed events in the rabbit embryonic shield. By marking different regions with neutral red dye, they have shown that some of the cell movements are not unlike those in other amniotes (Fig. 49), though it is not clear whether regression of the node takes place.

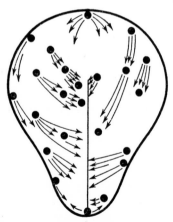

Figure 49. *Morphogenetic movements in the rabbit embryo during primitive streak formation. The black circles indicate the regions marked with neutral red. The arrows show the directions in which the marks were displaced. (From Daniel and Olson, 1966.)*

The notochord forms mainly from the head process and in many mammals it becomes fused with, and continuous with, the endoderm. Subsequently, the notochord rolls up, forming a rod. In some mammals, such as the mouse, a depression becomes enclosed in the rod, forming a lumen. This lumen is considered to be homologous with the archenteron of amphibian embryos and is known as the *chorda* canal. These developmental processes differ from those of the chick, where the notochord remains separate from the endoderm, and no *chorda* canal forms.

It will be recalled that there is some controversy among students of avian embryology as to whether the morphogenetic movements are related in any way to regions of high mitotic rate (p. 84). In the same way, evidence has been presented to support on the one hand the idea that localised areas of cell proliferation play a vital rôle in the gastrulation of mammals (Daniel and Olson, 1966) and on the other hand that they are not responsible for promoting and maintaining these movements (Corliss, 1953). Although Daniel and Olson studied the rabbit whilst Corliss made his observation on the rat, it seems unlikely that such a fundamental difference would really be present between the two species. Until further evidence is forthcoming it would perhaps be advisable to accept the findings of Corliss, since they are based on a more extensive study.

At a slightly older stage of development also, there appears to be a similarity between some of the morphogenetic processes in the mouse and those in the chick. For instance, Smith (1964), who transected mouse embryos, concluded from the distortion of the axis that there was a movement caudally of the already formed notochord and neural tube. A similar backward extension occurs in the chick embryo. An interesting difference is that if the rabbit embryo is transected at the primitive streak stage, the most posterior part is able to differentiate into neural tissue and somites (Waddington, 1937), whereas if the same experiment is carried out on the chick, differentiation ceases unless either the node or a substantial part of the node region remains. It would appear therefore that although regression movements do occur in mammalian embryos, they do not play such an essential controlling rôle as do comparable movements in the chick (Chapter 5).

Experiments that have been carried out on other aspects of morphogenesis of the early mammalian embryo at about the primitive streak stage also support the idea that the basic processes are similar to those of birds. Waddington (1937) cultured primitive streak stage rabbit embryos by the same techniques as he had so successfully used with the chick embryo, and was able to show that grafts of rabbit Hensen's node could induce a secondary axis if implanted into a chick host of comparable age (Fig. 50). Nicholas and Rudnick (1933), who isolated pieces of rat blastoderms on the chorioallantoic membrane of the chick, obtained comparable results to those which they got from similar experiments using the chick blastoderms.

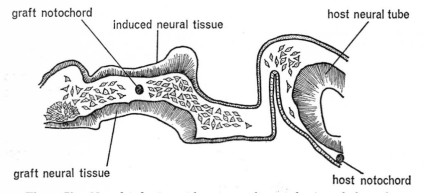

graft notochord

induced neural tissue

host neural tube

graft neural tissue

host notochord

Figure 50. *Neural induction with a mammalian graft. A graft from the anterior part of a rabbit embryo at the head fold stage was grafted into a chick embryo at the primitive streak stage and there induced a neural plate. (After Waddington, 1937.)*

(j) Development of the yolk sac placenta

About the time of implantation, the blastocyst cavity enlarges, probably by becoming distended with fluid. This is followed by a number of important changes in the composition of the blastocyst fluid; in the rabbit, at least, the concentration of various substances shifts until it is similar to that of the maternal blood plasma. For instance, soon after implantation there is a drop in bicarbonate concentration, but a rise in glucose, lactates and proteins in the blastocyst fluid (Lutwak-Mann, 1963).

Most of the endodermal cells that form in the embryonic shield will subsequently line the gut; these are known as the proximal endoderm cells. Some cells however now grow out from this endodermal layer and spread out to form a bag of extra-embryonic endoderm. These new cells are the distal endoderm cells; the bag is the yolk sac and it lines the blastocyst (Fig. 51).

In many mammals, such as man, the yolk sac appears to be a vestigial structure and it probably plays little part in the development of the

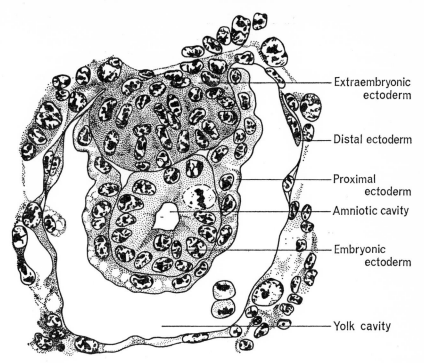

Extraembryonic ectoderm

Distal ectoderm

Proximal ectoderm

Amniotic cavity

Embryonic ectoderm

Yolk cavity

Figure 51. *Section of a mouse blastocyst soon after implantation, at about 5–6 days post fertilization. (From Snell and Stevens, 1966.)*

embryo proper. But in other mammals it becomes modified and adapted to form a yolk sac placenta.

In the eutherian mammals, yolk sac placentae are found among the rodents and insectivores and are always accompanied by a chorio-allantoic placenta (see below). The degree of erosion of the tissues varies in the different species. Yolk sac placentae are also present in most marsupials, and in some species this is the sole means of placentation, no chorioallantoic placenta developing at all. For a fuller account of yolk sac placentation in both eutherian and marsupial mammals see Amoroso (1952).

The walls of the blastocyst are known as the trophoblast (Figs. 46 and 52). Meanwhile the inner cell mass sinks down into the roof of the yolk sac to form the egg cylinder (Fig. 52). That part of the yolk sac pressed against the egg cylinder is known as the proximal or visceral yolk sac whilst that part lining the trophoblast is called the distal or *parietal yolk sac.* In the gap between the parietal yolk sac and the trophoblast is a granular layer of material called Reichert's membrane (see below). Subsequently, the proximal yolk sac endoderm becomes covered with a layer of mesoderm. Small blood vessels form in this mesoderm and link

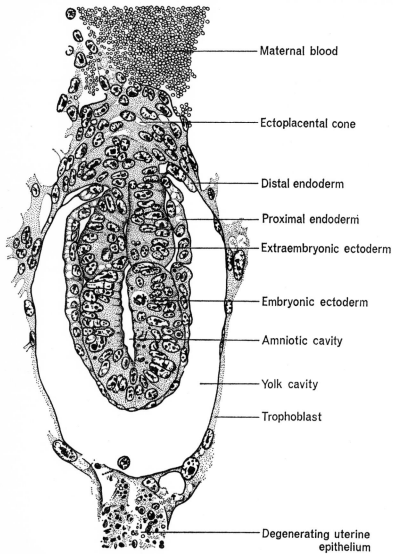

Maternal blood

Ectoplacental cone

Distal endoderm

Proximal endoderm

Extraembryonic ectoderm

Embryonic ectoderm

Amniotic cavity

Yolk cavity

Trophoblast

Degenerating uterine epithelium

Figure 52. *Section of a mouse embryo at a slightly older stage than in Fig. 51. The egg cylinder (i.e. the embryonic region together with the attached extraembryonic ectoderm) has lengthened. (After Snell and Stevens, 1966.)*

up with another to form the yolk sac circulation. Most of the parietal yolk sac now breaks down so that only the proximal part remains. This now becomes pressed against the uterine epithelium which is itself degenerating, and a region of close contact between embryonic and maternal tissue is formed. There is evidence that protein and other molecules may pass from the mother to the embryo at this point (Deren *et al.*, 1966a, b; Padykula *et al.*, 1966). For this reason, it is considered to be a placenta (although of a somewhat unusual kind in eutherian mammals, in that it is formed from the yolk sac).

The remnant of the yolk sac which lies wrapped around the embryo proper now presents what was originally an internal surface to the uterine wall. The effect is as if the yolk sac had been turned inside-out, and so we speak of the inverted yolk sac placenta.

Certain changes in the organisation of the endoplasmic reticulum and of certain apical granules in the mouse yolk sac suggest to us that during the first half of gestation there is a high level of synthetic activity; during the second half the emphasis is more on absorption (Galarco and Moyer, 1966). This is supported to some extent by autoradiographic studies on the uteri and embryos of pregnant mice where it was found that the DNA synthesis is extensive at days 10–12 of gestation, but is nearly completed by day 14 (Sacoman *et al.*, 1967).

Perhaps the most striking and well known histochemical feature of the yolk sac placenta of rodents is its high glycogen content (Wislocki *et al.*, 1946). This glycogen seems to be used as an energy source by the yolk sac. The amount of glycogen present in the yolk sac appears to be controlled by the cells themselves, for there is a cycle of storage between about 10 and 15 days, followed by glycogen depletion between about 15 and 21 days. This is found under normal conditions *in utero* and in pieces of the yolk sac explanted *in vitro* (Sorokin and Padykula, 1964). This high level of glycogen in the rat yolk sac is reminiscent of a similar high level in the yolk sac of the chick (Willier, 1968). The rat yolk sac also resembles that of the chick in possessing a high lipid content and in being well supplied with enzymes, such as alkaline phosphatase (Johnson and Spinuzzi, 1968). Other enzymes present in the rat yolk sac include succinic dehydrogenase, acid phosphatase and adenosine triphosphatase (Padykula, 1958).

Although it seems likely that many of these enzymes may be concerned with absorptive phenomena, it is still not clear exactly how substances are taken into the yolk sac placenta. Some authors who have examined the cells by electron microscopy have concluded that there are special canaliculi leading from the apex of each cell into large vesicles (Padykula *et al.*, 1966; Lambson, 1966). By contrast, Gascogne *et al.* (1966) concluded that substances were taken into these cells by pinocytotic vesicles.

The matter is evidently a complex one since a certain amount of selection can take place at the yolk sac splanchnopleure; this was shown by Hemmings (1956) who followed the distribution of I^{131}-labelled serum globulin into the intestine of the rabbit embryo. Similar results have been reported by others. For example, Anderson (1959), who treated rat embryos with labelled γ-globulins, reported that there was a barrier at about eleven days of gestation to its passage through the yolk sac, and that this barrier was probably located in the cells of the visceral yolk sac. Similarly, when pregnant rats were treated with trypan blue, the dye became concentrated in the yolk sac epithelium (Fern and Beaudoin, 1960). In a more recent study, Wild (1970) using fluorescent protein tracing techniques, has also concluded that selection takes place in this region, different proteins becoming localized in absorptive vesicles in the yolk sac endoderm.

Not all authors who have studied the location of the yolk sac believe that it is the endodermal cells that exert the selective action. Instead, they suggest that the effective barrier is provided by Reichert's membrane. The acquisition of passive immunity occurs in some animals before birth (Brambell *et al.*, 1951; Smith and Schechtman, 1962) by the absorption of maternal antibodies through the yolk sac. This does not occur until late in pregnancy and it has been suggested that it may not be possible until this time, because it is then that Reichert's membrane breaks down.

Reichert's membrane, which is about 15,000 Å thick in the guinea pig, is considered to be a highly specialised basement membrane (Petry and Kühnel, 1964). It is found associated with the yolk sac placenta of rodents (Plate VIII) and insectivores (Boyd and Hamilton, 1952). In an elegant study Pierce *et al.* (1964) showed that it was immunologically identical with the endoplasmic reticulum of the yolk sac epithelium. Furthermore, it had no antigens in common with erythrocytes, beta-globulins or various other tissue components.

This finding was of great interest from a general cytological point of view since basement membranes have often been considered to be formed by a condensation of inter-cellular matrix (Robertson, 1963).

Pierce and his collaboraters (Mukerjee *et al.*, 1965) brilliantly exploited the fact that a carcinoma of the parietal yolk sac of the mouse secretes large enough quantities of this basement membrane for biochemical analysis. This membrane is thought to be an exaggerated Reichert's membrane, though this theory does not appear to have been entirely proved (see Pierce *et al.*, 1962). It has been found to be a distinct muco-protein, and contains 75–80% protein, 12–14% carbohydrate and 2·5–3·0% sugars. Its amino acid content was found to be chemically distinct from that of connective tissue scleroproteins. Finally, it has an antigen in common with a variety of other basement membranes (Midgely and Pierce, 1963). Pierce has drawn attention to the fact that one would not expect such a similarity since basement membranes frequently act as the substrate for migrating cells; it is generally considered that the adhesion of cells to their substrate differs in different parts of the embryo, and that this plays a rôle in morphogenesis.

(k) *The amnion and chorion*

The blastocyst possesses two different types of ectoderm, the embryonic ectoderm which lies immediately above the blastocyst cavity, and the extra-embryonic ectoderm which is distal to it. It is in the latter that a cavity—the amniotic cavity—is formed at about five and a half days in the mouse embryo.

In some species, such as man, the amniotic cavity forms by a hollowing out of cells in the inner cell mass. In others, including the mouse and the rabbit, the amnion develops by a process of tissue folding similar to the process in the chick embryo. Some authorities regard the folding process as being more 'primitive' than the cavitation method (Mossman, 1937).

Shortly after the amnion has formed in the mouse the most distal of the extra-embryonic ectoderm cells proliferate rapidly to form the ecto-placental cone which extends toward the uterine lumen (Fig. 52). This is the primordium of the allantoic placenta, for lacunae develop in it and these become filled with circulating maternal blood. Meanwhile, the ectoplacental cone is also invaded by blood vessels of the allantois and the region formed is one in which exchange between the maternal and foetal circulations is possible, i.e. a placenta.

This type of chorioallantoic placenta is a highly specialised one and is characteristic of rodents. Other types of placenta are discussed in the next section.

(l) *The chorioallantoic placenta*

We can consider the mammalian chorioallantoic placenta in a number of ways: from the point of view of the gross histology, or morphology, or as a physiologically active organ carrying out complex exchanges between mother and foetus. The structure of the placenta varies so much from one group to another and is often so complicated that it would seem at first appraisal to be an ideal system for relating structure and function. Unfortunately, we do not know enough about the physiology of transport mechanisms to understand the precise relationship they may have to structural, or even ultra-structural, variations. Understandably, though disappointingly, 'placentologists' have often been reluctant to commit themselves in these matters and so the subject has tended to bifurcate into two main branches, the classifying of placentae on morphological bases, and highly specialised physiological investigations which though of extreme importance are generally carried out by investigators whose prime interest is in physiological processes rather than in embryonic development. For that reason they will not be discussed here. Excellent reviews have been given by Davies (1960), Huggett (1961) and Assali *et al.* (1968) and a recent review of placental hormones has been presented by Simmer (1968). A further difficulty is that no really satisfactory definition exists for the term 'placenta' though the best one is probably that of Mossman (1937) who stated that a placenta is 'an intimate apposition or fusion of the foetal organs to the maternal (or occasionally paternal) tissues for physiological exchange'. This definition is admirable in that it links structure and function, but in practice it has drawbacks, for we must often simply assume from the morphological appearance of a region of apposition that physiological exchange takes place there.

This definition is suitable for placenta of all types; that is, chorioallantoic placentae, yolk sac placentae and the specialised tissue relationships found in lower vertebrates such as the dogfish, *Mustelus* (Amoroso, 1952). Of these, most is known about the chorioallantoic placentae.

(m) Classification of chorioallantoic placentae

The earlier attempts to classify the different types of placentae were based on their gross morphology. For instance, T. H. Huxley (1864) distinguished between two groups, the deciduate (e.g. man) in which part of the uterine wall, the decidua, was shed together with the foetus at birth, and the non-deciduate (e.g. the pig) in which no maternal tissue was shed. A more elaborate scheme was that of Hubrecht (1894) who classified placentae into four groups according to the type of villous arrangements: diffuse (e.g. pig, horse) in which the villi were evenly distributed; cotyledonary (e.g. sheep, cow) had the villi arranged in clusters or cotyledons which were separated from one another by bare areas; the zonary (e.g. dog) in which the villi were arranged in a zone or band lying around the conceptus; discoidal (e.g. man) in which the villi were clustered together in one, or sometimes (as in platyrrhine and catarrhine monkeys) as two discs.

Both these schemes were considered to be insufficiently precise and, for this reason, when Grosser (1927) put forward an elaborate classification based on the histology of the placenta, it rapidly became accepted as the best so far available. It has been often criticised but most authorities are agreed that even with its faults it remains the most useful scheme we possess. Its basis is the supposed degree of intimacy that exists between the foetal and maternal circulation. In considering it we must ask ourselves two questions. Firstly, does the classification stand up to inspection by modern techniques of histology and electron microscopy? Secondly, does this scheme help us to understand the functioning of the placenta?

Grosser's classification is illustrated in Table I. He divided the placentae of eutherian mammals into four groups; the fifth, the haemo-endothelial, was contributed by Mossman (1937) who applied it to certain species that Grosser had included in group four, the haemochorial. In the first group, the epitheliochorial type, the embryo does not implant beneath the surface of the uterus but remains in the lumen. Consequently, there is practically no degeneration of uterine epithelium. This group corresponds with Hubrecht's non-deciduate placentae. But in all the other placentae certain layers of maternal or foetal tissues are eroded (see table) so that the maternal and foetal circulations are separated by fewer intervening layers.

Our task is now to consider briefly the type of criticism that has frequently been made about this classification. Firstly, Grosser implied that the fewer the layers, the greater the speed of diffusion of materials across the placenta, but there is no clear-cut evidence that this is so. Placental transfer probably depends to a large extent on cytochemical and cytological structure, but in any case the difference in actual width across the placenta in different types is less striking than the concept implies. Often the subject is discussed as if substances merely diffuse across the placenta, but this is too naive a concept since substances are often transported against a gradient. The only way to discover whether a membrane is a good transporter is to carry out the necessary physiological experiments on each substance for each species.

Secondly, even if there were a simple relationship, a placenta does not have a constant structure, as the table above implies, but its histology changes with age. Thus, initially, even a haemochorial placenta passes through an epitheliochorial stage, so the table is to be considered more as a guide to the histological state of the placenta in its maturity. In fact, it is

TABLE I

Grosser's scheme for the classification of the chorio allantoic placentae of eutherian mammals, together with Mossman's modification. The classification is based on the amount of erosion considered to have taken place at the foeto-maternal junction. Eroded tissues are shown as −, and non-eroded ones as +.

Grosser's categories	Maternal tissue			Foetal tissue			Examples
	Endothelium	Connective tissue	Epithelium	Trophoblast	Connective tissue	Endothelium	
Epitheliochorial	+	+	+	+	+	+	Pig, horse
Syndesmochorial	+	+	−	+	+	+	Sheep, cow
Endotheliochorial	+	−	−	+	+	+	Cat, dog
Haemochorial	−	−	−	+	+	+	Man, monkeys
Mossman's category							
Haemoendothelial	−	−	−	−	−	+	Guinea pig, mouse, rabbit

possible by careful scrutiny of most placentae to find many places where fewer layers of tissue intervene between the two circulations than one would expect. A striking example is that in places the trophoblast of the sow, which is a classical example of an epitheliochorial type, has even been seen to send processes between and past maternal epithelium into the region of the maternal capillaries, so that an endotheliochorial arrangement is present in those regions (Wislocki, 1955). More examples of this type are considered in the now classical review of Amoroso (1952) which were based on careful examination of sections by light microscopy. More recent work by electron microscopy has produced other criticisms of the Mossman scheme.

For instance, Hamilton, Harrison and Young (1952) using electron microscopy showed that in the cow and the fallow deer the epithelial lining cells are almost certainly maternal and not foetal as originally supposed; therefore these placentae were epitheliochorial rather than syndesmochorial in type. Similarly, electron microscopy of various other 'syndesmochorial' placentae, such as that of the camel and llama (Wynn, 1968 for references), has shown there to be an epitheliochorial condition.

Larsen (1963) who has studied the placenta of the rat by electron microscopy has concluded that trophoblast is present even at term—an amazing discovery for a so-called haemoendothelial placenta. He concluded that Mossman's modification of the Grosser scheme, in which he split off the haemoendothelial placentae, was unfounded. In Mossman's defence, however, we should appreciate that the gross morphological arrangements are indeed very different in this group from the haemochorial to which it has now been returned by Larsen. Similarly, thin layers of trophoblast have now been seen in other haemochorial placentae studied by electron microscopy (e.g. guinea pig, chipmunks, man, armadillo, rabbit) (Wynn, 1968). The actual thickness of the trophoblast is variable so that in 'haemoendothelial' types it is invisible by light microscopy. The number of layers of the trophoblast also vary, three layers being fairly common in rats, mice and voles (Enders, 1965). In each case the inner layer is cellular, the outer one being syncytial or possessing overlapping syncytium-like patches.

A third criticism of the Grosser scheme is that it deals only with the chorioallantoic placenta, and yet, as we have seen (p. 152), the yolk sac may also act as a placenta in some species. Furthermore, in certain animals a number of specialised regions are present that are thought to play a rôle in so-called histotrophic nutrition. For instance, certain haematomes are present in carnivores, areolae in various ungulates (Amoroso, 1952; Wynn, 1968). Although it is thought that maternal glandular secretions are taken up at these points by the foetus, their rôle is not covered by the Grosser classification.

For these reasons Grosser's scheme has been subject to many criticisms. But perhaps the fault lies not so much with the scheme as with our insatiable desire for tidiness in demanding one. The variations in placental types do not even fit comfortably into any phylogenetic plan, and attempts to consider them in this way have not been successful (Mossman, 1937).

Some recent electron microscopical work carried out on a variety of species, however, is suggestive for future helpful theories. First, many electron micrographs show that large numbers of microvilli are present at the absorbing surface and presumably increase the area (Terzakis, 1963). Secondly, pinocytotic-like vesicles are common. Thirdly, large numbers of mitochondria may provide enzymes that utilize the energy necessary for the passage of materials. Finally, the energy itself is probably provided by glycogen which has been known for many years to be present in both foetal and maternal tissues. In many animals the wall of the trophoblast consists of an outer syncytiotrophoblast and an inner cytotrophoblast. Galton (1962) has shown, by counting the mitoses and by estimating the DNA in the two regions, that in the human placenta the syncytial nuclei are derived from the cytotrophoblast. A similar conclusion was reached by Boyd and Hamilton (1966) who examined the human placenta by electron microscopy.

Until recently much of the electron microscopy of placentae has been overshadowed by historical preoccupations. What is now needed is electron microscopy combined with autoradiography or with labelled antibody work, or with similar fine techniques which can help us to study the transport and synthesis of materials in the placenta.

C. SUMMARY

In this chapter emphasis is laid on the ways in which the early development of mammals differs from that of birds. Apart from various types of foetal physiology (not dealt with here) experimental embryology of mammals is mainly confined to the pre-implantation stages, although it is possible to keep the operated embryos until birth by implanting them into the uterus of a suitable female. During the early period, the mammalian embryo is capable of extensive regulation; if two or more blastocysts are fused together they will form a single embryo. In this way it is possible to combine two individuals of different genetic constitution. This technique is valuable because it can be applied to a number of important problems of general biological interest, such as the formation of muscle syncytia, the immunological problem of self-recognition and the relationship between the primordial germ cells and the somatic cells of the gonad.

Some of the factors affecting implantation are discussed. The development of the yolk sac placenta is considered, special emphasis being paid to the structure and function of Reichert's membrane. A brief account is given of the different types of chorioallantoic placenta.

CHAPTER 8

SPECIAL ASPECTS OF THE EARLY STAGES OF THE DEVELOPMENT OF REPTILES

WE know less about the development of reptiles than about that of either of the other two classes of amniotes, at least as far as the earlier stages of development are concerned. This is largely due to the fact that few embryologists have had access to large supplies of reptilian embryos; furthermore, reptiles seldom breed in captivity with any regularity. Consequently there has been a tendency for embryologists to dismiss the reptiles by simply regarding their development as being almost identical with that of birds—which are of course more closely related to the existing reptiles, especially crocodiles, than they are to mammals. But, as we shall see, some important differences exist between the early development of birds and those reptiles which have been at all critically studied. Some of these differences may well repay further investigation.

The existing reptiles are a heterogeneous assemblage of diverse origins. Most of them—all the chelonians and crocodilians, the lone rhynchocephalian *Sphenodon* and the majority of the Squamata (lizards and snakes)—are oviparous. The fossilised eggs of certain extinct reptiles, notably dinosaurs, are also known. Presumably oviparity was the primitive mode of reproduction for the class, and there can be little doubt that the amniote egg with all its refinements was one of the innovations which led to the rapid success of the reptiles as the first truly terrestrial vertebrates. Romer (1966) has made the interesting suggestion that the adaptation of this type of egg to terrestrial conditions preceded that of the adult. The early reptiles may have laid their eggs on land (then perhaps a safer environment in terms of freedom from the hazards of seasonal drought than the water), while the adults were still more or less aquatic.

A. TECHNIQUES

Turtles are perhaps the most satisfactory material for research in reptilian embryology; a gravid female of *Chelydra serpentina*, the common snapping turtle of America, may produce as many as 30–50 eggs.

Techniques for their procurement, culture and use in experimental embryology have been described by Yntema (1964).

Techniques are also available for the culture of the eggs of the viviparous lizard, *Lacerta vivipara*, which is common in Western Europe. These involve removing the eggs from the mother and growing them in petri dishes whilst still surrounded by their own membranes and retaining their own yolk supply (Panigel, 1956) (Fig. 53). This technique, which has been used less successfully for the eggs of the slow-worm, *Anguis fragilis* (Maderson and Bellairs, 1962), generally allows normal development right up to hatching, and has been used for a variety of ablation and labelling experiments on the embryo (Holder and Bellairs, 1962). Another successful technique, devised by Raynaud (1959a, b) involves culturing slow-worm eggs in an elaborate system of glass cups which give support to each egg.

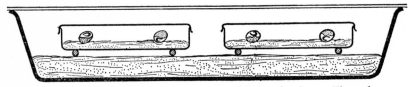

Figure 53. *Culture of the lizard,* Lacerta vivipara, in vitro. *The embryos in their membranes rest on moist cotton wool in small petri dishes. These small dishes are placed in larger ones, also containing moist cotton wool. (From Panigel, 1956.)*

An earlier attempt at culturing various lizard and snake embryos *in vitro*, which involved removing the blastoderm from the yolk and explanting it like a chick blastoderm on to a culture medium (Bellairs, 1951), was less successful, although survival for up to 20 days was achieved with some specimens. The growth of pieces of lizard embryos as grafts, either on the allantois of the same species (Dufaure, 1964) or on the chick chorioallantois, has also been reported (Nakao, 1939).

B. OÖGENESIS AND FERTILIZATION

The eggs of reptiles are large and yolky, like those of birds, and so far as is known no species has lost its yolk in the way that the eutherian mammals have done. Oögenesis apparently follows a similar pattern to that of birds and is therefore considered in Chapter 2, special points applicable to reptiles being mentioned. Unfortunately, we lack knowledge about the molecular biology not only of the oöcytes but also of the early embryos of any reptile. For instance, we are ignorant, as to whether masked maternal mRNA is stored in the oöcyte as in amphibians (p. 59); or whether the pattern of protein synthesis in these early stages corresponds with that in the mouse (p. 65).

C. CLEAVAGE AND GASTRULATION

Although relatively little experimental work has been carried out on these early stages, the morphological changes involved in cleavage and gastrulation have been described in detail for certain reptiles. These processes are similar in many ways to those in birds, though certain differences will be pointed out below. The same controversies that have dogged the student of avian embryology have also handicapped those interested in reptilian development. For instance, we still understand little about the early division of the embryo as it lies on the surface of the yolk, nor are we certain about the way that the endoderm forms.

The earliest cells are large and yolky and have been examined in various species by light microscopy, e.g. in *Lacerta vivipara* by Hubert (1962). For a general account of cleavage in reptiles see A. Brachet (1935), Pasteels (1937a) and Hubert (1962).

The orientation of the developing blastoderm on the yolk has been examined in various species. Just as in the chick embryo, the orientation of a reptile embryo is determined by gravity. This was shown by Clavert (1963) who found that if the mother lies on her back for a long period the

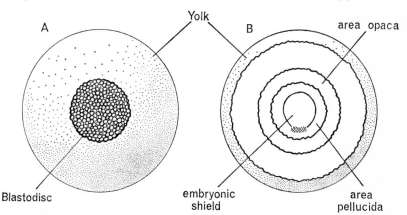

Figure 54. *Early development of the water tortoise,* Clemmys leprosa. *A. Cleavage: a flat disc of cells lies on the top of the yolk. B. Gastrulation: the embryo now covers a greater area of the yolk. The hatched region represents the archenteric (blastoporal) plate. (After Pasteels, 1937a.)*

egg rotates in the oviduct so that the animal pole (future dorsal surface in amniotes) is uppermost. Thus, the blastoderm leaves the normal meso-metrial side of the oviduct and migrates to the antimesometrial side. In general, reptiles also appear to follow the rule of von Baer (p. 72). In 93% of the embryos of *Lacerta vivipara* (Hubert, 1963), the head–tail axis was at right angles to the main axis of the egg. Similar percentages were found for *Anguis fragilis* (Raynaud, 1961) as well as for *Agama bibroni* (Bons, 1963).

In the water tortoise, *Clemmys leprosa*, cleavage takes place whilst the egg is still in the oviduct, and by the time it is laid a subgerminal cavity and an embryonic shield have developed. An area opaca is present, as in birds, but along the posterior edge of the shield a folded region appears (Fig. 54) and forms the 'primitive', 'blastoporal' or 'archenteric' plate (Ballowitz, 1901). It is here that a blastopore develops which is similar to the blastopore in amphibians. Such a structure is lacking in the chick and probably also in other birds. Sections through this region in *Clemmys* show that an archenteron (blastopore canal) extends forward from it beneath the ectoderm (Fig. 55). It will be recalled that some mammals have retained this structure in the form of the chorda canal (p. 151). Maps of the presumptive areas at this stage are available for a tortoise (Fig. 56).

As development proceeds, folds appear on the dorsal (ectodermal) surface of the reptilian blastoderm in the region of the blastopore and

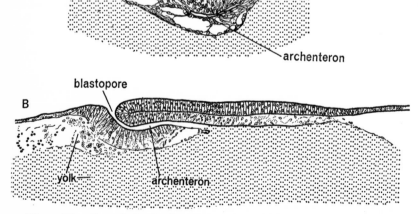

Figure 55. *Gastrulation in the water tortoise,* Clemmys leprosa. *Sections through the archenteron (blastoporal canal). A. Transverse section. B. Longitudinal section. (After Pasteels, 1937.)*

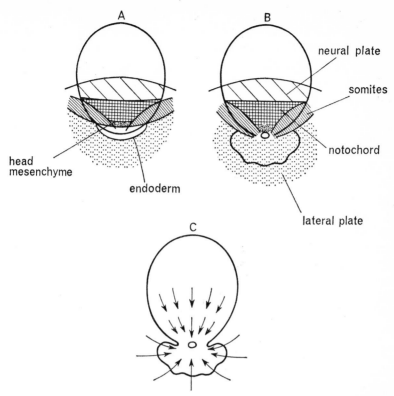

Figure 56. *Presumptive areas and morphogenetic movements in the water tortoise*, Clemmys leprosa. *A. Early gastrula before invagination of the endoderm. B. Mid gastrula. C. Morphogenetic movements during gastrulation.* (*From Pasteels, 1937a.*)

extend forward to form a structure composed of paired ridges (Fig. 57A and B). This structure is the developing neural plate, bounded by neural folds. In the region posterior to the blastopore a thickening develops which has been regarded by some authors as being a primitive streak homologous with that structure in birds and mammals. Pasteels (1940), who reviewed the subject, and Hubert (1962) who investigated *Lacerta vivipara*, have both pointed out that there is a confusion in the literature about the rôles of the blastopore and the so-called primitive streak in reptiles. It is particularly difficult to understand why both a blastopore and a primitive streak should be present, since in other classes of vertebrates these are alternative devices for invagination, the one being characteristic of amphibians, the other of birds and mammals. Pasteels

(1940) concludes on morphological considerations that invagination in reptiles is through the blastopore, but that the archenteron is not identical with that in amphibians and that the primitive streak of reptiles (if, indeed, it is a true primitive streak) is not identical with that of birds and mammals. Certainly, there seems to be no node regression corresponding with that in the chick embryo.

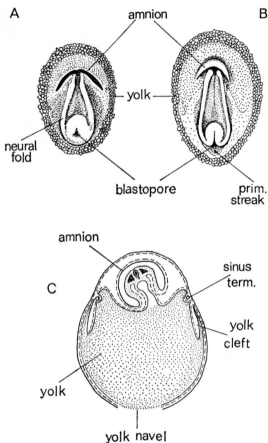

Figure 57. *Extra-embryonic membranes of reptiles. A and B. Two early stages in the development of* Crocodilus madagascariensis. *In A, the neural folds and amnion have started to form anterior to the blastopore. In B, the so-called primitive streak has developed posterior to the blastopore. (After Voeltzkow, 1902.) C. Section through the embryo of a lizard, the viviparous gecko,* Hoplodactylus maculatus. *The yolk cleft is formed by a downgrowth into the yolk of part of the yolk sac. The yolk navel is that part of the yolk that is not covered by yolk sac. (After Boyd, 1942.)*

As in birds, the problem of how the cells invaginate to form mesoderm is closely linked with that of how the endoderm develops. Some workers have maintained that it arises during cleavage by the delamination of cells from the lower layer of the blastula (Peter, 1934; Hubert, 1962), whereas others have suggested that it forms during gastrulation (Pasteels, 1937a) by a spreading out of cells from the base of the archenteron.

It appears, however, that both viewpoints are correct for Pasteels (1957b) concluded on the basis of a new study that in chelonians the endoderm forms by invagination during gastrulation, whereas in lacertilians it forms earlier by delamination during cleavage. Unfortunately, however, owing to the difficulty both of obtaining adequate supplies and of carrying out experiments on early embryos, all our information on endoderm formation in reptiles is based on the study of sectioned material. We have seen that in the chick embryo recent experimental work has led to a radical change in our views on formation of the endoderm (p. 81). It is advisable, therefore, whilst accepting these findings, to await possible experimental studies on the topic in reptiles with an open mind.

Unfortunately, with the exception of Pasteels' (1937a) own careful experimental work (Fig. 56), the morphogenetic movements of reptiles have scarcely been studied.

Our knowledge of the development of the early stages of crocodiles and alligators is based entirely on morphological studies, the most thorough being those of Voeltzkow (1902). The endoderm of the *Alligator mississippiensis* is already formed by the time the egg is laid and there appear to be no detailed descriptions of these earliest stages. According to Reese (1915) the mesoderm arises in two separate regions. Anterior to the blastopore it seems to form from the endoderm, whilst posterior to the blastopore it is derived from the ectoderm bordering the blastopore, the primitive streak.

D. REGULATION

Young reptile embryos probably all possess the same ability to regulate as do birds and mammals. If the blastoderm of *Lacerta vivipara* is cut into two parts during cleavage as it lies on the surface of the yolk, two separate embryos can form (Hubert, 1964). This experiment and its result are similar to that carried out in birds (p. 73).

Partial doubling of the axis is a phenomenon recorded for postnatal snakes discovered in the wild. This often takes the form of two separate heads on a single body (Strohl, 1925; Cunningham, 1937). These anomalies suggest that regulation is also a feature of the development of snakes for they are best explained by supposing that an initially single embryo has been partially divided into two.

The fact that regulation may take place gives us some confidence in our assumption that the embryos of reptiles undergo the same sort of processes as do the embryos of birds and mammals. Thus, we may presume that embryonic induction occurs in reptiles as it does in other amniotes. Further, we may expect that similar cell to cell communication takes place: that if a chick embryo scans itself continually during development (p. 14), then so does a reptile embryo. But in making these assumptions we must continually remind ourselves that these are assumptions, for the experimental analysis of reptile embryos has scarcely begun.

E. VIVIPARITY AND PLACENTATION

Not all reptiles lay eggs; a large number of lizards and snakes have adapted themselves to viviparity. The young are often born within their embryonic membranes, escaping from them shortly afterwards. These viviparous forms are found among different families and include many skinks, some anguids and the majority of vipers and sea snakes. Viviparity is advantageous to marine amniotes because the parents do not need to come ashore to nest, thereby exposing themselves and their offspring to great risk, as the egg-laying sea turtles do. Furthermore, the viviparous species are free to evolve extreme aquatic specialisation of body form. The extinct ichthyosaurs are thought to have been viviparous (Romer, 1966) and these fish-like reptiles must have been virtually helpless on land.

The advantages of viviparity to reptiles in general are probably also related to the problem of temperature. Unlike birds and mammals, reptiles are poikilothermic and tend to assume the temperature of their immediate surroundings. They are able to control their body temperature to some extent, however, by behavioural methods, basking in order to get warm and sheltering from excessive heat. It is true that the eggs of many oviparous reptiles are partially protected from extremes of temperature by being buried or laid in special sites such as heaps of rotting vegetation or termite mounds which act as natural incubators. Females of one species of *Python* at least (the Indian *P. molurus*) are believed to incubate their eggs, undergoing an increase in body temperature. A number of other lizards and snakes brood the eggs by coiling round them. Parental care of one sort or another is more widespread among reptiles than was formerly realised (see Bellairs, 1970). Nevertheless, it is reasonable to suppose that viviparous lizards and snakes often provide a more stable environment for their unborn young than the majority of oviparous ones which simply abandon their eggs after laying them. The viviparous mother can actively seek out optimal temperature conditions, besides providing better protection from dangers such as flooding and enemies. Viviparity in reptiles is often, though by no means invariably, associated with life in a relatively cold climate and it is well known that most, if not all, of the few reptiles which live under really severe climatic conditions are viviparous. The common lizard (*Lacerta vivipara*) and the adder (*Vipera berus*) whose ranges extend to the vicinity of the Arctic Circle are examples of this.

Viviparity in lizards and snakes differs in certain important respects from that in mammals. The eggs are almost always large and yolky like those of birds; in most viviparous forms the yolk is still the principal, if not the only, source of embryonic nourishment. The shell is lost or greatly reduced though a shell membrane may still develop. Weekes (1927) has shown that in some lizards it disappears when the chorion comes into contact with it and is perhaps partly digested by enzymatic activity. The oviduct also shows certain modifications (Boyd, 1942). The term ovoviviparity is sometimes applied to viviparity in reptiles, though some workers restrict its use to forms which do not have a well developed placenta. The ovoviviparous condition is exemplified in *Lacerta vivipara* where the placenta is rudimentary and the embryos can develop normally and hatch from their membranes when explanted from the mother and kept in a simple form of culture on a saline medium (Panigel, 1956).

Despite maternal behaviour, however, the embryos of many reptiles, both oviparous and viviparous, must be subject to greater variations in temperature than those of birds and mammals. This must be especially true of forms which inhabit temperate climes. There is much evidence that the embryos of some species, at least, can withstand a wide range of temperatures both under natural and experimental conditions and that their rate of development is on the whole accelerated by heat (below the lethal limit) and retarded by cold (A. d'A. Bellairs, 1970). Exposure to certain temperature changes during the earlier embryonic stages, however, has been shown to give rise to abnormalities in turtle (*Chelydra*) embryos (Yntema, 1960). The optimum temperatures for development in any species may well be related to the thermal preferences of the adult. Few adult reptiles can live for any length of time when their body temperatures are elevated above 45°C, and it is significant that Stephenson (1966) found that both adult and embryonic cells of certain Australian agamid lizards do not survive at 45° as well as avian cells. (For further remarks on the effects of temperature on development, see p. 206.)

F. FORMATION OF EXTRA-EMBRYONIC MEMBRANES

We have seen in Chapter 2 that, apart from certain specialisations, the extra-embryonic membranes of both oviparous and viviparous reptiles are basically similar to those of birds. Sufficient data is not available, unfortunately, for a critical comparative account of differences in the various reptilian groups. Conditions in crocodilia are especially poorly known.

In most reptiles, just as in birds, the amnion forms at about the time when the first somites appear but in *Sphenodon* and in *Chamaeleo chamaeleon* it develops earlier, at the time of gastrulation (Pasteels, 1956). In crocodiles it is starting to form when the neural folds first appear (Fig. 57). The main amniotic fold arises in front of the embryo and is continued as lateral folds which close over the embryo progressively from front to back until they meet another fold forming at the tail end. A peculiarity of some reptiles, however, is that the amnion becomes extended backwards over the tail region of the embryo as a tube, the posterior-amniotic tube, and it has been suggested that this forms a sero-amniotic cavity (discussion by Fisk and Tribe, 1949). It should be added that it is uncertain whether egg-white is present in the eggs of Squamata in any substantial quantities although it is certainly found in chelonian and crocodilian eggs. No reptile, so far as I know, produces its amnion by cavitation, as do certain mammals (p. 157).

There seems little doubt that the normal function of the amnion and its contained fluid is to prevent dehydration. An abnormal and deleterious activity of the amnion, however, has been shown to take place experimentally; that is, it may, if incised for the purpose of operating on the embryos, constrict around the tail to such an extent that this organ becomes amputated (Bryant and Bellairs, 1967). This experimental finding is of special interest since doubt has sometimes been thrown on the traditional idea that constriction by amniotic bands could be responsible for *in utero* amputation of parts of the limbs in human foetuses.

The allantois forms as in other amniotes by an outgrowth from the hindgut. In at least some reptiles, however, a modification has taken place in that the extra-embryonic allantoic blood vessels have become strung across the lumen of the allantois in a cytoplasmic bridge (Heimlich and Heimlich, 1950). It is thought that this is formed by a fold in the allantoic wall (Harrison and Weekes, 1925).

The yolk sac of reptiles forms as in birds from the periphery of the blastoderm and spreads over the surface of the yolk in much the same way. A peculiar feature in many, and perhaps in all, reptiles is that a superficial layer of yolk becomes separated by a cavity from the main body of the yolk (Fig. 57C). This is due to the downgrowth into the yolk of a part of the yolk sac. In oviparous forms, at least, this outer yolk layer subsequently becomes absorbed (Boyd, 1942). The significance of this event is not, as yet, understood.

Our knowledge of placentation in reptiles owes much to the work of Weekes (1927, 1935) on Australian skinks and the subject has been reviewed by Bauchot (1965), who lists the various species of lizards and snakes in which placentae have been described. Both yolk-sac and chorioallantoic placentae are found, often in the same species, but it is not clear whether one is ever present without the other.

Unfortunately, our knowledge of placental function in reptiles is based almost entirely on the idea that regions of placental exchange are those where highly vascularised maternal and foetal tissues are in close apposition. We have little direct knowledge to support such deductions. Panigel (1956), who injected radio-actively labelled salts into pregnant lizards (*Lacerta vivipara*, which has a poorly developed placenta), found very little radio-activity in the embryos.

Indirect evidence that substances pass from mother to foetus during the viviparous development of *Thamnophis s. sirtalis* has been derived by estimating the total contents of certain components of the egg and its embryo at different stages of development. Thus, Clark *et al.* (1955) showed that water, and probably amino acids, passed across the placenta to the foetus but that fats probably did not.

All the chorioallantoic placentae known in reptiles are of a relatively simple type, little or no erosion of the layers taking place, the two tissues merely lying in close contact. Weekes (1935) distinguished three types of chorioallantoic placentae which she arranged in a series of increasing complexity. At one extreme (Fig. 58A) the yolk content was high and the apposition of foetal and maternal tissues was a simple one. They lay in close contact with each other and were sometimes thinned out. At the other extreme (Fig. 58B) the yolk content was apparently reduced whilst the chorioallantoic placenta was more elaborate with extensive folding and vascularisation. She suggested that with a large yolk and yolk sac the main placental function was carried out by the yolk sac which formed

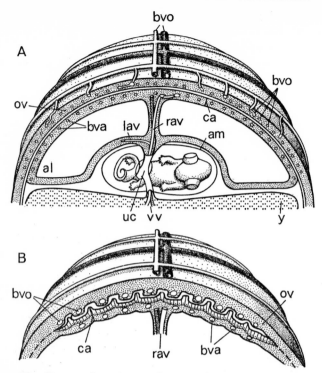

Figure 58. *Extra-embryonic membranes of certain viviparous lizards. Sections through the oviduct and extra-embryonic membranes. A. The simplest condition, where vascularised foetal and maternal tissues lie close together, but without any specialisation of the adjacent surfaces. A thin shell or shell membrane, (not shown), may lie between the surfaces in some species. B. The most advanced condition, in which foetal and maternal tissues are arranged in interlocking folds. In both figures, the right allantoic blood vessels are slung across the allantoic cavity in a cellular bridge (see text). al. allantoic cavity. am. amnion. bva. blood vessels of allantois. bvo. blood vessels of oviduct. ca. chorioallantois. lav. left allantoic blood vessels. ov. oviduct (cut wall). rav. right allantoic blood vessels. uc. umbilical cord. vv. vitelline vessels. y. yolk. (After A. d'A. Bellairs, 1970; based on Weekes, 1930.)*

an omphaloplacenta and which mainly imbibed water. With a decrease in yolk content, however, the respiratory and nutritive functions would take on more importance and these would be catered for by the chorio-allantoic placenta. These interesting suggestions still await experimental investigation.

The yolk sac placenta itself has been described in some detail for the New Zealand lizard, *Hoplodactylus maculatus*, and involves the enlargement of certain secretory cells in the wall of the uterus (Boyd, 1942). In the Australian reptiles described by Weekes, the remains of the yolk sac was withdrawn into the body at birth, in much the same way as the yolk sac of the chick is also withdrawn. In certain other reptiles (e.g. the *Hoplodactylus* geckonids of Boyd) it is left behind with the membranes and is eaten by the young lizards. In the viviparous English adder, *Vipera berus*, where the yolk sac is retracted into the body at birth, the infant snakes have only a few weeks of activity before hibernation begins. No feeding takes place during hibernation and the young snakes appear to be nourished before this period entirely by this yolk in the yolk sac (Bellairs *et al.*, 1955).

G. SUMMARY

Much less experimental work has been carried out on reptiles than on birds or mammals, although modern techniques now enable certain reptile embryos to be maintained *in vitro*. There is evidence that regulation can occur, as in other vertebrates. The cleavage and gastrulation of various reptilian species is compared with the same processes in birds. One of the main differences is that an archenteric canal forms in reptiles, though not in chicks, and probably not in most other birds. Although a structure called the primitive streak forms in reptiles, it is probably not homologous with the primitive streak of birds. We have no knowledge about the molecular biology of the early stages of reptile development. Not all reptiles are oviparous, many lizards and snakes being viviparous and possessing a placenta. This may be either a yolk sac placenta or a chorioallantoic one or both. Our knowledge of these placentae is almost entirely morphological, and we have little direct evidence about the substances that may pass from mother to embryo.

CHAPTER 9

ASPECTS OF THE BIOCHEMISTRY OF AMNIOTE DEVELOPMENT

THE present-day embryologist differs from his predecessor of even twenty years ago in that much of his inspiration comes from molecular biology. This is because he now sees development as a programme of protein synthesis; as the embryo grows it needs more protein and as it differentiates it needs different kinds of protein. Molecular biology is concerned principally with explaining how protein synthesis takes place. Much of this chapter will therefore be concerned with the molecular biology of amniote development.

Not all biochemical embryology is of course slanted toward molecular biology. Many workers have been concerned with collecting data which can be related directly to morphological processes. For example, investigators have studied the first appearance of collagen, or the specific effect of a teratological agent, or have enquired into which amino acids are necessary for differentiation. The accumulation of basic information of this type continues to provide a ground work for the more recent studies and will, therefore, also be considered in this and the following chapter. The earlier work of this nature will not be discussed since it is fully reviewed by Needham (1931, 1950) and Brachet (1950).

A. METHODS OF INVESTIGATION

Basically there are two methods of studying the biochemistry of an embryo: by direct chemical analysis of a lump of tissue, or by histochemical tests carried out on sections examined by microscopy (Deuchar, 1966 for discussion of the advantages and disadvantages of the two approaches). Analyses are frequently used as an end in themselves to obtain a description of the chemical events taking place in a developing embryo. They are also used to supplement other techniques.

The way in which a cell develops is essentially the result of an interaction between its genes and the environment (but see p. 11 *et seq.*). Many experiments that are carried out aim at either altering the DNA of the embryo (the genome) or at changing the environment. Examples of how

G

the genome may be altered are by nuclear transplants (p. 10), cell hybridisation studies (i.e. by fusing together two cells, p. 11, 185) or cell transformation (i.e. by the direct action of agents such as viruses on the DNA of the genome, p. 185). Examples of how the environment may be altered are by the action of drugs (p. 203 *et seq.*), hormones (p. 272) or anti-metabolites (e.g. p. 189, 198), the withholding of necessary precursors (p. 197), the action of other cells, either of the same or of different type (p. 107) or by general changes in the environment, such as temperature or amount of oxygen (p. 207).

B. MOLECULAR BIOLOGY OF DEVELOPMENT

The basic concepts of molecular biology, particularly the classical theory produced by Jacob and Monod to explain the control of protein synthesis in bacteria, are now so well known that it is unnecessary to discuss them in detail. Simple accounts are given on p. 30 *et seq.* Further information can be found in any advanced treatise of biochemical genetics.

Although it has gained wide acceptance, however, the Jacob and Monod scheme is not the only one possible. We shall briefly consider here three possible ways in which an embryo might control its protein synthesis. For authoritative discussions on the subject see Davidson (1968) and Denis (1968).

(a) *Control by altering the DNA*

A few examples are recorded in which the amount of DNA per cell becomes reduced during development. The best known is in the somatic cells of the annelid worm, *Ascaris*. In some insects there is a loss of certain chromosomes during development. Generally, however, it is thought that in most organisms the *amount of DNA* remains the same in all diploid nuclei present in the body throughout life, except during mitosis when it becomes temporarily doubled, and during meiosis when it becomes halved. The evidence in favour of this constancy is based largely on chemical analyses or ultra-violet microscopy.

The fact that the nucleotide sequences that make up the various DNA's seem to remain the same in different cells as an animal develops, has been taken to support the idea that the *structure* of the DNA also remains unchanged (see discussion by Denis, 1968). The evidence is based mainly on molecular hybridisation studies. These involve taking a strand of DNA from one cell and presenting it with RNA from another. If the two molecules pair and attach to one another, it is usually assumed that the arrangement of the nucleotides is the same. Thus, if DNA from an advanced embryo is able to link up with RNA from a young one of the same species, it is considered that the DNA has undergone no significant change in development. Recently, however, the basis of this argument has been questioned. Britten and Kohne (1968), who have examined the kinetics involved in reassociations of this type, have shown that the reactions occur at a much greater speed than would be expected if there were just a single copy of each gene present in the genome of each cell. They have shown that in a variety of animals they have studied, including *Xenopus*, thousands of copies of many genes are probably present; thus, a percentage of the genome of some animals probably consists of continually repeated sequences of the DNA. Not all this DNA will be used for transcription, and so some of it is considered to be 'redundant'. (Not all the genome is repeated, some 'single copy' DNA occurring only once.) The significance of this repetition is that during hybridisation studies RNA-DNA hybrids may form between any of the repeating genes: this means that the technique can no longer be regarded as a true test of small changes in the DNA.

Further evidence that the genome remains unchanged during differentiation, however, has been provided by experiments in which the nuclei of *Xenopus* eggs have been replaced by nuclei taken from gut cells of tadpole stages. These older nuclei have been able to direct development as well as if they had been nuclei of the fertilized egg (Gurdon, 1962).

A well-known 'modification' of the DNA that is thought to occur regularly in female mammals is the phenomenon known as 'inactive X' (Chapter 1). In this, one of the X-chromosomes becomes heteropycnotic (i.e. contracted into a densely-staining clump). It has been suggested that the inactivation of this X-chromosome is due to the fact that the chromatin has become physically condensed, thus preventing synthesis of mRNA (Hsu, 1962; Hsu *et al.*, 1964.)

(b) Control at the level of transcription

Even though the amount of DNA and its structure may remain unchanged during development, it is still possible that part of the DNA may be made *unavailable* at certain times and available at others. This is the theory of 'variable gene activity'. Apart from the special case of masked messenger RNA in the early stages of development (p. 59), most of the evidence is conducive to the idea that control of differentiation is brought about by controlling the transcription of the DNA. (For a summary of the evidence see Denis, 1968.)

The Jacob-Monod scheme, originally devised for bacteria, is based on this concept. The embryos of higher vertebrates are, however, considerably more complex than the bacteria. Even their DNA differs, being present largely in the chromosomes in the nucleus of the one, whereas there is no nuclear membrane in the other. The changing nature of protein synthesis in development is an elaborate one and few scientists would now suppose that a model devised for bacteria would be adequate to explain the situation as it stands in a complex multicellular organism.

We have seen that many of the sequences of DNA are repeated in the genome of animal cells. It seems likely that there is a continual change in the activity among these repeated sequences as differentiation proceeds. At present we do not know, at least in multicellular organisms, whether all genes are normally active unless specifically suppressed, or whether they are normally inactive unless specifically activated. Recently, Britten and Davidson (1969) have suggested that all the DNA sequences are inactive in transcription unless they are specifically activated. They propose that the activators are RNA molecules that are confined to the nucleus and are possibly produced by the 'redundant' DNA. Generally, it is considered that histones, and possibly other nucleo-proteins that form part of the chromosomes, play an important rôle in repressing gene activity.

(c) Control at the level of translation

According to this scheme, the entire information contained in the DNA is transcribed but a block occurs in the mRNA which prevents translation. This mechanism has been demonstrated to exist in the later stages of oögenesis, so that the messenger RNA is masked and only becomes unmasked at fertilization. We have also seen that other types of masking of mRNA may play a rôle in differentiation at later stages (p. 65.)

C. DISTRIBUTION AND AMOUNT OF NUCLEIC ACIDS IN AMNIOTE EMBRYOS

One of the generalisations that is frequently made is that the oöcyte contains almost enough DNA to last until the beginning of gastrulation. The total amount of DNA in the embryo, therefore, remains more or less constant until a new wave of synthesis starts at gastrulation (Chen, 1967). However, not all this DNA is present in the nuclei.

In the chick, DNA has also been found in the extra-embryonic yolk (Solomon, 1957b). It is probably also present in the intra-cellular yolk spheres since these have been found to take up H^3-thymidine (Emanuelsson and von Mecklenburg, 1968). A number of embryologists have suggested that in amphibians much of the cytoplasmic DNA is bound to the yolk, though this is not completely proved, for it is based mainly on staining by the Feulgen method; Williams (1965) has pointed out that this reaction may sometimes give a false positive result in the presence of lipid. As we have seen (p. 39), lipid is a major component of yolk.

Other possibilities are that the cytoplasmic DNA is derived from nucleolar material extruded from the nucleus (p. 34). It will be recalled that the nucleolus although rich in RNA also contains some DNA templates (p. 35). A further source of cytoplasmic DNA is the mito-chondria; one of the earliest demonstrations of mitochondrial DNA was in the chick embryo (Nass and Nass, 1963).

It used to be thought that there was no increase at all in the amount of DNA in the embryo until gastrulation. It now appears that some increase does take place in certain amphibians (Moore, 1959; Bristow and Deuchar, 1964), in chicks (Emanuelsson, 1961, 1965) and in the mouse (Mintz, 1965). In amphibians this new synthesis may not be enough to supply the additional requirements imposed on the embryo by the continual formation of new nuclei during cleavage. Various authors (e.g. Brachet, 1965) have therefore suggested that during cleavage some of the nuclear DNA is derived from the cytoplasmic reserve. Others, however (e.g. Dawid, 1965), have concluded that this cytoplasmic DNA cannot be used directly for replication of the chromosomes but must first be broken down and re-synthesised.

Our knowledge of the distribution and amount of ribonucleic acids at fertilization and cleavage has been discussed in Chapters 3 and 7. It will be recalled that amphibian embryos form practically no new RNA during cleavage, whilst mouse embryos start to synthesise it shortly after fertilization: chick embryos appear to occupy an intermediate position, and start to manufacture new rRNA in mid-cleavage. The composition of the rRNA (28S and 16S) stays fairly constant in the chick embryo throughout the first seven days after the primitive streak has formed (Lerner and Bell, 1963).

It has been suggested that the nucleic acids in the yolk (and also in avidin, a protein of the egg-white) form a reserve which may be drawn upon as required (Emanuelsson, 1966; Solomon, 1965).

As we shall see, the extracellular yolk is probably not used by the chick until about the second day of incubation, so that it might be expected that these reserves would not be tapped until then or even later. In fact, this reserve is not essential even at this stage for chick development for Solomon (1957b) demonstrated that entire chick blastoderms were able to synthesise both DNA and RNA when explanted at 5–10 somites on a synthetic medium in which the sole source of carbon was glucose. Under these conditions the total amounts of both DNA and RNA increased during the next 20 hours. The increase took place throughout the entire blastoderm but was greater in the embryo than in the extra-embryonic areas. This increase also occurs under normal *in ovo* conditions during this period (Solomon, 1957a, b; Emanuelsson, 1966). The factors which normally control the content of the nucleic acids are not known, but there is evidence that the total RNA of chick embryos in the later stages of development can be increased by growth hormone (Wang *et al.*, 1953). By contrast, the concentration of DNA does not seem to be affected by this hormone (Wang *et al.*, 1953).

After the initial stages, the picture becomes less clear, for estimates of total DNA or RNA have little meaning in a complex organism such as a chick of four days incubation or more. Now we must consider results based on histochemical or autoradiographic inspection of sectioned material, or on analysis of homogenates of dissected organs. For example, Gallera and Oprecht (1948) reported that at the start of incubation the chick blastoderm exhibited little basophilia but that as differentiation proceeded different regions, such as neural tissue and somites, became more basophilic. They attributed this basophilia to RNA. Gluck and Kulovich (1964) studied the distribution of nucleic acids in tissue sections of the developing chick by means of acridene orange, a fluorescent dye which binds nucleic acids, turning green with DNA and red with RNA; great diversity of distribution was found in different cell types. Similarly, Birge (1962), who found a high rate of RNA during differentiation of the nervous system, reported fluctuations in individual cells.

We have already seen that addition of RNA to chick explants may possibly affect differentiation (p. 121), and even induction. Emanuelsson (1962) achieved an increased rate of cell division when explanted chick blastoderms were treated with chick RNA. Similarly, Ranzi *et al.* (1962) reported that the ribonucleoproteins extracted from the organs of an adult fowl and introduced on to the chorioallantoic membrane of a young chick may bring about results that are specific according to the organ of origin. Thus, skeletal muscle ribonucleoprotein was found to induce the transformation of mesenchyme cells into muscular elements, whereas liver ribonucleoprotein induced the formation of typical glandular cells.

The converse experiment of treating embryos with RNAse has been carried out on chick embryos of 24–48 hours as they lay *in situ* within the shell (Lanot, 1963). The organs most affected were the somites, amnion and tail bud. Lanot suggested that the RNAse had interfered with cell movements. The effects of RNAse have also been investigated on amphibian eggs (Ledoux *et al.*, 1955) and on sea urchin eggs (Cormack, 1966) and in both groups abnormalities formed.

D. ATTEMPTS TO TRANSFORM VERTEBRATE EMBRYOS OR THEIR CELLS

Although there is a certain amount of evidence in the literature that transformations of vertebrate cells can be brought about in tissue culture (e.g. Bradley *et al.*, 1962) attempts to show that similar transformations can be obtained in the more complex system of an entire embryo have been very limited. One of the earliest experiments aimed at demonstrating that the genetic processes in vertebrates were biochemically similar to those in bacteria was that of Benoit and his colleagues, who injected DNA from Khaki Campbell ducks into Pekin ducklings. The ducklings grew to resemble the Khaki Campbells in certain ways and, even more significantly, these characteristics were thought to be handed on to their offspring (Benoit *et al.*, 1958). These exciting results could not, however, be repeated, even by the same authors, and the original procedure has since been criticised (for example, by Billett, Hamilton and Newth, 1964). Attempts to bring about genetic changes directly by treating the chick embryo with foreign DNA have also been disappointing (e.g. Martino-vitch *et al.*, 1962).

One of the ways in which the effective genome of a cell can be altered, however, is by infecting the cell with a virus. This is equivalent to introducing additional DNA into the cell. Most of this work has so far been concerned with the effects of viruses on populations of more or less identical cells in tissue culture (clones) (for discussion, Dulbecco, 1965; Ebert and Kaighn, 1966; Ebert, 1969).

We may expect that once the interaction between the virus and the cellular DNA is better understood, it will be possible to use viruses as tools for studying nucleic acid activity in embryos. An example of this type of investigation is the work of Taderera (1967) who showed that chick lung mesenchyme infected with Rous sarcoma virus loses its ability to induce the differentiation of mouse lung epithelium. Inactivated

Sendai virus has also proved a useful cytological tool in inducing the fusion of cells to one another in tissue culture to form cell hybrids (p. 11). Some recent work on cell hybrids of this type is described by Ebert (1969). The subject of cell transformation in general has been reviewed by Olenov (1968).

E. THE BLOCKING OF PROTEIN SYNTHESIS

A fruitful approach to our understanding of the control of protein synthesis has been that of blocking the activities of the DNA.

(a) Actinomycins

A number of workers have treated chick embryos with actinomycin A or D or with mitomycin C. These antibiotics, though apparently acting in different ways, all affect production of messenger RNA and hence of protein synthesis. The evidence suggests that actinomycin D selectively inhibits the production of nuclear RNA, especially messenger RNA, by combining with the DNA template. On the other hand, mitomycin C inhibits DNA synthesis from DNA precursors without affecting RNA (Schwartz et al., 1964).

Various authors who have treated embryos with actinomycin D have reported slightly different effects for different organisms and, not surprisingly, the results also varied according to the stage at which the embryo was treated (discussion by Brachet, 1965; Tiedemann, 1967a). McKenzie and Ebert (1960) also showed that different results were obtained if the chick blastoderm was treated from the ventral rather than the dorsal side. If explanted chick blastoderms are exposed to actinomycin D at about the head process stage, the neural tissue is generally affected (Brachet and Denis, 1963; Heilporn-Pohl, 1964; Ranzi, 1968) whereas if the actinomycin D is injected into the yolk sac at a later stage of development the main effect is on the skeleton (Pierro, 1961). By contrast, if mytomycin C is injected into the yolk sac, a variety of non-specific anomalies are formed (Kury and Craig, 1967). Heilporn-Pohl, however, found that the incorporation of H^3-uridine by the embryo was inhibited by at least 50% after treatment, indicating some disturbance of the RNA. But interesting and provocative though these results are, they cannot be regarded as conclusive proof of a direct interference with the DNA-RNA programme, for there is always the possibility that the antibiotic has had other effects on the embryo that are as yet unexpected.

(b) Histones

Another way of blocking the DNA-RNA information transfer is by treating the embryo with histones. There is now a considerable literature on micro-organisms to support the idea of Stedman and Stedman (1950) that histones are at least closely associated with inhibition of gene action, if not actually gene suppressors themselves (discussions by Busch *et al.*, 1964; Swanson and Young, 1965; Davidson, 1968). They appear to be a regular component of all cell nuclei, and probably act by affecting the RNA-polymerase (discussion by Davidson, 1968).

Moore (1963), working with amphibian embryos, argued that if histones are the normal DNA inhibitors then the nuclei of the early stages should be especially rich in them, for these nuclei are particularly inactive. Using histochemical methods, Moore demonstrated a high histone level in certain hybrids.

It might be expected that during differentiation there would be marked changes in the histones present. Kischer *et al.* (1966) carried out an amino acid analysis of the histones of chick embryos from the primitive streak stage up to seven days of incubation. They were unable, however, to find any essential qualitative or large quantitative differences which could be correlated with particular stages of development. It may be argued that although Kischer *et al.* were careful to remove the area opaca from the younger stages and all the extra-embryonic membranes from the older ones, nevertheless, the homogenisation procedure inevitably led to a number of very different organs becoming included in the same sample. Any differences in histone changes might in this way be masked. Against this argument is the fact that Lindsay (1964) and Kischer and Hnilica (1967) found that the composition of the histones in the chick embryo changed very little as the organs developed.

Few authors have studied the direct effects of histones on embryos, but Sherbet (1966) who treated chick blastoderms with calf thymus histones found similar effects to those produced by actinomycin D. In particular, damage seemed to be largely in the brain tissue if the blastoderm was treated at the primitive streak stage or head process stage. Malpoix and Emelinckx (1967) also similarly found that calf thymus histones resembled actinomycin in their action on the chick blastoderm in that they inhibited the development of the nervous system.

(c) Other blocking agents

In addition to the actinomycins and the histones, many other substances have also been used in the attempt to block protein synthesis.

For instance, Billett *et al.* (1965) treated young chick embryos with chloramphenicol which is considered to interfere primarily with messenger RNA. The main result was failure to close of the neural tube.

Perhaps it is not surprising that so many of these substances affecting DNA-RNA activity should show their greatest effects upon the neural tissue, for a great deal of new protein formation takes place in the development of that system. Unfortunately however, not every investigator has checked that protein synthesis is indeed affected by his experiments.

In considering these interesting results, moreover, it is important to bear in mind that all these experiments suffer from an overwhelming theoretical difficulty and one that is frequently glossed over; namely, that we can never be certain that some of the results are not due to side effects of the chemicals used. Although there is powerful evidence that these substances behave in a particular way in bacteria or under test-tube conditions, there is no guarantee that they do not have additional effects in multicellular organisms. We still cannot eliminate the possibility that they may also affect such factors as the active transport across the cells, the osmotic pressure or the carbohydrate metabolism. So far there seem to be no satisfactory control experiments for investigating these possibilities.

(d) Mitotic inhibitors

Another group of substances that interfere with protein synthesis is aimed specifically at attacking the cell when it is in the process of mitosis. For example, embryos may be treated with aminopterin or with amethopterin. These agents, which are well known mitotic inhibitors, act on the folic acid metabolism of the cell (discussion by Kihlman, 1966) causing a deficit in nucleic acid precursors and thus bringing about an inhibition of RNA and DNA synthesis.

When chick embryos are treated with folic acid antagonists the results vary according to the stage of the embryo at the time of treatment, the dosage of the drug and the method of application. Nevertheless, certain general conclusions can be drawn. When chick embryos are treated at four days of incubation or later (Karnovsky *et al.*, 1949; Cravens, 1952) with folic acid analogues, severe stunting results and a wide range of malformations is produced. In the earlier stages the main effects are on the development of the vascular system (Wagley and Morgan, 1948; O'Dell and McKenzie, 1963) as well as on the closure of the neural tube and the differentiation of the somites (Bellairs, 1954b; O'Dell and McKenzie, 1963). These results are somewhat similar to ones obtained in amphibian embryos (for example, Grant, 1965). In experiments performed *in ovo* the blastoderm continued to expand beneath the vitelline membrane in much the normal way (p. 21), for at fixation it covered an area comparable with that of the untreated controls; the area opaca of the treated embryos was present as an open meshwork and it was concluded that with the reduction in the mitotic rate the number of cells had been inadequate to cover the region, so that the expanding blastoderm had torn itself into holes (Bellairs, 1954b). Similar results were obtained by O'Dell and McKenzie (1963) who treated chick blastoderms *in vitro*.

We may now ask, what evidence exists that the folic acid antagonists affect solely the folic acid metabolism? The evidence is from 'protection' experiments; e.g. the deleterious effects of 2·5 μg of aminopterin on chick blastoderms can be overcome by simultaneously dosing the embryo with 34 μg of *Leuconostoc citrovorum* factor (Bellairs, 1954b). It is known from the study of micro-organisms or of cells growing in tissue culture that the effects of aminopterin and amethopterin can largely be overcome by simultaneous application of *Leuconostoc citrovorum* factor or folinic acid, provided that the dosage of the inhibitors is not excessively high. The mechanism of action of the *Leuconostoc citrovorum* factor is far from clear but it is likely that the aminopterin replaces folinic acid, *Leuconostoc citrovorum* factor or some similar compound when it enters a cell (Jacobson, 1954). At any rate, there appears to be some specific relationship between the two compounds. Such a relationship apparently exists in embryos also, and so it has been assumed by some investigators that aminopterin acts in the same way in the entire organism as in an isolated group of cells. This argument is generally considered to be supported by the fact that the same cytological events are visible, viz. a reduced mitotic rate with an arrest of cells in metaphase, together with occasional patches of necrotic cells. Thus it is often deduced that the effects of mitotic inhibitors of this type on embryos are entirely the result of mitotic inhibition.

Such a viewpoint seems somewhat naive, however, for embryos are complex developing organisms and not just aggregates of single cells. Indeed, organs with relatively low mitotic rates may be more badly affected than those with a high rate (Bellairs, 1954a). Further, other mitotic inhibitors, e.g. colchicine (p. 208), exist whose metabolic action is quite different (for a recent discussion on the effects of colchicine and other anti-mitotic substances on cells, see Deysson, 1968). These inhibitors all impose their own characteristic pattern on the embryo (Bellairs, 1954b). It seems probable, therefore, that each may produce additional effects to the ones we already know.

F. PROTEIN SYNTHESIS DURING DIFFERENTIATION

The major component of living systems is protein which may exist in a variety of forms or be bound with other materials such as with lipids in the lipoproteins of cell membranes, or with nucleic acids or with carbohydrates. In a chick blastoderm about 84% of the dry weight consists of proteins (Romanoff, 1967) and a significant proportion of these are enzymes.

All cells possess a number of essential proteins without which they cannot survive. But as they become differentiated they acquire specialised proteins. For instance, developing muscle cells (myoblasts) acquire myosin. Formation of these special proteins occurs before any visible differentiation of the cells or of the organ has begun. Thus, it is the *biochemical* changes which mark the first steps in differentiation of an organ. Possibly the most important problem in development is to find out what makes a cell begin to manufacture these special molecules.

A number of systems have now been studied in some detail. Two of these, the production of myosin by developing muscle cells, and of chondroitin sulphate by differentiating cartilage, are discussed on pp. 266 and 252 respectively. Two further examples, haemoglobin synthesis and the production of lens proteins will be considered here.

(a) Haemoglobin

A series of different kinds of haemoglobin are manufactured during the life cycle of an individual. For example, in the human, those that are present in the young embryo (embryonic haemoglobin) become replaced during the later stages of pre-natal development by foetal haemoglobin, and these in turn are exchanged for adult haemoglobin (Fig. 59) during the period just before and just after birth (Huehns and Shooter, 1965). The number of haemoglobins that an individual has during its life cycle depends on its species. Man appears to possess at least two types of adult haemoglobin and at least two embryonic ones (Huehns and Shooter, 1965); the fowl on the other hand probably has four adult and three embryonic ones (Hashimoto and Wilt, 1966). The haemoglobins of one species are distinctive and differ from those of another.

All these haemoglobins possess the same basic structure of four haem groups and four polypeptide chains. They differ from one another, however, in the amino acid composition of their globin (reviews by Solomon, 1965; Wilt, 1967).

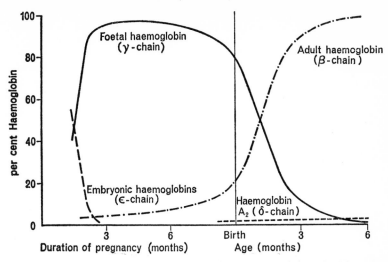

Figure 59. *The different types of haemoglobin present during the life cycle in humans. (From Huehns and Shooter, 1965.)*

The developing haemoglobin has proved to be one of the most useful systems available for studying biochemical differentiation. This is because the red colour of the haemoglobin provides a clear indication that the special protein has been synthesised. Technically, one of the easiest organs in which to study this process is the area vasculosa of the chick blastoderm (Fig. 21D). This is because the earliest (embryonic) haemoglobin appears in that region and this is easily accessible to the experimenter (p. 250 for description of blood islands). The area vasculosa can be readily extirpated and can be grown on its own or with other tissues *in vitro*, or on the chorioallantoic membrane. It may also be treated directly with various chemicals. Experiments of this sort have demonstrated that the potentiality to form haemoglobin is determined even before the primitive streak has formed (Fig. 60).

Much of our knowledge of haemoglobin synthesis comes from experiments in which the chick blastoderm has been treated with antimetabolites. These have led to the conclusion that, during differentiation, the control of haemoglobin synthesis takes place mainly at the level of translation.

The effects of the antimetabolites vary at different stages in development of the area vasculosa. If the blastoderm is treated at the head process stage or later with 5-Fu (5-deoxyfluoridine) or BUDR (bromodeoxyuridine), both of which affect the transcription of DNA to mRNA, the

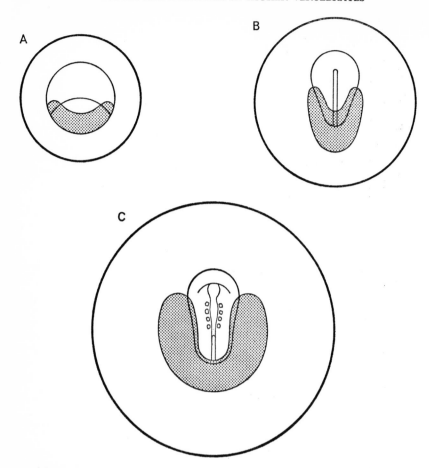

Figure 60.   *Haemoglobin formation. Pieces of the chick embryo at different stages of development were isolated on the chorioallantoic membrane and, after incubation, were examined for haemoglobin. Grafts from the shaded areas were found to be capable of giving rise to haemoglobin, even though the youngest ones had been prevented from taking part in the usual cell migrations. A. Pre-streak. B. Long primitive streak. C. 4–6 pairs of somites stage. It appears that the potentiality to form haemoglobin is determined at an early stage. (After Settle, 1954.)*

haemoglobin forms in the normal way. But if these agents are applied earlier than the head process stage, the haemoglobin formation is inhibited. It appears, therefore, that the mRNA specific for haemoglobin has already been formed on the DNA by the head process stage (Hell,

1964; Wilt, 1966). This idea is supported by the fact that actinomycin, which also affects the production of mRNA (p. 186), similarly fails to affect the initial stages of haemoglobin synthesis. Most authors, therefore, have concluded that mRNA is present many hours before it is used.

The final control mechanisms in haemoglobin synthesis are concerned with the metabolism of the haem component. If the embryo is treated with haem, this stimulates the synthesis of haemoglobin (Wilt, 1966).

(b) Lens proteins

The lens contains several highly specific proteins—the crystallins—which are found in no other organ. In mammals, α, β and γ crystallins are present, whilst in birds there are only α, β and Δ crystallins. The lens itself is formed from the epithelium overlying the eye-cup. It has long been known that it is induced by the eye-cup (for an account of the classical experiments on lens inductions, see Spemann, 1938). As the lens cells develop they begin to synthesise special proteins. In the chick embryo the Δ-crystallins can be detected when only a thickening of the epithelium (a 'placode') is visible (Zwaan and Ikeda, 1966). This thickening is the earliest visible sign of the developing lens.

In chick and mammalian embryos treated with tritiated thymidine, certain of the developing lens cells become labelled. The position occupied by these cells in the lens rudiment changes, however, as the organ develops. These experiments show that certain cells shift in position from the lens epithelium into the fibre region. Further, since cells will only take up tritiated thymidine during the S-phase of mitosis (p. 258), these experiments have also indicated that cell division stops in each cell immediately before it moves into the fibre area (discussed by Modak et al., 1968).

As the lens vesicle of the chick forms, however, the Δ-crystallins rapidly appear in all its cells. The α-crystallins can be detected slightly later and the β-crystallins soon after that. There is an interesting correlation between the structure of the cells in the different parts of the lens and the type of crystallin present in them. The distribution of both Δ and α-crystallins, though widespread at first, subsequently becomes more restricted; the Δ-variety becomes confined to the cells in the fibre area and the α restricts itself eventually more to the epithelial cells and less to the fibres. As the crystallins appear, the lens cells lose their nuclei (Stewart and Papaconstantinou, 1967); thus the further differentiation of specific proteins in these cells must depend on mRNA produced before the nuclei are lost.

G. ENZYME ACTIVITY DURING DEVELOPMENT

Until recently, embryologists interested in enzymes have concentrated more upon the acquisition of facts about the occurrence of enzymes than about their action in controlling the developmental processes. Despite all the technical difficulties involved in the work, an immense amount of information has been acquired about the enzymes of many types of embryos, much of it being concerned with the chick and a certain amount with mammalian embryos. It would be inappropriate to catalogue it all here, as several reviews exist by well-known experts (e.g. Herrmann and Tootle, 1964; Moog, 1965; Papaconstantinou, 1967; Wilt, 1967).

(a) Enzyme changes in development

The newly laid egg is poor in enzymes compared with the late embryo. Indeed it used to be thought that no enzymes were present in the earliest stages, at least not in an active state, though there is now evidence that even in oöcytes enzymes are present. For instance, glycolysis apparently takes place in some amphibian oöcytes (Williams, 1965). It is now generally considered that these enzymes are of maternal origin. Similarly, it is thought that maternal enzymes are present in birds' eggs, though we have no direct evidence as yet. Indirect evidence is, however, present in the fact that all oöcytes, and hence all fertilized eggs, contain large numbers of mitochondria. It is unlikely that these mitochondria would be so abnormal as to possess no enzymes. Some enzymes, e.g. liver hydroxylases, appear only late in development and then increase rapidly in activity, whilst others (e.g. cholinesterase in the central nervous system) appear early and increase slowly (Herrmann and Tootle, 1964). The activity of adenosine deaminase in the liver of the chick embryo reaches its peak at seven days of incubation and then remains unchanged throughout life (Solomon, 1960). The activity of lactic, malic, and glutamic dehydrogenases in this same organ increases two- to three-fold between the seventh and fifteenth days of incubation and then drops (Solomon, 1959). Changes of this type must be related to other events occurring within developing tissues. Some of these changes may be correlated with the general metabolic events experienced by any living organism, but many of the more dramatic ones that appear to occur as a regular event at a particular stage of development may be related to specific developmental changes. Tantalisingly, it is as yet seldom possible to make these correlations. An exception is the work of Ebert (1953), who was able to correlate the activity of myosin with its morphological appearance (p. 209).

During development, certain structural changes take place in the mitochondria. For instance, the mitochondria of the presumptive neural plate ectoderm of young chick embryos contain little matrix in sections examined by electron microscopicy and their cristae are easily disrupted, whilst in the differentiated neuroblast of the seventh day, the mitochondria are more electron opaque and their cristae are seldom damaged (Bellairs, 1959b). It seems probable that changes of this type might be associated with enzymatic development. Weber and Boell (1962) were able to show differences in enzyme activity of the mitochondria at various stages in the development of *Xenopus* embryos and suggested that these were correlated with structural changes (for a discussion of the rôle of mitochondria in development, Hermann and Tootle, 1964; Moog, 1965).

(b) Enzymes and protein synthesis

The building-up and breaking-down of proteins is not confined to developing tissues and occurs at times in most differentiated cells. For example, protein synthesis takes place during mitosis and also in the formation of many secretory products.

Each protein is composed of a number of amino acids. There are about twenty different amino acids in nature, and the ones that are used and the order in which they are arranged affect the properties of the protein molecule.

Before an amino acid can become incorporated into a protein it must be activated by an activating enzyme which joins it to the RNA. Each amino acid has its own activating enzyme. During activation, the amino acid reacts with a triphosphate molecule, such as ATP (adenosine triphosphate). The activated amino acid then becomes attached to the transfer (soluble) RNA and is carried to the ribosomes (p. 31). Here it is released and used by the cell to form protein (for further discussion, Staehelin, 1965).

Amino acid activation has been studied in a number of embryonic tissues. Deuchar (1961) showed that if leucine was presented to chick embryos it promoted greater activation in somite mesoderm than in other embryonic tissues.

(c) Enzymes and protein breakdown

As well as possessing enzymes that enable proteins to be formed, tissues also contain enzymes for breaking down proteins. The best-understood are perhaps the acid hydrolases (especially acid phosphatases) and the cathepsins. The latter play an important rôle in tail regression of *Xenopus* larvae (Weber, 1965, 1969). Acid phosphatases are a characteristic component of lysosomes (p. 267) and are therefore present in regions of the embryo where cell death occurs. They are also present in extra-embryonic tissues (Beck and Lloyd, 1966). It seems likely that the acid phosphatases in a degenerating tissue may not all be derived from the breaking down of lysosomes. In the involution of the tail in the larva of *Xenopus* there was a rise in activity of these enzymes which was too great to be attributed to lysosome breakdown (Weber, 1969). It appears that this rise is partly due to the synthesis of acid phosphatases by macrophages which invade the tail region.

In addition to the breakdown of cells, a considerable amount of non-cellular protein must be affected by enzymes during the utilisation of the yolk in reptiles and birds. It is significant, therefore, that the chick yolk is well supplied with proteases (Williams, 1967).

H. AMINO ACIDS AND OTHER PROTEIN PRECURSORS

Deuchar (1963a) has discussed the question of whether there is sufficient amino acid readily available at all times for synthesis of every type of protein, and has drawn attention to the importance of the 'free amino acid pool' in embryos. This pool is derived not only from the breakdown of yolk proteins but also from the autolysis of certain cells and tissues that occurs as a normal event during embryonic life. Since the breakdown of both yolk and of tissue proteins does not take place uniformly throughout development, the composition of the amino acid pool may be expected to vary. Deuchar (1962, 1963a), who has reviewed much of the work on amino acids in development, has shown that in *Xenopus* at least, several 'essential' amino acids are low in concentration at the beginning of development but become present in increasing amounts as gastrulation proceeds. These changes in *Xenopus* are probably correlated mainly with yolk breakdown; it is possible that the position in birds is similar.

In the chick embryo, nutrients are taken up from the yolk though there is little clear-cut evidence as to the form in which they enter the cells.

Walter and Mahler (1958) injected S^{-35} labelled proteins, peptides and amino acids into the yolk sac and subsequently concluded that proteins were taken up more readily than other substances. This work has, however, been criticised by Williams (1967) on the basis that it is difficult to draw conclusions from this type of experiment in view of our ignorance about the sizes of the amino acid and protein pools, and by Deuchar (1962) who points out that the proteins might have been broken down before being absorbed.

The very structure of the young chick embryo has, however, made it a most suitable system for studying the uptake of protein precursors *in vitro*. It is possible to explant the young embryo and supply it with controlled nutrients (e.g. agar gel fortified with known amounts of amino acids). This method of investigation was introduced by Spratt, who concluded from inspection of the embryos after incubation that glucose was the only essential additive required (Spratt, 1954). Hayashi and Herrmann (1959) showed further that a glucose-containing medium promoted a large increase in DNA and a small increase in protein glycogen.

The uptake of labelled amino acids from the medium has been studied *in vitro* by a number of workers using Spratt's method. Not surprisingly, it was found that there was a much higher rate of uptake of amino acids from explants placed with endoderm downwards than with dorsal side downwards (Deuchar and Herrmann, 1962). Furthermore, Hayashi and Herrmann (1959) demonstrated that a large area opaca was required for extensive uptake. Deuchar (1963a), however, points out that the addition of radioactive amino acids in excess of requirements is unlikely to give a true picture of normal utilisation, for even in chick embryos intracellular yolk inclusions are present (Bellairs, 1958).

Various authors have attempted to elucidate the basic requirements for protein synthesis. Britt and Herrmann (1959) and Herrmann and Marchok (1963) explanted blastoderms at the stage of 11–13 somites and measured the protein accumulation under various conditions. The protein content of the embryo increases most effectively on a medium with a high concentration of yolk and albumen (Britt and Herrmann, 1959; Klein *et al.*, 1962). Thus, the embryo probably utilises protein more efficiently than amino acids. Alternatively, as Williams suggested, there may be growth-stimulating substances in the yolk; he concluded that the embryo utilises only free amino acids for its protein synthesis. This view is supported by the fact that protein accumulation also occurs on chemically defined media lacking yolk or albumen. Leucine or lysine appear to be necessary at this stage, but not aspartic acid or proline (Klein *et al.*, 1962).

A number of workers have studied the effect of amino acid antagonists (or analogues) on development. These may be defined as substances whose chemical structure is closely related to that of a specific amino acid and which, if taken up by the tissues, will replace that amino acid; the tissues that take them up consequently tend to develop abnormally. It will be seen that amino acid antagonists are a useful tool for the embryologist, with the reservation that we can never be confident that all the abnormalities produced are the direct result of interference with normal amino acid metabolism.

Antagonists of leucine inhibit the segmentation of somites in explanted embryos (Rothfels, 1954; Herrmann et al., 1955; Deuchar, 1963a, c). This is particularly interesting since there is a higher level of leucine in somites than in unsegmented mesoderm (Deuchar, 1963). Further, the rate at which the mesoderm becomes segmented can be increased by treating chick embryos with additional leucine (Deuchar, 1960a). Thus, it appears that leucine is essential for the production of somites.

Leucine is, however, not the only amino acid necessary for somite segmentation in chicks; analogues of methionine (Herrmann et al., 1955) and of purine (Waddington et al., 1955; 1958) also lead to a reduction in segmentation.

The mode of action of the leucine analogues is not fully understood. Schultz and Herrmann (1958) and Deuchar (1960) suggested that in chick embryos, explanted at the stage of somite formation, bromoallyl-glycine (B.A.G.) acts by causing an increase in catheptic activity of the somite mesoderm. von Hahn and Herrmann (1962) however found that the formation of protein was inhibited in the somite mesoderm and neural tissue.

Other amino acid antagonists that have been used include: (1) α-methyl-norvaline which closely resembles both leucine and valine and has been found to inhibit haemoglobin synthesis in the area vasculosa (Deuchar and Dryland, 1965). (2) Ethionine (which inhibits methionine) and α-fluorophenylalanine (which inhibits glycine). Both these antagonists appear to affect the general growth of the embryo (Herrmann et al., 1955; Waddington and Perry, 1958). Feldman and Waddington (1955) showed that in the case of ethionine this was due to reduced uptake of methionine by the embryo. Herrmann and Marchok (1963) were able to show that fluorophenylalanine acted in a different way, since it did not affect the uptake of labelled glycine and phenylalanine into the embryo. Phenylalanine is a particularly interesting amino acid, since urodele ectoderm treated with it will become transformed into neural crest (Wilde, 1955).

The rôle of other amino acids and protein precursors in embryonic development is discussed in some detail by Grant (1965).

I. SUMMARY

In this chapter development has been considered mainly as a programme of protein synthesis. Our knowledge of the biochemical ways in which the embryo controls this programme are discussed. The distribution and amount of nucleic acids and of various protein precursors are considered. An account is given of the effects on the embryo of substances known to interfere with protein synthesis, and it is shown how some of these substances may be used as tools to analyse the way in which cells synthesise special proteins during their differentiation.

CHAPTER 10

TERATOLOGY AND IMMUNOLOGY

IN the previous chapter development was considered as a continually changing pattern of protein synthesis. In this one, two further aspects of the biochemistry of amniote embryology will be discussed, the teratological and the immunological.

A. GENERAL TERATOLOGICAL STUDIES ON AMNIOTE EMBRYOS

Embryos tend to be more susceptible to treatment by obnoxious substances than are adults; many chemicals that appear to have no apparent effect on the adult may cause havoc with the embryo. For instance, pregnant women treated with thalidomide experienced little, if any, permanent damage, yet their foetuses frequently suffered extensive abnormalities of development. Even a mild attack of rubella (German measles) in the mother may lead to severe interference of development in the foetus, often resulting in blindness, deafness and heart disease (Rubella Symposium, 1965). The reason for this excessive sensitivity of embryos is not far to seek. The obnoxious substances frequently interfere with protein synthesis at the crucial stage of differentiation of certain organs. The rubella virus may lead to blindness in the embryo if it attacks when the lens is in the process of developing, but will have no effect on this organ if it does not reach it until after this period. It is easy to understand that until differentiation of the organs has taken place during the early part of development, embryos are more sensitive than they are later. Even during this early phase, however, certain periods of exceptional sensitivity have been noted (Hamilton, 1952b). To make matters more complicated, it has been shown in chicks, that what would normally be harmless amounts of potentially teratogenic substances may, if combined with small quantities of similar drugs, act synergistically. That is, the effect of each drug is so enhanced that there is a high incidence of malformations (Landauer and Clark, 1964).

(a) Drug testing

The practical importance of this extreme sensitivity of the early embryo was emphasised in 1962 by the thalidomide tragedy, for it then became apparent that any drug administered to pregnant women was potentially dangerous to the foetus. Efforts have since been made in many countries to devise reliable techniques for detecting possible teratogenic effects of drugs. Drug testing for teratogens is usually carried out on pregnant rats, mice or rabbits and emphasis is laid on the earliest stages of development. These animals are especially convenient since they are cheap, can be obtained in large numbers and are easily maintained in the laboratory. Compared with larger domestic animals such as the cow, they have big and frequent litters combined with a short gestation period. Many teratologists have acquired a good deal of faith in these tests, and it is claimed that no substance found to be teratogenic in man has, when belatedly investigated, failed to produce effects in those laboratory test animals (WHO report 1967). Nevertheless, few embryologists can feel entirely happy at the thought of relying on animals that differ so much from man, for there is an important fundamental difference between these particular species and our own: that is, in the possession of very different types of placentae. Both man and the test animals possess chorioallantoic placentae, but that of man is the haemochorial one, consisting essentially of foetal villi hanging in a pool of maternal blood. The chorioallantoic placenta of rodents and lagomorphs (rabbits) is the complex haemoendothelial type which consists of closely juxtaposed and highly modified foetal and maternal cells tunnelled through by a labyrinth of blood sinuses (p. 159). An even more significant difference between the two types is that whereas man has only the chorioallantoic placenta, the rodents and the lagomorphs have a yolk-sac placenta as well. In man, once the chorioallantoic placenta has formed, most if not all drugs reach it by this route. In the test animals the presence of two types of placentae in the same animal means that the drugs might reach the embryo in two different ways, and it seems probable that in the earliest, most vulnerable, stage the drugs that enter the embryo do so predominantly via the yolk-sac placenta.

Unfortunately, we do not know how significant these differences between man and his test animals are, but it might be that some substances that are broken down and stored by the yolk-sac placenta pass quite easily through the human placenta. It is also possible that some drugs affect the foetus indirectly by interfering with the metabolism of the placenta (Ginsburg, 1968). It is reassuring to see that a recommendation that pregnant primates be used for test animals is included in the WHO report, though, in view of the expense and difficulty involved in maintaining adequate numbers of monkeys in the laboratory, the possibility that they might replace rats and mice seems a remote one.

(b) Thalidomide

An account of the symptoms produced by thalidomide and of its clinical history has been given by Smithells (1966). Since 1962 a considerable body of research has been devoted to trying to elucidate the mechanism of action of thalidomide on developing tissues. Many theories have been advanced to explain its action, one of the most popular being that thalidomide is an antagonist of glutamine or glutamic acid (Faigle et al., 1962; discussion by Smithells, 1966). Like many other teratogenic substances, thalidomide seems to affect mainly mesodermal structures (Jurand, 1966) though such an analysis is perhaps not very meaningful since mesenchyme is the most omnipresent tissue in the embryo.

(c) Lithium ions

It has been known since 1893 (Herbst) that if sea urchin eggs are treated with lithium ions the embryo fails to develop an animal (or dorsal) half properly, but overdevelops the vegetal (or ventral) half. Similar vegetalisation occurs in lithium-treated gastropods (Raven, 1952). In some embryos, e.g. Loligo (a cephalopod), certain fish, and amphibians, lithium often leads to the production of cyclopia (Ranzi, 1957). In addition, lithium may affect the inductive activity of tissues though the results are far from clear cut, some authors considering that certain types of inductive response are enhanced in amphibians (discussion in Tiedemann, 1966) and others that they are reduced (Lallier, 1954).

The effect of lithium in chick embryos is also somewhat controversial. This is probably because if it is injected into the egg it sometimes fails to reach the embryo. Most investigators have found that neural tissues are especially affected, though a variety of anomalies including cyclopia have been reported (Naz and Rulon, 1946; Rogers, 1963; Nicolet, 1965a; Noto, 1967).

The way in which lithium ions act on cells is not understood though most theories suppose them to interfere with protein synthesis. For example, Ranzi (1962) reported that in amphibians lithium chloride affected the physical properties of the proteins. According to Tiedemann (1966) the lithium ion is a more highly hydrated one than that of potassium or sodium, and by substituting for them, lithium can change the shapes of proteins and the interactions between the proteins. Some of the effects of the lithium ions, however, may be produced by direct action on the DNA (Ross and Scruggs, 1964). They may also be due to an inhibition of ribosome synthesis (De Bernardi *et al.*, 1968).

Rogers (1963, 1964) who treated chick blastoderms *in vitro* with lithium chloride, included in the culture medium various radioactively-labelled protein and nucleic acid precursors. In the presence of lithium chloride these precursors were not incorporated into the proteins and there was an inhibition of protein synthesis which was especially marked in the brain and optic vesicles. Nicolet (1965b) who treated young chick embryos *in vitro* with lithium chloride concluded that the main effects were brought about by changes affecting the morphogenetic movements and the mitotic rhythm.

(d) Trypan blue

The earliest extensive investigation of the effects of trypan blue on development was that of Gillman *et al.* (1948) who carried out experiments on pregnant rats. The most common defects were *hydrocephalus*, *spina bifida* and various eye and tail anomalies. Similar results have since been obtained for other species of rodents as well as for rabbits, but attempts by Beck and Lloyd (1966) to produce anomalies in mammals other than those possessing a yolk-sac placenta were unsuccessful. Various authors (e.g. Mulherkar, 1960; Stéphan and Sutter, 1961) who treated chick blastoderms *in vitro* or *in ovo*, respectively, with trypan blue, have obtained a number of malformations, especially of the somites. Beaudoin (1961) who injected trypan blue into the hen's egg obtained various anomalies, including *spina bifida*. Unfortunately, it is difficult to compare Beaudoin's results with those obtained from experiments on the chick embryo *in vitro* since he allowed his embryos to develop for eight-and-a-half days after treatment before examination; that is, they were then a total of ten days old, whereas the *in vitro* specimens were maintained for only about twenty-four hours after treatment.

Beck and Lloyd (1966), in an authoritative review of the work carried out on the teratogenic activity of this and other azo dyes, drew attention to the fact that many commercial samples of trypan blue are not pure, so that considerable doubts now attach to the results of many otherwise admirable pieces of work by reputable scientists. (See also, Lloyd *et al.*, 1968).

A useful characteristic of trypan blue is that because of its blue colour it is possible to identify it in any tissue. It is thus of especial interest that affected embryos seldom appear blue in colour, but that the yolk-sac placenta in mammals and the yolk-sac of the chick, if it is treated *in ovo*, stain heavily. Similarly, if a radioactive label is attached to the dye the radioactivity is subsequently found predominantly in the same sites. It appears, therefore, that the dye is absorbed or phagocytosed in some manner by the cells of the yolk-sac. Recently, Beck *et al.* (1967) have demonstrated that the dye becomes broken up by the lysosomes of the proximal yolk-sac. They suggest that the embryo may develop specific malformations because it lacks enzymes necessary to digest nutritional elements, and so it is deprived of essential substances at the early stages of development. This hypothesis is supported by the findings of Berry (1970) who showed that trypan blue caused a retardation of growth in rat foetuses. An alternative interpretation that might be worth pursuing is that, having been broken down by the lysosomes, an unlabelled component of the trypan blue is taken up by the embryo and has a deleterious action on certain organs. Certainly, the yolk-sac is probably not the only site of action by trypan blue. Davis and Gunberg (1968) have seen it also in the gut epithelium. It also appears that in addition to affecting the lysosomes of the yolk-sac, trypan blue leads to a significant increase in the absorption of certain ions (Kernis and Johnson, 1969).

(e) Variations in temperature

(i) MAMMALS

Except in situations of extreme heat or cold, mammals are able to maintain a relatively constant body temperature. Consequently, their embryos are seldom affected by abnormal temperature conditions. In experiments in which pregnant mice were kept at environmental temperatures of $-30°C$, the deep body temperature was apparently maintained even though the animals were in a state of permanent stress. The embryos were able to develop at their normal temperature although the size of the litters was reduced (Barnett, 1965).

When unfertilized rat eggs were directly exposed to colder temperatures than normal, they sometimes failed to eject their second polar body and occasionally underwent a few parthenogenetic divisions (Austin and Braden, 1954).

When early mammalian embryos were exposed directly to abnormally high temperatures they showed the same type of vulnerability as do other amniote embryos. For instance, when fertilized rabbit ova were maintained for six hours at an elevated temperature of 40°C before being transplanted to a recipient uterus, there was an increase in the mortality rate. It is interesting to note that the effects of increased temperature were not always apparent until later stages (Alliston *et al.*, 1965). Similarly, if the entire uterine horn of a pregnant rat was exteriorized and put into saline at 40–41°C for 40 minutes a number of malformations resulted (Skreb and Frank, 1963).

(*ii*) REPTILES AND BIRDS

The embryos of reptiles and birds, unlike those of mammals, are frequently at risk from temperature variations. Many reptiles resemble mammals in being viviparous but are unlike them in that their body temperature undergoes fluctuations depending on the external temperature. The situation is, however, less erratic than was formerly supposed, since it is now known that certain reptiles at least can regulate their body temperature by such behavioural patterns as changes in bodily posture, or by moving into or out of the warmer spots in the environment (see A. d'A. Bellairs, 1970). Similarly, although it might be expected that eggs laid by oviparous reptiles would be in considerable hazard from the temperature changes in the environment, in many cases, behavioural instincts of the parents lead to the eggs being laid in a relatively stable environment. For example, certain monitor lizards (*Varanus*) utilise the nests of termites as natural incubators for their eggs.

The effects of abnormal incubation temperatures for hens' eggs have been investigated by a number of workers; the normal incubation temperature is 38·5°C. The chief effect of incubating eggs at excessively low temperatures such as 25°C is that the blastoderms are not only slow to form a primitive streak, taking about twenty times as long as normal, but the streak seldom reaches the normal length and by the tenth day has regressed and degenerated (Harrison, 1957). If the eggs were kept at a low temperature for less than ten days and then transferred to a normal temperature, at least 90% were found to develop satisfactorily. If the transference was delayed until about the twelfth day, however, only about 8% developed well. These results are especially interesting, since in nature the eggs of many birds must be routinely subjected to short periods of cooling when the brooding bird leaves the nest.

By contrast, chick embryos seem less well able to tolerate high temperatures. Deuchar (1952) found that morphogenetic movements were affected in embryos subjected to temperatures at 45·5°C for brief periods. Rather less high temperatures (about 40–42°C), whether applied for all or merely part of the incubation period, tended to accelerate growth, especially of the viscera which frequently herniated through the abdominal wall (Delphia and Elliott, 1965) and of myocardial muscle (Shelley, 1961). Perhaps the most frequently reported effect of high temperature, however, is on the blood vessels, which are said to be enlarged.

(f) The gaseous environment

Variations in the gaseous environment are likely to occur only in oviparous eggs and are, perhaps, mainly a hazard for those eggs artificially incubated. To some extent oxygen consumption in chick eggs is dependant on temperature; the higher the temperature the greater the oxygen consumption. At normal temperatures, however, the oxygen taken up by the red blood corpuscles decreases steadily during incubation (O'Connor, 1952). To some extent, this may be due to the change-over from one type of haemoglobin to another (p. 190), though it has been suggested that it may also be the result of a reduction in metabolic rate as the corpuscles mature (Boyer, 1950). If the temperature was normal, growth (as measured by protein nitrogen) of the explanted chick embryo was enhanced by culturing under high levels of oxygen, although the developing yolk sac circulation was poor and the proportion of red blood corpuscles was lower than normal (Klein *et al.*, 1964). Curiously, these results could not be obtained for chicks developing *in ovo*.

Reduction of oxygen levels was found by Grabowski (1966) to lead to an increase in body fluids in five day chick embryos and this was accompanied by changes in the composition of the serum. In particular, lactic acid, free amino acids, potassium and carbon dioxide increased, whilst sodium, calcium, chloride and glucose decreased. Further, marked differences were found in the concentration of the electrolytes in the blood stream and in the fluids surrounding the extra-embryonic vessels. Grabowski concluded that some extra-embryonic organ, possibly the yolk-sac, normally has a controlling effect on the osmo-regulatory system. Heussner and Zahnd (1963) were unable to obtain any evidence for a diurnal rhythm in oxygen uptake in chick embryos. By contrast, a diurnal (Circadian) rhythm has been reported in the oxygen consumption in the embryos of five species of snakes (Dmi'el, 1969).

The levels of carbon dioxide present in the hen's egg undergo some fluctuations since this gas is given off, not only by the embryo and the membranes, but also by the shell as the calcium carbonate becomes dissolved out of the mammillary layer during incubation (Simkiss, 1967). It is possible that a certain level of carbon dioxide is essential for normal development. Decreasing the amount of carbon dioxide available for explanted chick blastoderms inhibited differentiation of the central nervous system, and if this gas was almost completely removed degeneration of the entire blastoderm tended to occur (Spratt, 1949).

The normal requirements for oxygen and carbon dioxide are discussed by Romanoff (1967), but somewhat higher levels of carbon dioxide can be tolerated under experimental conditions (Taylor and Kreutziger, 1966).

(g) Radiation damage in the amniote embryo

One of the main effects of X-irradiation on cells is to interfere with mitosis in the late prophase stage and to cause necrosis and cell death. As with all other teratogenic influences, the results are more drastic the younger the embryo, even very small doses—ones that would not hurt adult tissues—leading to malformation of vertebrate embryos. Some of the effects of irradiating later stages of mammalian embryos are discussed by Hicks and D'Amato (1966). They include skeletal anomalies, anophthalmia, microcephaly, malformations of snout and nostrils, harelip and spina bifida.

X-rays have also been used as a tool in embryological research, Wolff probably being the greatest exponent of the technique. For instance, in 1936 he carried out a systematic survey of the effects of selectively destroying different regions of the chick blastoderm with localised X-rays. In classical genetical studies also, X-rays have been frequently used as a tool for producing mutations since they act directly on the genome.

(h) Other teratological agents

Innumerable teratological agents have been used on amniote embryos and it is impossible to mention more than a few. Colchicine is known to affect mitosis, probably by its action on the sulphydryl groups of the spindle proteins (Diwan, 1966). But according to Overton (1958), who treated chick blastoderms with colchicine in ovo, the results of colchicine treatment are not primarily due to the effect on the spindle, for a pronounced shrinkage occurred throughout the blastoderm and its normal elasticity was destroyed. She concluded that the main action was on the tension (p. 21) in the blastoderm. Diwan (1966), however, concluded that the main effects were on morphogenetic movements and induction. It may be noted that these conclusions are not really at variance since tension plays a large part in the ability of the embryo to carry out normal morphogenetic movements (Bellairs et al., 1967). Nitrogen mustard—another mitotic poison—has also been used upon the chick embryo

(Jurand, 1963; Salzgeber, 1966). (For an account of the effect of many further chemical teratogens on vertebrate embryos, see Ancel, 1950.)

B. AMNIOTE EMBRYOS AND IMMUNITY REACTIONS

It is fortunate for the embryologist that early embryos show little or no immunological activity. By this we mean that if any antigen, such as a protein from another animal, is introduced into the embryo, either as an extract or as a tissue graft, it does not evoke an immunological response; it does not stimulate the embryo to produce antibodies to destroy the foreign antigen. It is only the absence of immunity reactions in young embryos that has made possible many of the experimental investigations such as tissue grafting which have taught us so much about development. Without this work we would know very little about the ways in which such processes as embryonic induction occur, or of the other ways in which cells co-operate with one another.

Many investigators have used immunological techniques for studying the development of proteins specific to certain organs (Ebert, 1958). They cleverly utilise the antibodies of adult tissues against the embryo. An early example of the way such an investigation is carried out is the now classical study of Ebert (1953) on the development of cardiac myosin in the chick embryo. Ebert first prepared an antiserum to the myosin of the heart of an adult bird and then treated chick blastoderms with this substance at the primitive streak stage; as soon as the cardiac myosin of the embryo developed, it became destroyed by the antibodies in the adult antiserum. The effects of this destruction were visible in the cells of the embryo. Thus it was possible to see where the cardiac myosin was localised. Ebert thus concluded that cardiac myosin was widely distributed in the chick epiblast at the primitive streak stage, but became restricted to the presumptive heart region during the head process stage.

Nowadays, in experiments of this type, the antibody is generally labelled with a fluorescent dye or with a radioactive material. The label is chemically attached to the antibody. If the label can subsequently be recognised in the embryo, this is taken as evidence that the antibody has also been tracked down, and that the embryo has produced specific antigens. Unfortunately, we cannot always guard against the possibility that the label has become detached from the antibody during the experiment.

Many other experiments of this type have been carried out on early embryos. One of the most widely investigated topics is that of the development of the lens proteins (reviewed by Solomon, 1965; also p. 193).

These immunological techniques are perhaps the most sensitive approach that we possess to tracing the development of specific proteins in the embryo, but unfortunately they suffer from a number of pitfalls and theoretical disadvantages. For instance, it is often necessary to use adult antigens since it is difficult to obtain a sufficiently large amount of embryonic ones for experimental purposes. This means that an experiment is based on the assumption that the adult and embryonic proteins are identical. In many cases this is so, but in some situations (e.g. haemoglobins and serum proteins) the embryonic and adult forms are definitely not identical, whether examined immunologically, electrophoretically, or by oxygen affinity (Solomon, 1965). (For a review of immunological techniques useful in embryology, see Wolff, 1964.)

(a) Onset of immunity reactions

Returning now to the onset of immunity reactions in the embryo itself, we can see that this is also related to the ability to produce special proteins; it is not possible for it to take place until a certain amount of development has occurred. The stage at which antibodies can first be found differs in different species. In the chick they are present from about 15 days of incubation (Solomon, 1965) or possibly even nine days (McCallion and Trott, 1964). In some mammals such as man, they are found before birth, whilst in others, such as cattle, they are not detectable until after birth. These first antibodies are, however, not manufactured by the embryo itself but are acquired 'passively' from the mother. In the chick, they are derived from the yolk where they were deposited by the hen during oögenesis, and they enter the embryo via the yolk sac (Brierley and Hemmings, 1956; Ebert and DeLanney, 1959). In those mammals which receive their passive immunity before birth, they enter the embryo via the placenta; in those that receive it after birth, however, they are taken in as immunoglobulins with the highly specialised thick milk—the colostrum—that the mother produces in the first few days after giving birth. This 'passive' immunity serves the new-born animal until its own 'active' immunity has had time to develop, an occurrence which also varies greatly in its time of onset according to the species (Brambell, 1958). Of great interest is the fact that the intestine of these new-born animals is only capable of absorbing these antibodies for a short period and that the length of this period is dependent on the type of food ingested by the young animal. If new-born piglets are fed on tea alone, the intestine remains capable of absorbing antibodies for a longer period than if they are fed on proteinaceous food (Lecce *et al.*, 1964).

(b) *Mechanisms of immunological reaction*

Our knowledge of the mechanisms involved in the immunological reactions of embryos is based on what is known about similar reactions in adult tissues. This is such a rapidly advancing field of study that the reader is recommended to consult recent reviews on the subject. Nisonoff and Inman (1965) give an account of the chemical basis of the specificity of antibodies, and further aspects of the subject with especial relevance to differentiation are considered in the symposium edited by Warren, 1968. Formerly it was believed that antibodies were produced by macrophages since these become very plentiful in inflamed or wounded tissues. But it now seems generally to be agreed that certain specialised plasma cells are the actual producers of the antibodies. These cells can even produce antibodies in culture as a response to antigen. The process in a living animal that has already been sensitised (i.e. has previously suffered exposure to the same foreign protein) is that, as soon as the antigen has been taken in by the body, plasma cells begin to form in large numbers in the lymph nodes and bone marrow and subsequently move to the infected part of the body where they then immobilise the antigen. The actual biochemical mechanism by which they are able to produce the appropriate antibody is not entirely clear.

The plasma cells themselves are derived initially from lymphoid cells, and these are produced in the young animal principally by the thymus or, in the case of birds, by the bursa of Fabricius (Good and Papermaster, 1964), a diverticulum from the cloaca. If these organs are extirpated at birth immunological ability is greatly reduced (Miller, 1964), although if thymectomy is not carried out until after the animal has become immunologically competent, there are usually no deleterious effects (Sterzl and Silverstein, 1967). This is probably because a sufficient population of competent lymphoid cells is now available elsewhere in the body. Similarly, if the bursa of Fabricius is extirpated in chick embryos the ability to form antibodies is reduced (Warner and Szenberg, 1964).

In addition to producing lymphoid cells the thymus also appears to manufacture a hormone, which is thought to travel in the blood stream to other lymphoid tissues and stimulate them into becoming antibody producing cells. This has been shown by experiments in which thymus cells have been re-introduced into a thymectomised mouse in such a way that cells have been unable to escape from the implant, although humoral

H

substances have been able to do so. The thymus cells were enclosed in a plastic capsule with a pore size of less than $\frac{1}{2}$ micron (Levey, 1964). Mice treated in this way did not lack immunological potential.

During embryonic development many of the cells of the lymphoid tissues differentiate slowly and it is only when the animal becomes exposed to a different environment at birth that there is a marked increase in lymphoid and plasma cells. In the so-called 'germ-free' animal which has been protected from all infections during and after birth, the lymphoid organs remain in their foetal condition. Conversely, if the foetus is immunised *in utero* the lymphoid tissue is stimulated to develop prematurely (for full discussion, Sterzl and Silverstein, 1967).

(c) Development of immuno-competent tissues

Our knowledge of the development of the immuno-competent cells has recently been reviewed by Auerbach (1967). The thymus itself has both epithelial and mesenchymal components, the former giving rise to thymic lymphocytes and the latter to the stroma in which they can form. Each component influences the other during differentiation. Similarly, experiments carried out *in vitro* have shown that the thymus can affect the development of the bone-marrow and *vice versa*. For instance, pieces of thymus have been shown to stimulate the differentiation of bone marrow in tissue culture, and conversely the bone marrow can initiate the development of lymphocytes in a thymus that lacks them.

In the same way, interactions between thymus and spleen, and between bone marrow and spleen have been demonstrated in the developing embryo.

In addition to the thymus, immuno-competent cells may be formed in embryos from bone marrow, and possibly also in the embryonic liver and yolk sac (discussion by Auerbach, 1967).

(d) Self-recognition

An important aspect of any immunological system is that the organism shall be able to recognise its own proteins so that it does not react against them. We still know relatively little about how self-recognition takes place or indeed about how it develops. It appears, however, that if two cells of different genetical type are closely associated during early embryonic development, and providing that they remain closely associated, they will subsequently never recognise each other as foreign. The evidence for the tolerance, postulated by Burnett and Fenner (1949) and demonstrated by Billingham *et al.* (1953), has been elegantly supplemented by Mintz and Palm (1965, 1969) and Mintz and Silvers (1967). They found that adult mosaic (allophenic) mice which had been produced experimentally by aggregating embryos of two different genetic strains (p. 137) possessed two different populations of cells but that these were mutually compatable. Moreover, these mosaic adults were also immunologically tolerant of red blood cells of the two original strains of mice. On the basis of these experiments Mintz and Silvers concluded that immunological self-recognition does not arise in embryonic life until the post-blastocyst stage.

(e) *The pregnant mammal*

One of the most tantalising aspects of the development of immunity is the question of why pregnant mammals tolerate their young since these are organisms possessing a different genetic constitution. A related problem is the converse one of why the foetus tolerates its mother.

A number of theories have been put forward to explain this phenomenon. They are as follows.

(1) That the foetus is antigenically immature.
(2) That the mother's immunological activities are reduced or suspended during pregnancy.
(3) That the uterus is an immunologically privileged site.
(4) That an immunological barrier exists at the placenta.

The evidence has been considered in several recent reviews (Brent, 1966; Kirby, 1968; Currie, 1968) and will not be discussed in detail here. Most authorities now agree that the first theory should be discarded since transplantation antigens have been demonstrated in the embryos of the mouse (Edidin, 1964) and of other species.

Some evidence exists to support the second theory. Anderson (1965) showed that if the skin from new-born rats was grafted on to the chest walls of their mothers, it took and flourished, but if it was grafted on to another, unrelated female it was generally rejected. This immunological inertia disappeared soon after birth and it seemed possible that it might have been constantly reinforced during pregnancy by continual leakages of antigenic material across the placenta. But evidence based on the activity of neo-natal skin grafts has been questioned by Billingham (1964), who has suggested that the physical structure of the skin of the new-born infant may prevent the ready passage of transplantation antigens. Indeed, other experiments strongly imply that although some reduction probably does occur in the maternal immunological activities, this is not enough to explain the tolerance of the foetus. If the foetus is removed from a pregnant rat or rabbit and implanted into the flank muscle of the same animal for instance, it becomes subjected to an immunological reaction by the mother (Woodruff, 1958).

It seems unlikely that the third possibility is correct. If suitable antibodies are injected through the vaginal route into the uterus of the guinea pig they produce an antigenic response (Behrman and Otani, 1963). This shows that the uterus is not immunologically inert. Further, occasionally an embryo implants in the abdominal cavity (ectopic pregnancy) and survives to an advanced state of pregnancy. This would not be possible if the uterus were the only privileged site in the body.

The fourth theory is the one supported by the greatest weight of evidence, though not all investigators agree on the exact location of the immunological barrier. One possibility has been that it was located in the fibrinoid material that is deposited in most chorioallantoic placentae. Fibrinoid becomes located at the place where the trophoblast cells abut the maternal tissue (Kirby *et al.*, 1964; Kirby, 1968; Billington *et al.*, 1968). It has been pointed out that the fibrinoid is thicker in the placentae of hybrid foetuses than in those where mother and foetus are from the same strain and are genetically very similar. Kirby suggested that the fibrinoid acts by 'masking' the antigens of the trophoblast.

Currie (1968), however, whilst supporting the idea of the fibrinoid barrier, proposed that the sialo-mucin at the fibrinoid surface of the trophoblast cells might affect the surface properties of the cells.

Certain difficulties exist in accepting the idea that the immunological barrier is provided by the fibrinoid layer. One is the fact that cells are known to pass across the fibrinoid barrier into 'hostile' territory and to survive. One of the most striking examples is the 'deportation' of small clusters of trophoblast cells into the maternal circulation of humans. These cells have been found to pass into the maternal lung (Attwood and Park, 1961). In the human, foetal blood cells have been discovered in the maternal circulation in as many as 54% of pregnancies (McClary and Fish, 1966). Conversely, there is evidence that cells may also pass from mother to foetus (discussion by Kirby, 1968; Assali *et al.*, 1968). Some of these maternal cells may become lodged in foetal organs (Tuffrey *et al.*, 1969).

A further argument against the fibrinoid theory is that this material is not a continuous layer between the maternal and foetal tissues, e.g. it is not present around the villi in the human placenta (Lister, 1969). Thus, if the fibrinoid concept were correct, the trophoblastic cells would themselves be at risk in these unprotected regions; an alternative hypothesis is that the trophoblastic cells themselves are the site of the barrier.

Evidence to support this concept has been presented by Curzen (1968, 1969) who, using fluorescent antibody techniques, obtained a strong reaction in the cytoplasm of human trophoblast to an antiserum produced in mice in response to innoculation by homogenised human placenta. This concept has certain theoretical attractions. Unlike the fibrinoid, the trophoblast cells form a continuous, unbroken layer around each foetus. They are also present in all types of chorioallantoic placentae.

Although the exact immunological mechanism remains to be established, it has been suggested that 'immunological enhancement' plays a rôle (Curzen, 1968; Currie, 1969; Lanman, 1969). In this process an antibody is produced that is relatively harmless to its target cell; in a mammalian pregnancy the target cells are probably trophoblastic, since these are the only ones in continuous and direct contact with maternal cells. These relatively harmless antibodies now attach to all the available antigenic sites on the target cells, which in this way are then protected from more powerful antibodies. Thus in the mammalian embryo, the cellular immune reactions at the trophoblast become inhibited.

The degree of tolerance is not the same in all species. Kirby (1962) found that mouse blastocysts transplanted to rat kidney capsule implanted and formed a healthy trophoblast far more frequently than when rat blastocysts were transferred to the comparable site in mice. He suggested that the mouse trophoblast was not only antigenically weak but that it might also be capable of absorbing antitrophoblastic antibodies.

One of the best known examples of the failure of immunological tolerance in pregnant mammals is the effect of Rhesus (or Rh) antigen in human red blood cells. Some individuals are Rh-positive; others are Rh-negative. The situation is that when an Rh-negative mother is gestating an Rh-positive foetus, the mother becomes immunised against the foetus, perhaps by the leakage of foetal blood corpuscles carrying Rh antigens into the maternal blood stream. The mother then produces antibodies against the foetal antigen and if these get back to the foetus they are likely to react with the foetal red blood corpuscles and destroy them. The extent of the damage may be only slight in the first pregnancy, but with subsequent pregnancies the already sensitised mother becomes more efficient at producing antibodies, and if the foetus is Rh-positive it becomes liable to severe damage.

One of the potentially interesting aspects of the impact of immunology on development is that it may be possible to induce sterility in women by immunising them against sperm. There is some possibility that certain habitual aborters may have become sensitised to the sperm or seminal fluid of their husbands. But the experimental results obtained so far are somewhat controversial (discussed by Brent, 1966; Kirby, 1968).

C. SUMMARY

Development of an embryo may be affected by certain drugs or by unsuitable environmental conditions. Little is known about the ways in which some of these teratogenic agents affect the tissues, though it seems probable that they often interfere with protein synthesis.

The absence of immune reactions in the early embryo is considered, and the problem of how a mammalian embryo avoids being immunologically destroyed by its mother is discussed.

CHAPTER 11

SELECTED ORGAN SYSTEMS

WE have seen that the location of a particular organ rudiment may be determined by the processes of morphogenetic movements and of embryonic inductions, and that any mistakes made in the early stages may often be rectified by the regulative mechanisms available. Once the organs have been laid down, however, this regulative ability is rapidly lost and in the higher vertebrates regeneration is restricted to a few specialised regions.

As its size increases, the whole embryo must co-ordinate and control the development of the various organs. At the same time most of these organs must begin to function, for the embryo is faced with many of the same physiological problems as the adult; for instance, it needs to respire and excrete. Simultaneously, however, it must prepare itself for some totally new situations that will confront it at birth and afterwards. In particular, it must develop a respiratory system that, without any practice, can be ready to function immediately after birth. The way in which the embryo co-ordinates its development is not well understood but some advances in our knowledge have been made in the last twenty years.

It is not possible to do more here than select a number of topics that illustrate various principles, and inevitably the selection is a biased one. The developmental anatomy of the organ systems will not be described in detail for it is covered for the human embryo in specialised text-books such as Patten (1953a) and Hamilton, Boyd and Mossman (1962); for the pig by Patten (1953b); for the chick by Patten (1953c), Hamilton (1952b) and Romanoff (1960); and general accounts of certain organ systems are included in the recent work edited by DeHaan and Ursprung (1965). There appears to be no text dealing fully with the development of any reptile, though certain aspects are described by Voeltzkow (1902), Goodrich (1930) and others. Of course, many of the processes of organ formation are comparable in quite different amniotes, but it is important to guard against assuming that this is inevitably the case. Sometimes there are extensive modifications, the reasons for which are easy to understand; for instance, the development of the skeleton in most birds is modified by the penetration of air sacs into the bone (pneumatisation) so that the bones become lighter in weight; this helps the bird to fly (A. d'A.

Bellairs and Jenkin, 1960). Many specific modifications are, however, probably overlooked, or if we are aware of them we have difficulty in understanding their importance.

The problems we shall be considering here are those which face the embryo after the basic embryonic body axis has been laid down. We have seen (Chapter 5) that once gastrulation has been completed a simple embryonic axis forms (Fig. 34), consisting of neural tube, notochord, somites, lateral plate mesoderm and endoderm. One of the main problems now faced by the embryo is that of ensuring that each organ starts to develop in the appropriate region of the body. Most of the examples considered in this chapter will be drawn from experiments carried out on the chick embryo.

A. MOVEMENT OF CELLS TO THE APPROPRIATE REGION

The cells of some organs have already reached their destination by the time a simple embryonic axis is formed (Fig. 61). Perhaps the most obvious are the cells of the central nervous system which have already

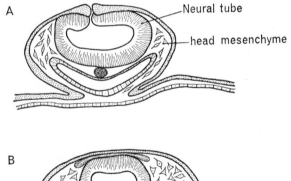

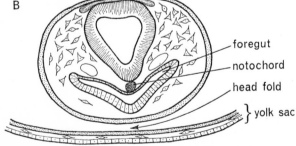

Figure 61. *Morphogenetic movements to form the foregut. A, B, C. Three sections through regions progressively further and further back in the chick embryo. In A and B the foregut is closed off.*

begun to differentiate into a neural tube. Most cells, however, undergo further migrations and displacements. The way in which they move varies. Some cells (such as the gut wall) migrate in sheets, others move as individual cells (e.g. primordial germ cells and neural crest cells). Although it has been possible to trace the paths taken by some of the migrating cells, we know relatively little about the forces controlling these movements.

(a) The movements associated with the gut and heart

The ventral body wall is formed by the folding of the tissues anterior, posterior and lateral to the central axis of the body. The mechanism

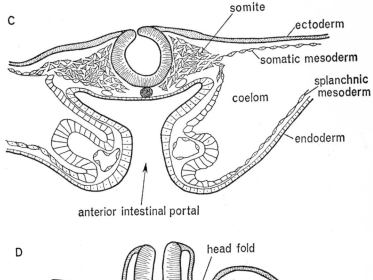

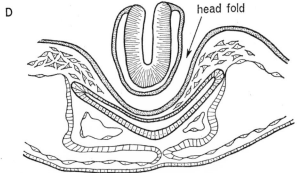

Figure 61 (continued). *In C the foregut is still open ventrally at the anterior intestinal portal. D. Section through an aberrant embryo in which the head fold has passed dorsal to the foregut. This shows that the foregut is not formed by being mechanically scooped up in the head fold movement.*

appears to be much the same in the different classes of amniotes though slight differences are recorded (Nelsen, 1953). The principal tissues involved are the somatic mesoderm and the ectoderm.

Many of these cell and tissue movements were described by 19th century embryologists who were able to make their deductions from studying serial sections of embryos at progressively older stages. Others have only been worked out clearly by the use of modern cell marking techniques (p. 80). Some of these movements (e.g. those involved in the folding off of the head and the body) are accompanied by cholinesterase activity (Kussäther *et al.*, 1968) though its purpose in these tissues is unknown.

One of the earliest organs to appear is the foregut which can be seen as a pocket on the ventral side of the chick embryo (Fig. 61). The movement of the endoderm cells from which it is formed (Fig. 30) were traced by carbon marking experiments (Bellairs, 1953), and the conclusions obtained were later confirmed by experiments in which certain regions

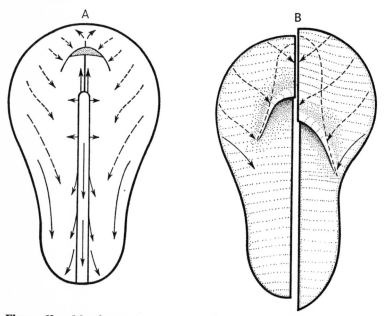

Figure 62. *Morphogenetic movements during foregut formation. When labelled cells were followed in the endoderm of the chick embryo, two types of cell movement were found. (a) '2-dimensional', which stayed in the original plane: black arrows; (b) '3-dimensional', which moved ventrally: dotted arrows. A. headfold stage. B. two later stages. Both A and B are viewed from the ventral (endodermal) side. (From Bellairs, 1953.)*

were destroyed by localized X-rays (Le Douarin, 1961). These movements are of two kinds (Fig. 62), one staying more or less in the original plane of the flattened embryo ('2-dimensional') whilst the other kind moves ventrally in a different plane ('3-dimensional'). We do not know what initiates these movements. It is often implied that the foregut is formed by being scooped up in the headfold (see below) but this seems improbable since the foregut can develop even if the head fold fails to form (Fig. 61D). Stalsberg and DeHaan (1968), who have recently re-investigated the problem, have suggested that the regression movements

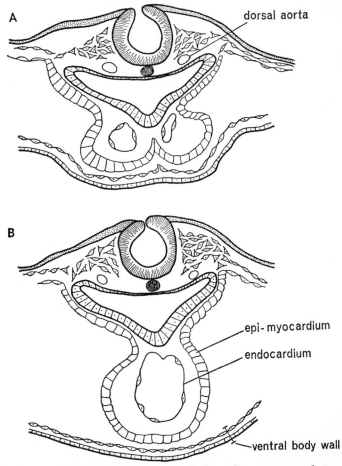

Figure 63. *Morphogenetic aspects of the cell movements during the formation of the heart. The heart is formed from the splanchnic mesoderm which has moved ventro-medially (cf Fig. 61).*

(described on p. 93) lead to a posteriorly directed tension on the endoderm. They have produced a theoretical model to show how, as this tension is increased, the endoderm on either side of the mid-line is pulled ventrally and medially. In this way the foregut increases in length.

At the same time as the foregut forms, the closely applied head mesenchyme and the splanchnic mesoderm move ventrally. The splanchnic mesoderm gives rise to the muscles and connective tissue of the gut and its derivatives, and to part of the lining of the coelomic cavity, but it also forms the heart (Figs. 61C and 63).

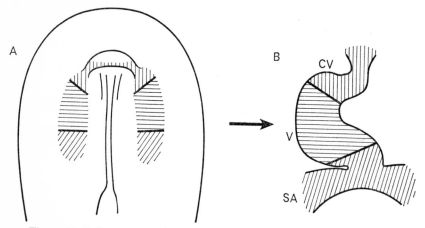

Figure 64. *Cell movements during the formation of the heart. A. Presumptive map showing different regions of the undifferentiated splanchnic mesoderm. When cells in these regions were marked with particles of iron oxide, it was found that they gave rise to the tissues shown in Fig. B. cv. cono-ventricular tissue; v. ventricle; sa. sinu-atrial tissue. (From DeHaan, 1965.)*

The movement of the presumptive heart cells into their correct situation has been studied by DeHaan (1963, 1965). It has been shown by various workers (Willier and Rawles, 1935; Ebert, 1953; Rosenquist and DeHaan, 1966) that the presumptive heart mesoderm lies as two separate areas on either side of the head process of the chick. DeHaan, by using time-lapse cinematography, demonstrated that the mesoderm in each of these areas consisted mainly of many discrete clusters of cells. Some of these clusters migrated away from the heart-forming area, though most became part of the heart. To some extent the clusters were found to be carried passively on the endoderm as it folded ventrally, but each cluster also appeared actively to move over the endodermal substrate. We can compare their movement with that of a man walking slowly down a rapidly descending escalator. We can also take our analogy further, for just as the path of the man is directed by the steps of the escalator, so the path of the pre-cardiac mesoderm appears to depend on the structure of the underlying endoderm. DeHaan (1965) found that at a certain stage, individual endoderm cells in a crescent-shaped area (Fig. 64) change in shape from being squamous and polygonal to become spindle-shaped; he suggested that they formed a track directing the movement of the mesoderm. Subsequently, however, Stalsberg and DeHaan (1969) have suggested that the movements of the splanchnic mesoderm are also affected by the overlying ectoderm. They conclude 'It is as if the movements of the splanchnic mesoderm were the resultant of relative movements of the layers above and below, one pulling the splanchnic layer forward, the other drawing it backward'. Another interesting correlation between the two tissues is that the ability of the pre-cardiac mesoderm to regulate seems to be reduced at the time the endoderm cells change from squamous to polygonal (Orts Llorca and Collado, 1967). As the mesoderm itself migrates it retains its integrity as a sheet and does not break up into cells migrating separately (Manasek, 1968; Stalsberg and DeHaan, 1969). DeHaan (1968) has suggested that there is an increase in intercellular adhesiveness in the pre-myocardial mesoderm which causes it to condense and thicken as it moves into position.

(b) Cell movements associated with somites and somatic plate mesoderm

In the chick embryo, the somites increase in number until about sixty pairs are present. It has long been known that soon after they have formed (p. 101) the somites undergo modifications and become differentiated into three regions (Figs. 65 and 66). These are, (1) the sclerotome which gives rise to much of the axial skeleton, (2) the myotome which forms part of the muscular system and (3) the dermatome which is said to differentiate into the dermis of the skin, at least of the dorsal body wall. It is not clear whether it forms the dermis of the entire body. Langman and Nelson (1968), who examined the formation of the various components of the somites by means of tritiated thymidine labelling, found that the myotome was formed mainly from the division of cells in the dermatome. On either side of the somites are the somatic and splanchnic layers of the lateral plate mesoderm. The somatic mesoderm gives rise to much of the body wall as well as to the muscles of the wing bud and probably of the leg bud too. It is also continuous with the extra-embryonic mesoderm that covers the amnion and lines the chorion (Fig. 66).

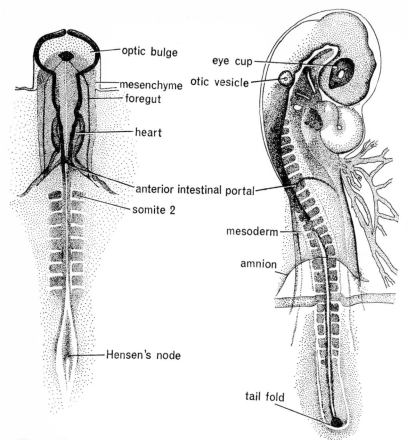

optic bulge

eye cup

mesenchyme   otic vesicle

foregut

heart

anterior intestinal portal

somite 2

mesoderm

amnion

Hensen's node

tail fold

Figure 65. *Further development of the embryonic axis of the chick (cf Fig. 34). Left. At about 35 hours of incubation (about stage 10 of Hamburger and Hamilton, 1951). Right. At about 55 hours of incubation (about stage 15 of Hamburger and Hamilton, 1951).*

The muscles of the limbs and of the body wall originate individually from either the myotomes or from the somatic plate. The relative distribution of these two types of mesoderm has not been easy to determine, largely because of the extensive migrations that occur. The few experimental analyses that have been made have all involved either marking different regions (with carbon particles or with tritiated thymidine) or studying the effects of extirpating pieces of tissue.

Formerly, it was thought on morphological grounds that the muscles and skeleton of the limbs in amniotes were derived solely from the somites, but experiments (Saunders, 1948; Seno, 1961) have now shown that they are formed from lateral plate. There does, however, appear to be some contribution to the lateral plate itself from the somites (Seno, 1961) at an early stage.

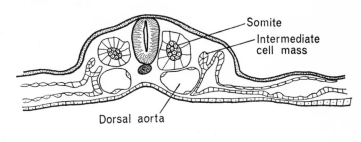

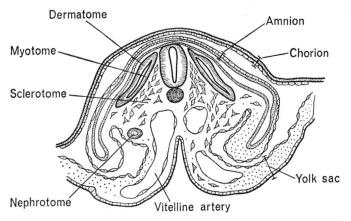

Figure 66. *Differentiation of somites. Top. Transverse section at the start of differentiation. Bottom. Differentiation into dermatome, myotome and sclerotome.*

The relative distribution of the somitic and lateral plate mesoderm to the body wall is less well established. On morphological grounds it has often been stated that the trunk muscles (and ribs) are formed entirely from somites, but the position seems to be more complicated than that. Straus and Rawles (1953) concluded from their experiments that in the chick the somites form only those structures in the dorsal third of the body wall, whereas the somatic (lateral plate) mesoderm forms the remainder. Seno (1961), however, using similar, though perhaps more refined, marking concluded that the somites formed the entire axial skeleton and the musculature of the abdominal wall whilst the sternum and pectoralis muscle were derived from somatic mesoderm. Further experiments are perhaps needed before we can be certain of the details. Meanwhile, however, much of the older work, based solely on morphological studies, has turned out to be reliable. It is a well known principle, for instance, that the origin of muscles is indicated by their innervation. In an interesting analysis of the tongue muscles of the chick embryo, Deuchar (1958) was able to show that in at least one example this line of reasoning was correct. When carbon marks were placed in the occipital somites they were subsequently found in the mandibular arch and then in the foregut, thus confirming the anatomical deduction that, since the tongue was innervated by nerve XII, its muscles must have originated from the occipital somites.

From all the experiments mentioned here we can conclude that extensive migrations of the myotomes and of the lateral plate mesoderm cells occur before histological differentiation of the muscles takes place. Similarly, the cells of the dermatome (which gives rise to the dermis of the trunk) and of the sclerotomes (which form the vertebrae) also undergo migration. The dermis of the limbs, and possibly of the ventral body wall (Rawles, 1955), forms from somatic mesoderm and the dermis of the head is probably derived from the head mesenchyme.

By the time that the developing muscle cells have reached their final destination they have already undergone considerable biochemical differentiation. Various workers who have used immuno-fluorescent antibody techniques (p. 209) have detected muscle proteins, especially actin and myosin, at about stages 17–18 in the chick or even earlier (i.e. when about 30–35 somites are present) (Holtzer et al., 1957; Ikeda et al., 1968). Shortly after this stage, the spindle-shaped myoblasts become fused together to form the syncytial myotubes (p. 137).

At one time it appeared from a variety of experiments that this progressive differentiation from myotome cells, to myoblasts, to myotubes, was dependent on some inductive influence emanating from the neural tube (Lash et al., 1957). Subsequently, however, it was found it was possible to obtain muscle cells directly from somites growing in culture without the intervention of any other tissue. Ellison, Ambrose and Easty (1969b), who have studied the effect of different culture conditions on the somites, have concluded that this early phase of development is one where those cells which already have a tendency to form muscle become 'stabilised' to do so by the environment. The same authors reached a similar conclusion about chondrocyte differentiation (p. 252). In the embryo proper the stabilisation probably takes place during the period of cell migration.

(c) Movements of individual cells: neural crest, primordial germ cells and axons

The examples of cell migration considered so far have been concerned largely with the movements of cells in tissue masses. The material from which some organs are formed, however, reaches its destination in a different way, by migrating as single cells which either pass through or around other tissues. The best-known examples are probably the cells of the neural crest, the primordial germ cells and the developing axons of the nerves.

(i) THE NEURAL CREST

This initially arises as a double band of cells lying one on each side of the mid-dorsal region of the neural tube just beneath the ectoderm (Fig. 67). These cells migrate ventrally through the mesoderm and give rise to a variety of structures. The early experimental work on the neural crest was carried out mainly on amphibians (for review, see Hörstadius, 1950). Experimental work on the neural crest of the chick has shown that it forms much the same structures as does its counterpart in amphibians, that is, the pigment cells (Dorris, 1940), certain nerve ganglia (Strudel, 1955; Yntema and Hammond, 1955) and some of the visceral skeleton (Hammond and Yntema, 1964). If the neural crest is extirpated, these structures do not develop.

Dorsal roots of the spinal nerve ganglia, which are arranged one pair per segment of the adult body, come from the neural crest. The ventral roots grow out from the spinal cord itself. The motor axons penetrate into new somites and as these differentiate into muscles and move into new positions the nerves go with them. The Schwann cells, from which the sheaths that cover the nerves are formed, are possibly also derived from the neural crest, though there is some disagreement about this. Among recent workers, Weston (1963) suggested that at least some of the Schwann cells come from the nervous system, although Johnston (1966), who used a similar autoradiographic technique, found extensive contributions to the Schwann cells from the neural crest.

No direct experimental work on the neural crest of mammals has so far been carried out, but morphological studies suggest that it gives rise to much the same structures as in other classes. However, some interesting results about one of the neural crest derivatives, melanocytes, have been obtained indirectly from experiments in which mosaic (i.e. 'chimaeric' or 'allophenic') mice have been produced (p. 136). When mouse blastocysts from two differently pigmented strains were combined, the individual that developed was often 'dramatically striped' with light and dark bands of fur (Mintz, 1967). Presumably the light fur was derived from one individual, the dark from the other. Mintz suggested that this banding occurred because there were initially only a small number of primordial melanoblasts, and that each of these gave rise to a clone (for a definition of 'clone', see p. 124). Seventeen bands of colour were normally present on each side of every experimental animal. Mintz concluded, therefore, that thirty-four primordial melanoblasts were initially derived from the neural crest and that each of these formed a clone of pigmented cells. Mintz suggested that a similar cloning also occurred in normal mice. Other workers who have investigated mosaic mice have described the coat colour as mottled, rather than striped (Mystowska and Tarkowski, 1968). Further analysis therefore appears to be necessary before we can apply Mintz' theory to normal development.

Many workers have been fascinated by the way in which the neural crest cells reach their destination, especially since a certain amount of regulatory behaviour seems possible. Thus, if the crest is extirpated on one side of the body only, neural crest cells from the other side will migrate into the deficient region.

Various theories have been put forward to explain what factors control the movement of these cells and what directs them so that they eventually arrive at their correct destination. The first possibility is that there is a chemotactic control, the cells moving into a chemically more favourable environment either by attraction to their destination or by being repulsed from their site of origin. No evidence in favour of this hypothesis seems to be available. An alternative explanation is that the migration of the cells is guided by the substrate on which they crawl. Weston (1963), using cells labelled with tritiated thymidine, found that in the chick, migration occurred as two streams of individual cells, one moving dorsolaterally into the superficial ectoderm and the other passing ventrally into the mesenchyme between the neural tube and the developing myotome (Fig. 67). The direction of the migration appeared to be independent of the layout of the mesenchyme, for when the labelled neural tube was inverted *in situ* the neural crest cells still migrated as two streams. The direction in which they moved appeared to be controlled by the neural tube which they were leaving, rather than by the mesenchyme into which they were passing. Weston (1963) suggested that both contact guidance and contact inhibition (p. 20, 21) play a rôle in directing the path of these migrating cells. He could find no evidence that any of the neural crest

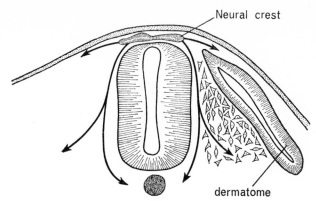

Figure 67. *Neural crest. Direction in which labelled cells were found to migrate in the chick embryo. (After Weston, 1963.)*

cells were predetermined in any way to pass to specific sites in the embryo. He tested this possibility by grafting neural crest from donors to younger hosts. These experiments had the effect of delaying the time at which individual cells left the neural crest, and so affected the chronological order in which they migrated out. The experiments showed that the cells of the neural crest were still capable of giving rise to the full range of distal and proximal derivatives (Weston and Butler, 1966). However, when young neural crest was grafted into older hosts the migration of the crest cells was gradually attenuated, and Weston and Butler concluded that the changing embryonic environment had affected the extent of the migration.

(ii) THE PRIMORDIAL GERM CELLS

Like the cells of the neural crest, the primordial germ cells also migrate at the appropriate stage of development from one part of the body to another. There is good evidence that in several species of amphibians they are derived directly from a specialised region of the fertilized egg which is already set aside by the two-celled stage. This region is the vegetal pole cytoplasm, for if it is irradiated in *Rana pipiens* with an ultra-violet beam, the embryo develops normally except that its germ cells are absent (L. D. Smith, 1966). Similar regions may be set aside early in the embryos of all species of vertebrates. We can feel confident that it is from such a region that all the primordial germ cells and so all the germ cells will arise.

The primordial germ cells are first visible in the yolk sac in most groups of vertebrates, including not only the amniotes but also the anurans, although there is some possibility that in urodeles they form in the splanchnic mesoderm (Nieuwkoop, 1950). In the chick, they have now been identified as early as the stage when the endoderm first forms (Dubois, 1969). Their target organ in all animals is the gonad and having reached it, they differentiate into either oöcytes or spermatocytes (p. 27).

Much of the work on the primordial germ cells was formerly influenced by the controversy surrounding the theory of the 'Continuity of the Germ Plasm' (well reviewed by Everett, 1945). This controversy was concerned with whether the germ cells proper (oöcytes, spermatocytes) are derived directly by a series of cell divisions from the primordial germ cells, or whether they are budded off from the so-called germinal epithelium of the gonad, i.e. the peritoneal covering of that organ. The theory of continuity suggests that a special area of cytoplasm is set aside in the fertilized ovum, that it gives rise to the primordial germ cells, and that these, by multiplication, ultimately give rise to the entire population of ova and sperm produced by the adult throughout its entire life.

The reason for the controversy was mainly that critical methods of identifying primordial germ cells were not available until recently. Happily, however, mouse primordial germ cells have been found to stain selectively with Gomori's technique for alkaline phosphatase (Chiquoine, 1954) and rabbit primordial germ cells with glychemalum-eosin (Chrétien, 1966); chick primordial germ cells could not be stained in this way but were found to have a characteristically high glycogen content (Clawson and Domm, 1963a, b).

In all the three classes of amniotes, there is now clear evidence that the germ cells develop from the primordial germ cells. In a classical investigation, Mintz and Russell (1957), taking advantage of the alkaline phosphatase techniques, were able to show that the embryos of a mutant mouse strain which carried a gene for sterility possessed very small numbers of primordial germ cells compared with normal embryos (Fig. 68). They demonstrated, moreover, that in the normal embryo there is a greater increase in the numbers of these cells during their period of migration.

Similarly, Simon (1960), using a special glycogen stain for the chick, was able to demonstrate the necessity of the primordial germ cells for gonocyte development. The presumptive cell area lies at the anterior border of the area opaca adjacent to the area pellucida at about the head process or early somite stage, not only in the chick (Willier, 1937) but also in the duck (Fargeix, 1966; Rogulska, 1968). If this region is

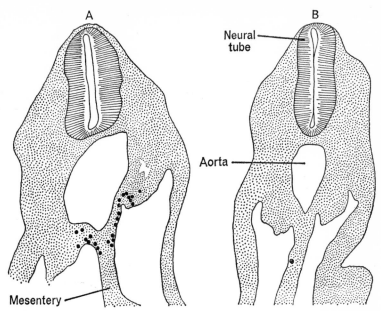

Figure 68. *Primordial germ cells. A. Section through normal mouse embryo with many primordial germ cells (shown in black) passing up the mesentery. B. Section through sterile mutant mouse embryo with only one primordial germ cell visible. (After Mintz, 1957.)*

extirpated, a sterile gonad is formed (Willier, 1937). In her experiments Simon cut small holes in the shells of two fertilized eggs which were then pressed together in such a way that their chorioallantoic membranes were able to fuse and the two circulations came into parabiotic union. She found that even if the presumptive primordial germ cell area was extirpated from one embryo, a chick with a normal gonad was nevertheless formed, apparently because cells had migrated in from the other egg.

In a different series of experiments, Reynaud (1969) injected the primordial germ cells of turkey embryos into the blood stream of chick embryos in which the germinal crescent had previously been destroyed. He found that the primordial germ cells of the turkey were morphologically recognisable, and that they had colonised the gonads of the host and proliferated. In this way, a normal population of germ cells had been restored.

The path taken by the migrating primordial germ cells is of especial interest. In the mouse it has been clearly shown by Mintz and Russell (1957) that there is a migration of the cells from the yolk-sac up through

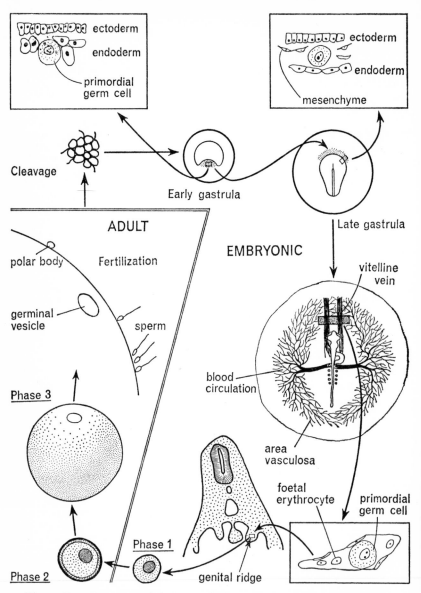

Figure 69. *Continuity of the germ plasm in the chick. The germ plasm is probably already established in the cleavage stage (though we have as yet no direct evidence for this statement in birds). It has been located in the early gastrula endoderm at the posterior end. In the late gastrula it has moved to the anterior border of the area pellucida and is situated in the mesoderm*

the spanchnopleure of the yolk sac stalk and then around the gut to the region of the gonadal ridge. There is also evidence, though of a more general morphological type, that a similar track is followed by the primordial germ cells of other mammals. Mintz and Russell found no evidence of any vascular transport in mice and suggested that the cells moved by amoeboid activity. Subsequently, Blandau *et al.* (1963), who filmed the primordial germ cells of mice, showed directly that they were indeed capable of moving in an amoeboid manner.

By contrast the experiments of Simon and of Reynaud on the chick embryo strongly suggest that in birds the passage of the primordial germ cells (Fig. 69) is primarily a passive one, although it does seem probable that the cells may be capable of active migration when they enter and leave the blood vessels. We have already seen (p. 82) that the forerunners of these cells actively migrate at an earlier stage of development from the posterior to the anterior germ wall (Dubois, 1967).

This amoeboid movement continues at a later stage when they appear to migrate into the blood vessels; in fact, the cells are capable of migrating even in the absence of the blood vessels (Dubois, 1969). This author has shown that if the region of the chick embryo containing the primordial germ cells is combined *in vitro* with various other tissues, such as skin, liver, and lung, the primordial germ cells will even invade them. He concluded that the primordial germ cells are not passively attracted into the blood vessels. On the contrary, there is evidence that these cells leave the blood vessels because the gonad produces a chemical substance that attracts them.

Formerly, it was thought that the primordial germ cells possibly collected in the gonad because of some peculiarity of the vascular network in that organ which resulted in the primordial germ cells becoming physically trapped; it will be remembered that characteristically these are large cells. However, the work of Dubois (1964, 1968) makes this less likely. This author associated *in vitro* the gonadal region of a sterilised chick embryo (i.e. one deprived of its presumptive primordial germ cells)

---

where it differentiates into the primordial germ cells. Each of these subsequently enters the circulation and passes to the genital ridge, there undergoing its first phase of development as an oöcyte. Here it remains with very little further development until the embryo hatches and grows to adulthood. Then it increases in size, is ovulated, fertilized and begins to undergo cleavage. Rectangular boxes show enlarged views of the primordial germ cells at three stages. Oögenesis may be considered as consisting of three phases: 1. embryonic. 2. phase of little growth. 3. phase of rapid growth due to yolk deposition just prior to ovulation.

with fragments of the germ crescent (presumptive primordial germ cells) of a normal embryo. Under these conditions it was not possible for the primordial germ cells to be transported in the blood stream. He again found that the primordial germ cells migrated actively to the sterile gonad. The attraction was even exerted across a permeable barrier, such as egg vitelline membrane.

A similar attraction probably exists in certain amphibians where Gipouloux (1964a, b) has shown that if the dorsal axial organs are extirpated, the germ cells fail to migrate, but that if a tiny piece of somite is retained, the primordial germ cells move toward it.

These examples of a chemotactic stimulus are extremely interesting, for with the exception of the findings of Niu and Twitty (1953) (p. 119), there is little other direct evidence that such a mechanism plays much part in morphogenesis. It is especially important to see if further experiments elaborate and confirm the findings of Dubois.

According to Rogulska (1968) there is a considerable individual variation in the numbers of primordial germ cells in the early stages of normal duck embryos. The numbers appear to be adjustable to some extent, since this author found that if an unincubated duck blastoderm was transected in such a way that two embryos formed (p. 73), the total number of primordial germ cells in the entire blastoderm became about twice as high as in a normal embryo.

Our knowledge of the primordial germ cells in reptiles, and of the routes which they take during migration, is based entirely on cytological studies. Pasteels (1953, 1962) reviewed the body of work then available. In the water tortoise, *Sternotherus*, the movement appears to be by an active migration through the tissues, as in mammals. In *Sphenodon* the journey is partially by active migration and partially by being carried passively in the vascular system. Lizards and snakes resemble the birds in using a vascular route. Recently, Hubert (1970) has suggested that differences in the mode of migration exist even among the lizards and snakes.

The factors controlling the differentiation of the primordial germ cells once they have reached the developing gonad are little understood. It is clear that the environment exerts some effect on them, since they fail to differentiate properly if they settle down in other regions of the body. Primordial germ cells also failed to differentiate normally when pieces of hind gut (containing primordial germ cells) were taken from nine-day mice embryos and transplanted to the anterior chamber of the eye of adult mice (Ożdżenski, 1969). The primordial germ cells frequently migrated from the gut epithelium to the surrounding mesenchyme but failed to undergo mitotic activity, and many of them degenerated. It seems probable that the study of tumours derived from germ cells may throw light on the problem (for a review of teratomas, i.e. tumours containing differentiated tissues, Stevens, 1967).

(iii) AXON GROWTH

The outgrowth of the nerve axons from the neuroblasts of the central nervous system differs somewhat from the two preceding examples, for here part of the cell remains *in situ* and only its extremity migrates. Unlike the neural crest and primordial germ cells, the axon has already undergone a considerable amount of cytological differentiation by the time it starts on its journey (Bellairs, 1959b; Lyser, 1968).

Several theories have been put forward in the past to explain how the nerve fibre grows out in the right direction and makes contact with its proper end organ (for review, Hughes, 1968). Today, the most widely accepted view is that each nerve fibre follows the track of existing structures in the body. Generally these are blood vessels; indeed, nerves frequently lie close alongside blood vessels in many vertebrates. This idea, that the nerves follow other structures by a type of contact guidance (p. 20), is based largely on the results of growing embryonic nerve cells in tissue culture. In a classical series of experiments, Weiss (1934, 1945) found that the direction in which nerve fibres grew out from an explant was largely determined by the structure and orientation of the substrate on which they were migrating. The alternative hypothesis to contact guidance is that of neurotropism, initially put forward by Cajal (1929). Essentially, this supposes that the nerves are attracted to their end organs by reacting to a chemical influence. It is possible that such a chemical attraction plays a rôle in directing the nerve fibre toward the end organ in the terminal stage of its journey. The evidence for both theories is discussed fully by Hughes (1968), who points out, however, that neither theory is adequate to explain the pattern of the peripheral nervous system or 'the enormously more intricate plan of the fully differentiated neural tube'.

B. INVASIVE CELL MOVEMENT AS A FACTOR IN ORGAN GROWTH

Many organs in the body are composed of more than one tissue and in some of these the two components remain discrete and are merely arranged in close association with one another. For example, the skin consists of both dermis and epidermis lying one beneath the other. In certain other organs, however, the two components are more intimately associated with an intermingling of their cells. A special type of cell migration occurs in the formation of these composite organs in which the cells of one type of tissue actively infiltrate among the others. The placenta is a special case, resulting in complex cellular arrangements, often accompanied by tissue breakdown (p. 160). In the composite organs of the embryo itself infiltration is usually accompanied by some sort of inductive relationship.

One of the most interesting of these composite organs is the *liver*, which consists of a mesenchymal matrix into which epithelial cells of endodermal origin penetrate and then proliferate to form cords. The presumptive liver area in the chick blastoderm has been located as

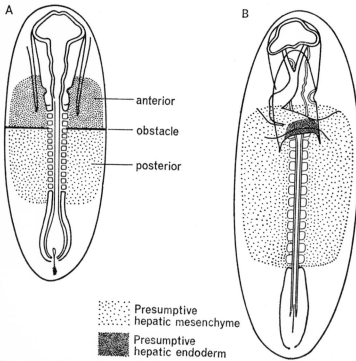

A

B

— anterior

— obstacle

— posterior

Presumptive hepatic mesenchyme

Presumptive hepatic endoderm

Figure 70. *Development of the liver in the chick embryo. A. An obstacle is inserted in the chick embryo so that the presumptive liver area is divided into two. Only the anterior part of the region forms liver. B. The endoderm and mesodermal components from which the liver forms (see key). (From Croisille and Le Douarin, 1965.)*

early as the head process stage. If this region is extirpated and grown on the chorioallantoic membrane it is capable of differentiating into liver (Rudnick, 1935; Rawles, 1936). However, if the presumptive area is divided *in situ* by a transverse slit, which is then kept open by inserting some foreign body (Le Douarin, 1964), liver will form only in the anterior part (Fig. 70). The posterior part is not capable of differentiating into liver; nor does it form liver if it is isolated on its own on the chorioallantoic membrane, but develops into loose mesenchymal masses (corresponding in some ways with the right and left lobes of the liver) which are totally lacking in epithelial cords. (The differentiation of the liver is discussed further on p. 248.)

A similar invasion of one tissue by another occurs in the formation of the lung (Sorokin, 1965).

The factors which initiate or control this type of migratory activity are not known, but some hints may be obtained if we consider the behaviour of cells under certain artificial conditions in tissue culture. There is some evidence that contact inhibition (p. 20) is less between cells of different type than between cells of the same type. Thus, when certain epithelial and mesenchymal cells are grown together *in vitro*, some of the mesenchyme cells are able to insert themselves between the epithelial sheet and the substrate (Abercrombie and Middleton, 1968). This kind of behaviour is not possible between cells of the same type because of powerful contact inhibition. It seems possible that a similar reduction in contact inhibition occurs in the embryo proper, enabling one tissue to infiltrate into another.

C. DIFFERENTIATION OF THE TISSUES OF THE ORGANS

Most organs do not begin to differentiate histologically until their component parts are in the appropriate situation in the body, although this is not always the case, since tissues can sometimes undergo differentiation in abnormal positions. For instance, if the two halves of the heart are prevented from fusing in the mid-ventral line, two separate lateral hearts will develop (diplocardia); most organs that have a bilateral origin behave in the same way. Some organs undergo further movements in the body after formation. For example, although the heart develops at the most cranial end of the thorax, it subsequently migrates further caudally.

At this point it is perhaps necessary to decide what we mean by *differentiation* as far as an organ is concerned. To some extent, we mean the development of an organ that has a gross morphology that we can recognise as characteristic for that organ even though it may be somewhat malformed. Such an interpretation is not enough, however, since an organ of composite origin may look morphologically correct whilst being deficient in certain histological components. A normal-looking liver may develop from the mesenchymal component but be histologically abnormal in that it lacks the endodermal tissues (Le Douarin, 1964). A morphologically normal-looking gonad will form even if the primordial germ cells never reach it, although such an organ is histologically abnormal in that it lacks gonocytes. In considering whether an organ is differentiated, therefore, we must consider its histology as well as its morphology. We should also remember that all histological differentiation must itself be preceded by biochemical changes in the individual cells (p. 190).

Some cells will differentiate autonomously into one particular organ by the time they reach their destination in the body. For example, the cells that grow out from the brain to form the optic cup will usually differentiate into this organ even if placed in an abnormal position in the body or if they are grafted on to the chorioallantoic membrane. Similarly, myotomes may develop into muscle if isolated from their normal environment. In most organs, however, an interaction takes place between neighbouring tissues and this is often of an inductive nature, and is known as a secondary induction to distinguish it from primary induction (p. 106). This inductive relationship is not surprising, since it is as important for the organs to develop in an appropriate relationship to one another at these stages as it was in the earlier ones.

(a) Ectodermal/mesodermal relationships

The skins of amniotes have become modified into a variety of structures, such as hair, fur, feathers, scales, hooves, claws, nails and horns. In each case, an interaction between the ectoderm and the underlying mesoderm appears to play an essential rôle in differentiation.

The first indication of the development of these derivatives is usually that condensations of cells appear in the mesenchyme at intervals beneath the epidermis (Fig. 71A). Such condensations are not an invariable accompaniment of epithelial differentiation; for instance, they do not occur as a prelude to scale development in certain lizards (Flaxman et al., 1968).

It is possible to think of several ways in which condensations of this sort might arise. The cells might actively move into clumps, or localised contractions might take place which would serve to pack the cells more closely, or there might be areas of localised cell proliferation. The problem has been investigated by Wessels (1964, 1965) using the skin of chick embryos prior to feather formation. He found that before the condensation of the mesenchyme, tritiated thymidine became incorporated in the mesenchyme in a random way, but that in older skin which was beginning to form dermal (mesenchymal) condensations, the DNA-precursor became localised in the condensations. These condensations apparently came into being as a result of increased mitotic activity in localised areas. Hale (1956) showed that in the chick eye, where similar condensations form, which eventually take part in the formation of the scleral papillae, these also arise by enhanced mitosis.

The relationship between these mesenchymal condensations and the overlying epidermis has been studied by a number of investigators. McLoughlin (1961a, b), took the skin from 5-day old chick embryos and separated it into epithelial and dermal components. Various techniques exist for performing separations of this type (p. 23). She grew the epidermis separately in tissue culture and also in combination with mesenchyme taken from various regions such as the gizzard, the proventriculus and the heart. She found that contact with mesenchymal connective tissue of some sort was essential if mitosis were to occur in the epidermis, but that each type of connective tissue 'exerted a different and characteristic effect on the differentiation, spreading, and possibly mitotic activity of the epithelium'. For instance, epidermis taken from the back underlain by gizzard mesenchyme became ciliated and secreted mucus, whilst if it was underlain by proventriculus, it first secreted mucus and then became keratinised.

It will be recalled that chick epidermis may also be persuaded to develop into a mucous or keratinised variety by either decreasing or increasing the vitamin A available to it (Fell and Mellanby, 1953) (p. 7). It seems unlikely that this effect is brought about in the same way as the mesenchymal one, however, since chick epidermis that has been separated from mesenchyme will also become keratinised in the presence of excess vitamin A (Dodson, 1967). The effects of vitamin A on the skin also seem to vary with the species. Mouse skin treated with excess vitamin A sometimes undergoes metaplasia and becomes abnormal (Hardy, 1968).

Separation of the epidermal and mesodermal components followed by recombinations of various sorts have been carried out on most of the epidermal derivatives. In general, we can conclude from this work that there is usually an interacting relationship between ectoderm and mesoderm, though, as we shall also see, the precise type of structure that develops is also subject to outside influences. We shall concentrate our attention on two systems—the feather and scale system in birds, and the development of the limb buds in birds. A number of similar systems have recently been reviewed in the symposium edited by Fleischmajer and Billingham (1968).

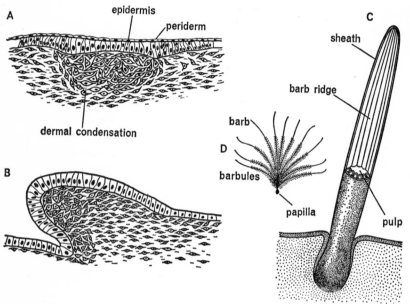

Figure 71. *Feather formation. A. Feather germ showing condensation of mesoderm beneath the epidermis. B. Feather germ at a later stage. C. Down feather still enclosed in a sheath (shown in partial section). D. Down feather after the sheath has split. (C and D after Rawles, 1960.)*

(i) THE FEATHER AND SCALE SYSTEM

The morphological events in the development of the down feather of the chick embryo are illustrated in Fig. 71. The feather tracts appear as papillae at about seven to eight days incubation and gradually extend out from the surface of the body as epidermal cylinders, around a mesenchymal pulp. The ectoderm undergoes modification into a series of barbs. As the surface of the young chick dries after hatching, the epithelial sheet ruptures, releasing the barbs as down. The papilla at the base of each down feather remains as a mass of dermal cells covered by a thin layer of epidermis, and it is from this papilla that all succeeding feathers arise. The epidermal component gives rise to all parts of the feather proper and is re-formed with each successive new feather, but the dermal component is permanent and is essential for feather formation.

In the adult bird different kinds of feather are found in different regions of the body; it is generally considered that the dermal papilla induces the overlying epidermis to become a feather, but that the specific type of feather formed depends on the epidermis and not on the dermis. The evidence is mainly derived from the experiments by Lillie and Wang (1943) who took samples of epidermis from different parts of the body and combined them with mesoderm from other regions. For example, if a papilla from the saddle region was stripped of its epidermis and transplanted beneath breast epidermis, it induced a breast-type feather.

This ability of the adult epidermis to control the type of feather that develops was shown by Cairns and Saunders (1954) to be absent early in embryonic life. For instance, they found that if they removed the mesoderm from a wing bud and replaced it with mesoderm from the thigh, the feathers which developed were typical of the thigh region.

Several investigators have followed up these studies. Rawles (1960) has been especially interested in the relationship between scales and feathers, and in particular why scales should form in some regions of a bird and feathers in another. The relationship is a curious one since scales which bear feathers are found in some breeds of fowls, e.g. Silver Campines and White Leghorns.

Both scales and feathers are first visible as mesenchymal condensations. Rawles' experiments consisted of combining ectoderm and mesoderm from different regions of the embryo and growing them as explants on the chorioallantoic membrane. She took her grafts from the *back*, which normally develops feathers, the *leg*, which normally develops scales, and from the *beak*, which normally becomes keratinised. She deduced that whether or not the mesenchyme could elicit a response in the epidermis with which she combined it depended on two factors—the age of the tissues at combination and the 'strength' of the inductor.

For instance, the dermis taken from the back at six days incubation was able to induce feathers from all types of epidermis, but dermis from the leg did not acquire a capacity to induce scales until the thirteenth day. The age of the epidermis appeared to be less important, for it was able to respond to any of the dermal inductors when tested between five and eight and a half days. The type of structure that formed often depended on the dermis. Thus, beak dermis could induce a beak from back epidermis (which would normally have given rise to feathers). Not all dermis, however, was so 'strong' an inductor. That from the thirteen day foot (which would normally induce scales) could only stimulate eight day back epidermis sufficiently to produce its normal derivative, feathers. It appears, therefore, that the dermis of the embryo resembles that of the adult fowl in being necessary for the development of the epidermal structures, such as feathers or scales, but that it differs from the adult in being able to determine the type of structure that forms.

In addition to the effect of the dermis on the epidermis there also appears to be a reciprocal effect of the epidermis on the dermis. The evidence is as follows. Firstly, if strips of epidermis are removed and replaced *in situ* with reversed orientation, normal feathers develop, although they individually lie pointing toward the head rather than toward the tail of the animal (Sengel, 1964). Thus, the epidermis has imposed its orientation on the underlying dermis and has affected the differentiation of the dermal cells.

Secondly, if dermis is taken from the skin of a strain of chicks that lack scales (the mutant strain, 'scaleless') and combined with epidermis from the leg of a normal chick, then scales will develop. Epidermis from a 'scaleless' embryo is, however, incapable of developing scales even if combined with normal dermis. It appears, therefore, that in this mutant the epidermis imposes its deficiency on the dermis.

A similar effect of the epidermis upon the dermis has also been discovered in other systems, such as the development of the scleral ossicles (Coulombre *et al.*, 1962), and it also plays an important rôle in limb bud formation.

It is not surprising that, since the development of the skin has been investigated experimentally by so many workers, there should be great interest in the chemical changes that take place during differentiation. There are three main lines of evidence:

Inhibition. One of the striking things is that although the presumptive feather or scale is able to respond to a variety of different mesenchymes, its reactivity can be completely suppressed under certain conditions. Moscona and Moscona (1965) found that although disaggregated skin cells from 8-day chick embryos could reaggregate and give rise to feathers, this reaction was completely inhibited if the disaggregated skin cells were mixed with other types of cell. If they were mixed with liver, heart or kidney cells from the same embryo, the disaggregated skin cells collaborated with them to form liver, heart or kidney respectively. Similarly, if they were mixed with skin cells from an older chick embryo, or even from a comparatively older mouse embryo (15 days), feather formation was suppressed. Skin cells from young mouse embryos

(13 days), however, did not have this ability to suppress feathers and even participated in forming them. The analysis of these effects is still in an early stage but it appears that as certain types of cells lose the ability to participate in feather formation they acquire the property of being able to suppress it (Moscona and Garber, 1968). It will be interesting to see if this effect is related to changes in adhesiveness; Ishii (1967) has shown that there is a loss of adhesiveness in disaggregated chick skin cells with increasing age.

Growth-stimulating effect. The differentiation and development of the chick epidermis can be stimulated by an 'epidermal growth factor' (S. Cohen, 1965) which is apparently produced by the sub-maxillary gland of the mouse. Reminiscent of the 'nerve growth factor' (p. 272), this substance is of great interest, though at present it is difficult to assess what relevance it has to normal feather development.

Biochemical analysis. Bell and his collaborators have been able to demonstrate that there is a synthesis of a new set of proteins during feather formation (Bell, 1963, 1965) and that three new antigens appear at about six days of incubation that are probably correlated with feather induction (Ben-Or and Bell, 1965).

Whatever the inductive agent may be it appears to be capable of traversing a millipore filter (Wessels, 1965).

Before leaving the epidermal derivatives, it should be noted that some workers believe that the basement membrane plays an important rôle in their differentiation (Grobstein, 1968). There is some evidence that the basement membrane is itself formed, at least in part, from the epidermis. This was shown by Kallman and Grobstein (1966), who labelled salivary epithelium with tritiated glucosamine and then combined it with un-labelled mesenchyme, the two being separated by a millipore filter. The label was subsequently found to appear in the basement membrane. (Further remarks on the origin of the basement membrane are made on p. 156.)

To summarise our knowledge of the development of epidermal derivatives, there is probably in most of these tissues an inductive action of the mesoderm upon the epidermis. There may even be a transfer of small protein molecules. The mesoderm generally appears to determine the type of reaction that occurs, but there are exceptions to this rule. Apparently there is also a reciprocal effect from the epidermis upon the mesoderm.

I

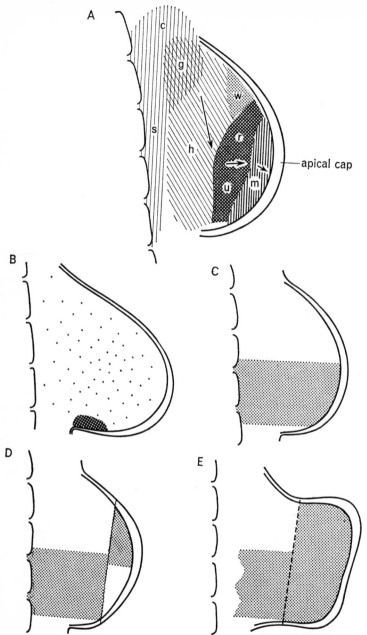

Figure 72. *Development of the limb buds. A. The presumptive areas in the wing bud of the chick embryo at about 2½ days of incubation (stage 17 of*

(ii) DEVELOPMENT OF THE LIMB BUDS

The limb rudiments in terrestrial amniotes are first seen as paired thickenings on the thorax and abdomen. Gradually these thickenings bulge out into buds, each of which consists of a mass of mesenchyme covered with ectoderm. At the apex of the bud the ectoderm is thickened and known as the apical cap (Fig. 72). The presumptive areas of the chick wing bud were mapped out by Saunders (1948) who marked individual groups of cells with carbon particles and followed their subsequent fate. He was able to establish in this way that the wing buds arose from the somatic mesoderm of the lateral plate, and not, as was previously thought, from the somites.

There is little information on the factors that cause the limb buds to develop in their specific regions, though in amphibian embryos they can be induced either by ear vesicles or by nasal placodes (Balinsky, 1925).

More attention has been given to the problem of the factors controlling the development of the limb bud into a limb. Formerly it was thought that a 'mosaic limb field' was present (p. 18), though post-war work has shown this to be too simple a concept. At present there are two main theories about limb bud development in birds. The situation in amphibians will not be discussed here.

The theory of Saunders and Zwilling. These authors suggest, on the basis of their experiments, that mesoderm induces the overlying ectoderm to form the ectodermal ridge, as a result of which the ectoderm now influences the mesoderm to grow out and form limb structures. Their main evidence (most of which is summarised by Zwilling (1960a) is: firstly, that if the ectoderm is removed the mesoderm fails to develop (Saunders, 1948); secondly, in a homozygous, wingless mutant, wing buds form but fail to develop further, and this is associated with regression and disappearance of the apical cap (Zwilling, 1949); thirdly, if the

---

Hamburger and Hamilton). c: coracoid; g: glenoid; h: humerus; w: interdigital web; u: ulna; m: manus; r: radius; s: scapula. B. Location of the 'posterior necrotic zone' in the wing bud at about four days of incubation (stage 23). C. Distribution of 'maintenance factor' in the chick wing bud at stage 21. D. Experiment in which the tip of the wing bud is cut off and then replaced with reversed orientation so that the location of the maintenance factor is changed. E. An early stage in the duplication of the limb which results from the above experiment. The new distribution of the maintenance factor is hypothetical. (A and B after Saunders, Gasseling and Saunders, 1962; C, D and E after Saunders and Gasseling, 1963.)

ectoderm is removed from limb bud mesoderm and replaced with ectoderm from the flank of the embryo, the limb bud does not develop into a
limb; fourthly, mesodermal cells taken from a limb bud, allowed to reaggregate and then implanted into the chorioallantoic membrane, do
not form a limb but if they are wrapped in limb ectoderm before
explantation on to the chorioallantoic membrane they will form a limb
(Zwilling, 1964b). Ectoderm from another region of the body, however,
does not have this effect. A further piece of evidence is that the transient
limb buds seen in the embryos of the limbless lizard, *Anguis*, do not
possess an apical cap (Raynaud, 1962).

The theory of Saunders and of Zwilling also holds that this interrelationship of the ectoderm and the mesoderm continues as the limb
grows out; for, although those structures already present will continue
to develop if the ectoderm is removed from a partly developed bud, no
new structures will form. Thus, the ectoderm continues to be necessary.
They propose that the mesoderm also continues to affect the ectoderm
by producing a so-called 'maintenance factor', which is distributed
asymmetrically, being mainly in the post-axial portion (Fig. 72c) and
which passes from the proximal to the distal part of the bud. Their
argument is based on a variety of experiments. For instance, reversal of
the tip of the bud affects the development of the limb, causing it to
double, which they believe to be due to a redistribution of maintenance
factor (Fig. 72d). If a Millipore filter is inserted between the tip of the
bud and the more proximal region, the effect that is obtained varies with
the pore size of the filter; thus, the smaller the pore, the fewer limb buds
that developed a double tip (Saunders and Gasseling, 1963). Earlier
experiments by Zwilling (1956) and Zwilling and Hansborough (1956)
were also interpreted in terms of maintenance factor. (Further information on the maintenance factor is given on p. 247.)

The theory of Amprino. The difference between this theory and the last is
that Amprino does not believe that the apical ectoderm induces the
mesoderm, but suggests rather that the apical outgrowth of the mesoderm is an expression of the inherent growth pattern of the mesoderm.
He states (1965) that 'individuation, determination, growth and differentiation of the various limb territories depend on the propagation of
organogenetic influences from the proximal limb regions'. To some
extent the difficulty is a semantic one, Amprino objecting to the particular usage of the term 'induction' made by the American authors. The
main point, however, is that Amprino is of the opinion that all the
results obtained so far can be explained on other bases. For instance, he

has suggested (1965) that the reason why the limb bud ectoderm is necessary is that it forms a biological boundary to the mesoderm and that it also has a modelling effect upon the latter. The force of this argument is perhaps lost, however, as a result of experiments (mentioned above) in which disaggregated mesoderm reacted differently according to the type of ectoderm.

Amprino (1965) has also complained that the time relationships of development do not support the theory of the American workers. His argument is that if the outgrowth of the mesoderm is stimulated by the inductive activity of the apical ectoderm, and if the thickness of the ridge is an indication of its activity, then this thickening should occur before the outgrowth of the mesoderm both in normal and experimental conditions. The reverse is true, however; the mesodermal thickening starts to take place before the apical ectoderm has thickened. This difficulty does not yet appear to have been explained by Zwilling and Saunders and their collaborators.

Camosso and Racanelli (1962) showed that there is some correlation between the mitotic rate of the mesoderm and its differentiation. (For a general discussion on the rôle of mitosis in differentiation, p. 259.)

Finally, Amprino (1965) has suggested that when the apical ectoderm is removed the limb may become arrested because of a high pycnotic rate, but he is impressed by the fact that even in the absence of the apical ectoderm, differentiation sometimes takes place. In answer to this, Zwilling (1964b) has suggested that these special cases may perhaps be accounted for by the regeneration of the ectoderm.

In considering the relative merits of the two main theories, it would appear that there is more concrete evidence in favour of the first though as yet not all the criticisms of Amprino have been met. Amprino's own theory by contrast suffers from a certain vagueness, which makes it difficult to devise critical experiments to test it, so that in the absence of other evidence, the Zwilling and Saunders hypothesis is generally accepted.

Recent work on the maintenance factor has shown that it is absent in one specialised region of the posterior edge of the wing. This region is known as the *posterior necrotic zone* (Fig. 72b); it is characterised by the fact that most of the cells in it die at a certain stage of development (p. 266). Saunders and Gasseling (1968) showed that if the mesoderm from the posterior necrotic zone was inserted beneath the apical ectoderm, the latter flattened out and no further outgrowth occurred at the wing tip. On the contrary, two separate and more or less complete wings grew out, one on either side of the graft.

In further experiments Saunders and Gasseling demonstrated that wherever the graft of posterior necrotic zone tissue was placed in the host wing bud, two sets of wing parts developed. One was that of the host but the other was an additional set produced by influence of the graft. (The only exception was when the graft was placed near the host's own posterior necrotic zone.) These experiments indicated that although the posterior necrotic zone itself lacks maintenance factor, it somehow stimulates the formation of maintenance factor from the tissue adjacent to it.

The orientation of the limbs is also related to the interactions that take place between epithelium and mesenchyme (Saunders and Gasseling, 1968). The proximo-distal relationships appear to be controlled by the epithelium; this is shown by the fact that the parts of the limb are laid down in a sequence in the mesoderm under the action of the apical ectoderm. The antero-posterior axis, however, is apparently determined by the mesoderm of the posterior necrotic zone. This was demonstrated by Saunders and Gasseling in the experiment outlined above. They found that supernumerary limb buds tended to be orientated so that their antero-posterior axis was always directed toward the posterior necrotic zone.

(b) *Endodermal/mesodermal relationships*

Many of the visceral organs (e.g. the liver, pancreas, lungs) have a composite nature, being fashioned from both mesodermal and endodermal components. Other organs, whilst being formed entirely of mesodermal tissues, are influenced in their development by the endoderm. As an example of the first group we will consider the liver and of the second group, the heart and blood vessels.

(i) DEVELOPMENT OF THE LIVER

The differentiation of this organ has been investigated especially by Le Douarin in the chick embryo (see p. 235 *et seq.*). The liver has a bilateral origin, the presumptive areas being traceable as early as the head process stage. If both the mesoderm and endoderm of these areas are explanted together on the chorioallantoic membrane, they will give rise to liver tissue, the mesodermal part forming the matrix and the endodermal part developing into the epithelial cords as in the normal embryo. At this stage, any part of the hepatic region is able to develop into liver if it is grown on its own in tissue culture or on the chorio-allantois. By the time the embryo has reached the 9–15 somite stage only a restricted region is capable of forming liver. This is the region lying near the anterior intestinal portal (Fig. 70) and Le Douarin has been able to show by a series of experiments that this is because it is the only region at this time that possesses both the endodermal and mesodermal components of the organ. She found that the posterior part of each presumptive liver area does not differentiate properly if isolated from the anterior region; this isolation can be done by inserting a foreign body between the anterior and posterior regions. In these experiments the posterior part forms loose mesenchymal masses which roughly correspond with the lobes of the liver but lack the endodermal component (epithelial cords). The anterior region, however, develops into a histologically normal liver containing both the mesodermal and endodermal components. Similar results were obtained when the regions were isolated on the chorioallantois.

Le Douarin, who has investigated the interaction of the endoderm and the mesoderm, has concluded that several steps are involved in liver differentiation. First, there is a primary induction of the endoderm by the mesoderm of the cardiac area. This step is apparently essential if the next is to take place. Secondly, there is another induction of the endoderm by mesoderm, this time the mesenchyme being that of the hepatic area itself.

The first induction occurs at about the 4-somite stage. Before this time the endoderm is incapable of differentiating into liver cells if it is separated from the mesenchyme. Even if it is recombined with mesenchyme from the hepatic region proper it will not form hepatic tissue. After the 5-somite stage the endoderm has apparently received some induction since, if it is now combined with hepatic mesenchyme, it can form hepatic cords. It appears that this combination with the hepatic mesenchyme is also essential for development, since without it the endoderm will still not differentiate. This secondary induction of the mesoderm upon the endoderm is now followed by a reciprocal effect in which the endoderm exerts an effect upon the development of the mesenchymal component of the liver. The evidence is discussed in detail by Croisille and Le Douarin (1965).

(ii) DIFFERENTIATION OF HEART AND BLOOD VESSELS

We have seen that the endoderm plays a rôle in directing the cell migrations that lead to the formation of the primitive heart (p. 223). But it appears that the endoderm also influences the development of the earliest blood islands that form—those in the area vasculosa of the yolk sac in the chick. It is generally considered that these blood islands are developed in the splanchnic mesenchyme at about stage 8 (about 26–29 hours). Bremer (1958a, b, 1962) has shown that there is a correlation in early development between the extension of the mesoderm in the area opaca and the shape, volume and density of the endoderm cells that lie beneath it. These facts led him to conclude that the two tissues influence each other's development. Electron microscope studies indicate that they lie very close to one another (Mato *et al.*, 1964). One aspect in which they do not appear to be synchronised, however, is in their mitotic rates (Bremer, 1960).

The classical morphological accounts of the development of the blood islands were given by Dantchakoff (1934), Sabin (1917) and Murray (1932). The general conclusions have been that the blood islands develop from certain cells of the extra-embryonic splanchnic mesoderm, which according to Murray are either predetermined or 'strongly biased' to differentiate in this way.

The first stage in the formation of the blood islands is the clumping together of groups of certain cells which then become known as haemangioblasts. The outermost cells undergo certain changes and now become known as angioblasts, whilst the innermost cells either become blood corpuscles or disintegrate to form plasma. Cords of syncytial cells grow out to connect one blood island with another, and when these cords become channelled a continuous circulation forms.

These classical ideas are, however, beginning to undergo modifications as a result of recent studies. In particular, Edmonds (1966) has shown that the angioblastic cords are not syncytial as was supposed, although the earliest red blood corpuscles have desmosome-like attachments. Cattaneo (1963) has suggested that the yolk inclusions of the haemangioblasts are important in that they provide the developing blood cells with ferritin for the production of haemoglobin.

If the endoderm is removed from the area opaca, the mesoderm cells do not usually clump together or form angioblasts (Wilt, 1965). The blood islands develop more readily when the mesoderm is combined with the endoderm, even if the two tissues are separated by a Millipore filter (Miura and Wilt, 1969). Since some development of the mesoderm is possible in the absence of the endoderm, it appears that the interaction between the endoderm and mesoderm is not an inductive one. Miura and Wilt suggest, rather, that the situation is comparable to that in the development of chondrocytes (p. 252). The endoderm probably assists the mesoderm by providing more favourable conditions for the development of the blood islands; the mesoderm itself responds readily because it already has a tendency to differentiate in this way.

By contrast, if the ectoderm overlying the presumptive heart region at stages 5–8 is extirpated, the heart development continues although its position and curvature may be affected (Orts-Llorca, 1964). The rôle of the endoderm in influencing the heart continues into late morphogenesis for, although the bulbo-ventricular part of the heart has been found to possess the capacity to differentiate from the heart tube at stages 12–14 (about 50 hours), the sinus and atrium can only develop if the anterior intestinal portal is present (Orts-Llorca, 1963).

It would be misleading to leave this discussion without pointing out the important rôle played by some of the other factors in the development of the heart and blood vessels. One of these is the effect of the circulating blood. By the time the heart has formed sufficiently to promote blood flow, the pattern of the vessels has already been laid down, but the fine details are moulded by the passage of the circulating blood (Hughes, 1937). For instance, the path by which blood flows in the arterial arches of the chick embryo can be modified by ligaturing certain vessels, so that the aortic arch develops on the left side, as in mammals, and not on the right (Stéphan, 1949). Similarly, in the differentiation of the heart itself, some morphogenesis will take place even if no blood flows through it, as for instance if the heart tube is grafted into an intracoelomic site. Good differentiation of the endocardium and the myo-epicardium are, however, dependent on a flow of blood (Orts-Llorca and Gil, 1967).

The development of the heart beat has naturally received much attention. This is a problem that is related to the antero-posterior polarity of the organ as well as to the differentiation of the myo-filaments. G. Le Douarin et al. (1966) who isolated different parts of the presumptive cardiac area, however, concluded that the atrial, ventricular and bulbar regions were already determined. The determination of polarity has also been studied by Orts-Llorca and Collado (1967), who have cut out and reversed strips of presumptive heart region and sometimes obtained reversal of the heart. They concluded that early determination is labile rather than determined. They found also that determination coincided with the differentiation of the endodermal cells from polygonal to spindle-shaped, as described by DeHaan (p. 223).

All parts of the heart increase in spontaneous contractility with age and each part, if isolated, is capable of contracting (DeHaan, 1963; G. Le Douarin et al., 1966). The first visible contractions occur at about the stage of ten pairs of somites, and this corresponds with the period when the myofibrils have first been laid down (Croisille and Le Douarin, 1965;

Olivo *et al.*, 1964). (For an authoritative review of our knowledge of form and function in the developing heart of the chick embryo, see DeHaan, 1968.)

(c) *Mesodermal/mesodermal relationships*

(i) CARTILAGE FORMATION

For many years it was accepted that the vertebrae, which are first laid down as cartilage and then later ossified, were formed as a result of induction by the spinal cord and notochord. This idea was based on a variety of experiments (Avery *et al.*, 1956; Strudel, 1962; Cooper, 1965; Grobstein and Holtzer, 1955). For instance, using tissue taken from chick embryos, Avery, Chow and Holtzer (1956) found that if they grew somite cells in the absence of spinal cord or notochord, they were unable to obtain cartilage. If either of these two tissues was present, however, vertebral cartilage differentiated. The only other tissue that appeared to be capable of inducing cartilate from somites was kidney (summary of experiments in Holtzer, 1961). This is particularly interesting, since in the normal embryo the neural tube and notochord lie to one side of each somite and the intermediate cell mass (from which the kidney forms) lies on the other (Fig. 66).

The initial discovery of the inductive relationship was followed by an attempt to isolate an inductor (review by Holtzer, 1961). It was found that inductive activity could take place through a Millipore filter, which suggested that a chemical inductor was active. Subsequently, Lash *et al.* (1962) obtained inductions with a nucleotide-containing fraction of the spinal cord and notochord of the chick.

This basic idea that the somites can form cartilage only after induction has, however, recently had to be modified. This is because several investigators, using improved tissue culture techniques, have obtained synthesis of chondroitin sulphate from somites that have been grown on their own (Lash, 1963; Strudel, 1963; Holtzer, 1964; Ellison *et al.*, 1969a). Chondroitin sulphate is not only a necessary component of cartilage but it is also specific to it. Consequently, its appearance in a cell means that differentiation into a chondrocyte has already begun. Earlier experiments had seemed to show that chondroitin sulphate was only manufactured by somite cells when they were induced. It now appears that even when the chick embryo is so young that it has only about seven pairs of somites these have already acquired a strong bias to develop into chondrocytes (Ellison *et al.*, 1969a). It has been suggested that after having acquired this biochemical bias to form cartilage, the cells must now go through a period when this tendency is 'enhanced' (Lash, 1968) or 'stabilised' (Ellison *et al.*, 1969a) before cartilage as such becomes histologically differentiated. Whether or not cartilage develops at the end of this period depends on the environment. The factors that are necessary appear to be present in the 'natural' inductors (i.e. notochord and neural tissue) but are also found in the enhanced culture medium of Ellison *et al.* (1969a).

All the mesodermal tissues of early limb buds of the chick embryos begin to synthesise chondroitin sulphate irrespective of whether they will subsequently develop into muscle or cartilage cells (discussion by Zwilling, 1968). It appears, therefore, that the synthesis of muscle proteins may begin later than that of chondroitin sulphate, or that it may be overshadowed by the synthesis of cartilage.

The extent to which the cartilage cells are permanently fixed to be cartilage has been studied by Coon (1966). Cartilage cells taken from eight-day limbs or 12–14 day sterna of chick embryos were grown for many generations in culture until they had apparently de-differentiated and lost their cartilaginous characteristics. They were, nevertheless, still capable of reverting to cartilage if placed in a favourable environment (discussion on modulation, p. 7).

(*ii*) THE INTERRELATIONSHIPS OF TISSUES IN THE DEVELOPING URINARY SYSTEM

The vertebrate kidney and its excretory duct are formed from the intermediate cell mass (Fig. 73). It is generally considered that during the course of embryonic development each amniote passes through three stages of kidney formation in which it develops firstly a pair of pronephric kidneys, which are then replaced by a pair of mesonephric kidneys, which in their turn are finally supplanted by the metanephric or adult-type kidneys (Fig. 73). Each pronephric kidney empties into a duct—the pronephric duct. Later, when the mesonephros appears, this duct is taken over by that kidney and its name is changed to the 'mesonephric duct'. The mesonephros subsequently degenerates as the metanephros forms and a new duct develops as an outgrowth from the base of the mesonephric duct. This outgrowth is the ureteric or metanephric bud. But a number of investigators have recently pointed out that the pronephros scarcely appears in amniote embryos and that the mesonephros shades almost imperceptibly into the metanephros at its

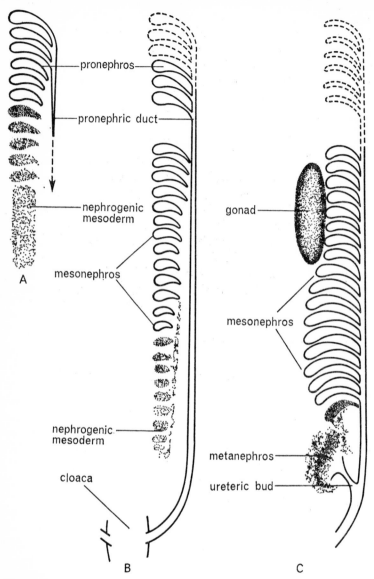

Figure 73. *Plan of the relationships of the pronephros, mesonephros and metanephros. (Redrawn from Burns, R. K. in Willier, Weiss and Hamburger: Analysis of Development. Philadelphia, W. B. Saunders Company, 1956.)*

caudal end (discussion by Torrey, 1965). It seems possible, therefore, that these terms together with the concept of a sequence of kidneys forming during development, may soon be abandoned, at least by embryologists.

The interrelationships between these developing organs have been studied experimentally in amphibian and chick embryos (reviews by Burns, 1956; Torrey, 1965).

The pronephric duct is usually thought to act as an inductor of the mesonephros since, in early experiments, it was found that the latter would not differentiate properly in the absence of the duct (Gruenwald, 1937; Waddington, 1938b; Boyden, 1927). This conclusion has since been questioned by Gruenwald (1942) and by Calame (1962) who have now found that some self-differentiation can take place. It may be that the mesonephros is very sensitive to the environment. It is perhaps significant that it is capable of forming not only renal tubules but also cartilage, depending on the conditions under which it is explanted in tissue culture (Strudel and Pinot, 1965).

There is firmer evidence that the metanephric kidney is induced by the ureteric bud. If the pronephric duct fails to grow caudally, either as a naturally-occurring malformation or as a result of experimental inter-ference, the ureteric bud cannot form and this always results in a failure of the prospective metanephric mesoderm to differentiate (Boyden, 1927; Gruenwald, 1937; Gluecksohn-Schoenheimer, 1949). Similarly, if, in man, a double kidney forms as a malformation, this is always associated with the presence of two ducts (Wharton, 1949)—a fact that supports the idea that a close relationship exists between the two.

The relationship has been experimentally studied by Grobstein (1955) using the metanephros of the mouse embryo. Essentially, this organ forms from two components—the epithelial (i.e. ureteric bud) and the mesen-chymal. The former gives rise to the collecting ducts, and the latter to the uriniferous tubules. If the rudiment of the organ is isolated *in vitro*, these structures develop fairly normally. If they are separated, however, with trypsin neither can differentiate properly, suggesting that each part induces the other. The development of the ureter can only occur in the presence of metanephric mesenchyme. Thus the reaction is thought to be highly specific. But the reciprocal induction of the mesenchyme can be brought about by other epithelia, such as salivary gland epithelium (Grobstein, 1953) or embryonic spinal cord (Grobstein, 1955).

Experiments in which the two tissues were separated with Millipore filters showed that close contact was not essential but could take place across a distance of about 80 μ. Only in filters with a large pore size was it possible to obtain inductions, and this may be taken to imply that it was carried out by large molecules.

D. SUMMARY

This chapter has been concerned mainly with the ways in which tissues interact with one another during organ formation. Many of these interactions cannot take place until the cells have migrated into new positions in the body. The reactions that occur frequently appear to be of

Experiments in which the two tissues were separated with Millipore an inductive nature. Recent work with some tissues has led to the idea that some cells may be 'committed' to develop along a certain path and that induction is merely a process which enables them to carry out this commitment.

CHAPTER 12

GROWTH CONTROL OF THE ORGANS

ONE of the most remarkable features of development is that the individual organs grow in an orderly manner, so that not only do they form in the correct anatomical relationship to one another but each also develops to the size and shape appropriate for the body's requirements. This chapter will be concerned with some of the factors involved in the control of growth and differentiation.

A. PROCESSES OF ORGAN GROWTH

(a) Mitosis

The most important growth process is by cell division and this continues not only throughout the embryonic period but also during the juvenile stage, and in some organs (e.g. the hair and nails) even throughout life. It is not appropriate here to discuss the cytological events of mitosis, especially as modern accounts are readily available (for example, Mazia, 1961). The presence of tubule-like spindle filaments is now well established by electron microscopical studies (Allenspach or Roth, 1967), but it is interesting to note that there appear to be differences in the fine structure and arrangements of the spindle filaments between the epithelial and the mesenchymal cells of young chick embryos. These may be correlated with the fact that the epithelial cells divide by shortening the 'chromosome-to-pole' distance, whereas the mesenchyme cells undergo extensive spindle elongation. They suggest that differences in cleavage patterns are influenced by the fact that epithelial cells are fixed to one another at their free edge by terminal bars or desmosomes, so that they are unable to undergo the spindle elongation of the mesenchymal cells.

In preparation for mitosis, the number of chromosomes in the cell becomes doubled and so the DNA in the cell must also double. This extra DNA is synthesised in the 'interphase' stage (Fig. 74). Formerly it was considered that interphase was a 'resting' stage for the cell, but there is now ample evidence (Mazia, 1961; Howard and Pelc, 1953) that cells of many types will incorporate tritiated thymidine in the middle (S) period of the interphase and that this is when active synthesis of DNA is taking place (Fig. 74). Both before and after the S period, that is, in the G_1 and G_2 periods, the cell fails to incorporate the thymidine. There is some evidence that certain cells even within the same population may differ in their mitotic rhythms. For instance, in epidermal cells of mouse ear, not all cells pass regularly through mitosis; some may pause in the G_1 phase whereas others stop in the G_2 phase for as long as two days (Gelfant, 1963). The rate at which mitosis proceeds in a tissue may help us to understand the rôle that cell division plays in the moulding and formation of organs.

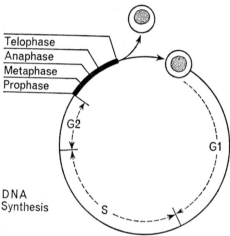

Figure 74. *Mitosis. DNA synthesis takes place in the S-period. (From Sisken and Kinosita, 1961.)*

In adult tissues, the presence of diurnal rhythms in mitosis has been convincingly illustrated. In the epidermis of both man and mouse there is a higher rate of cell division during periods of rest or sleep (Bullough, 1965); it might seem unlikely that embryos would have such well-marked rhythms since their activity is more spasmodic. In comparing the mitotic rates in different adults, attempts are usually made to fix the tissues at the same state of diurnal rhythm. It is, however, seldom possible to take such precautions with embryos. One of the difficulties in assessing the mitotic rate in a particular tissue and the rôle that mitosis plays in the moulding and formation of an organ, is that the assessment of the mitotic rate is a somewhat unsatisfactory process. This is because not only is there disagreement on the best way of making the assessment but there is also a strong subjective element in the results obtained by different individuals. The usual method is to make serial sections of the tissue and then, in sections selected at regular intervals throughout the series, to count the numbers of both the mitotic and non-mitotic cells. The results for the tissue as a whole are then expressed as *mitotic rate*, this being the percentage of cells in mitosis. Difficulties arise in recognising some of the earliest phases of prophase and consequently these are often left out of the counts by some investigators.

The counting process is a tedious one especially as large samples need to be obtained to provide enough data for analysis. For this reason, some investigators use a modification of the technique which increases the numbers of cells seen in mitosis. This involves treating the embryo with a mitotic inhibitor (p. 188) of which colchicine or its derivative, colcemid, is the most widely used. This drug does not prevent cells from entering mitosis but inhibits them from completing it. If the embryo has been treated several hours before it is sacrificed, then all the cells that have embarked on division during that time will be arrested in metaphase; the relative mitotic rate can then be compared in two tissues, or at different stages in the same tissue. By increasing the actual numbers of dividing cells visible in each section the counting process becomes easier, but this gain must be weighed against the disadvantage that the colchicine may have side-effects on the cells which distort the results (p. 189, 208). One possible side result is on the cell surface; if the adhesiveness between cells changes, the tissue is likely to become physically distorted.

Another method of determining the mitotic rate is to estimate the rate of DNA synthesis in the growing tissue since the mitotic division of each cell is preceded by the synthesis of new DNA. Various modifications of this technique are critically assessed by Wimber (1963).

(b) Mitosis and differentiation

The relationship between mitosis and differentiation is not clear and it seems possible that it varies in different tissues. Mitosis, of course, takes place in all the cells of the early cleavage stage embryo and continues throughout gastrulation. After this early period, however, there is a general decline in the rate of mitosis throughout the organism (Dondua et al., 1966; Laird, 1966). This fall is not a steady one and there is evidence that in many tissues there is a rapid burst of cell division just before differentiation begins. This happens in the skin of the chick just before the first stages of feather formation (Wessels, 1965).

With the onset of differentiation this rapid mitosis ceases. This has been demonstrated in the development of the normal lens of the chick

(Modak *et al.*, 1968) as well as during Wolffian lens regeneration (i.e. regeneration of lens from iris) (Yamada, 1966). Cell division also drops abruptly at the onset of differentiation of the myoblasts (which is generally judged to begin when myosin formation starts) (Okazaki and Holtzer, 1966) as well as with the start of haemoglobin synthesis by erythroblasts (Marks and Kovach, 1966).

Findings of this type have led to two further ideas: (1) that differentiation cannot take place during periods of rapid cell division; that is, the two processes are incompatible and (2) that rapid cell division is an essential step through which a tissue must pass before it can begin to differentiate.

It is not easy to test either of these hypotheses *in vivo*, since most organs consist of a mixed population of dividing and non-dividing cells, as well as frequently containing several different types of cells. For this reason much of this work has been carried out in tissue culture using clones of cells (p. 124).

Chondrocyte cells growing in tissue culture have proved especially useful, since they manufacture chondroitin sulphate as they begin to differentiate. The production of this substance can be taken as an indication that differentiation has begun. Abbot and Holtzer (1965) found that dividing chondrocytes in culture were unable to produce chondroitin sulphate, and though negative, these results appeared to support the first idea. Subsequently, however, Cahn and Lasher (1967) showed that with somewhat different culture conditions they were able to do so and therefore the hypothesis became less tenable. Other tissues whose differentiation can be measured by their production of specialised molecules include the pancreas (Rutter *et al.*, 1968) and cardiac muscle cells (Manasek, 1968). In these tissues also it appears that, provided the tissue culture conditions are adequate, some differentiation can take place in cells that are also rapidly dividing. Further, in the chick, the regenerating lens cells (i.e. cells formed by the process of Wolffian regeneration) were found to form γ-crystallin during mitosis (Yamada, 1966). It is apparent therefore that in some cells, at least, mitosis and differentiation cannot now be regarded as incompatible phenomena. How far these results are typical of all dividing cells is not clear. Certainly there are many cells which do not appear to show this dual activity.

There is perhaps more evidence for the second idea—that some mitotic activity takes place that is important for differentiation. For example, the virgin mammary gland will respond to treatment by prolactin and hydrocortisone, not only morphologically but also by producing casein (Turkington, 1968). The response will only occur, however, if the epithelium has previously undergone a burst of mitosis. Thus, not only structural changes but also biochemical ones are dependent on cell proliferation. If the tissue is treated with mitotic inhibitors such as colchicine, the production of casein is inhibited (Stockdale and Topper, 1966). Similarly, pancreatic cells will not differentiate if exposed to 5-fluorodeoxyuridine (FUDR), an agent that blocks DNA synthesis and consequently also inhibits mitosis (Wessels, 1968). This type of experiment has led to a further suggestion that the cells become reprogrammed during their final mitotic division before differentiation. Thus, their final division is seen as a critical one.

Although there is as yet no direct evidence that this is so, it would seem probable that the decision as to which is to be the last division prior to differentiation is taken as a result of some change in the environment. It is well known that subtle changes in the conditions around a cell will affect differentiation in tissue culture. These include not only the composition of the culture medium which itself undergoes change but such factors as the density and the types of cells in the culture dish. In the embryo itself there is a continual change in the composition of the environment of each cell. We have already seen that in some way each cell must be able to assess these changes (Chapter 1).

(c) Mitosis as a factor in pattern formation

One of the interesting aspects of mitosis is that it can give rise to areas of localised growth. For example, localised regions of cell division play a rôle in shaping the cardiac tube of the chick embryo (Sissman, 1966; Stalsberg, 1969). Another concept, however, is that an orderly sequence of cell division may play an important rôle in laying down a pattern of different cell types within an organ.

In the retina of many vertebrates several different types of receptors are present and there is evidence that these are arranged in a regular mosaic pattern. In some fishes the arrangement of cones is very regular and various schemes have been put forward to show that this pattern may be the result of an orderly series of mitotic divisions (Müller, 1952; Engstrom, 1963). Amongst amniotes the pattern in the arrangement of receptors has been described in a gecko (Dunn, 1966) and in the chick (Morris, in press), and although less regular than in some fishes the pattern in the chick could arise as the result of an orderly sequence of mitoses (Fig. 75).

The division of cells in an epithelium such as the neural plate or the retina differs from that in loosely arranged mesenchymal cells. This is

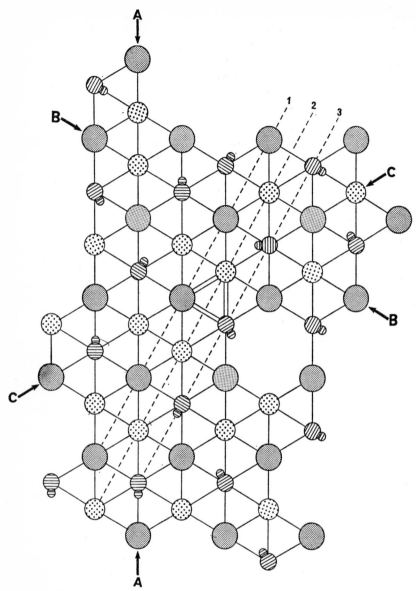

Figure 75. *Idealised diagram to illustrate the pattern of the receptors in the peripheral retina of the chick. Rows of receptors (eg AA, BB and CC) intersect at angles of 60°. Each of these rows is composed of repeating sequences of one rod and two other types of receptors. Another pattern is indicated along the broken lines: broken line 1 is a row of rods, broken line 2*

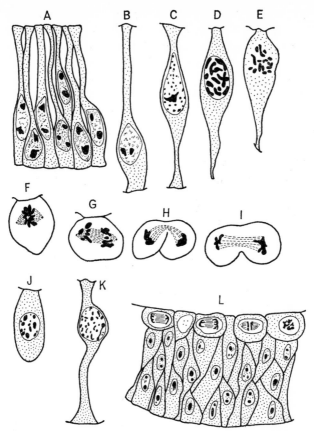

Figure 76. *Interkinetic migration. A. In an epithelium each cell is attached to its neighbours at the free (luminal) edge. B, C and D. In preparation for mitosis, the nucleus moves to the free edge and the cell shortens. E, F, G, H and I. Mitosis takes place at the free edge. J, and K. The new cell lengthens. L. Epithelium with mitotic figures at the free edge, the so-called 'germinal layer'. (After Sauer, 1935, 1936.)*

is a row of single cones, broken line 3 is a row of double cones. A third regular pattern which is visible throughout the lattice is a triangle of receptors consisting of one rod, one single cone, and one double cone; an example is indicated with double lines. The triangle can be regarded as the developmental unit from which the overall pattern is constructed. Thus, this unit is probably derived from a stem cell which underwent an orderly sequence of mitoses. At the first division the stem cell might give rise to a rod, at the second division to a single cone, and at the third division it might convert itself into a double cone. Rods: fine stipple; single cones: coarse stipple; double cones: hatched. (After Morris, in press.)

probably related to the fact that an epithelium must always retain continuity as a sheet of tissue. It had long been apparent that the mitotic figures in an epithelium were always arranged at the free, luminal edge. For this reason, in the neural tube the free edge is known as the germinal layer (or ependymal layer). Formerly, it was thought that the cells in this region either migrated up to the free border to divide, or were perhaps a special population of non-differentiated cells (stem cells) which were continually undergoing mitosis. There now appears, however, to be strong evidence for the theory of *interkinetic migration* (Figs. 76 and 77).

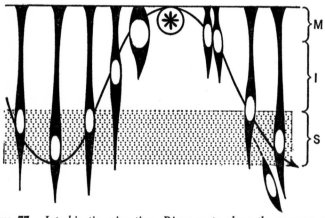

Figure 77. *Interkinetic migration. Diagram to show the passage of an individual cell (black) and its nucleus (white) to the free border. DNA synthesis takes place in the S-phase (stippled). M. mitotic zone. I. intermediate zone. (After Fujita, 1964.)*

F. C. Sauer, using light microscopy (1935), and M. E. Sauer and Walker (1959), using autoradiography, showed that in the early neural plate each cell is firmly attached to its neighbours where it abuts the lumen. During mitosis the cell migrates up to this free edge. After mitosis, the nuclei of the two new daughter cells migrate away from the lumen. The cells themselves, however, do not lose contact with this region even though they must undergo a change in shape to retain their 'foothold' on the lumen. Generally, these cells become elongated when not in mitosis.

This concept of interkinetic migration has been confirmed by many investigators using both light- and electron-microscopy (reviewed by Watterson, 1965). Interkinetic migration is, however, not restricted to the neural plate and has been reported in other epithelia, e.g., the lens rudiment of the chick (Zwaan *et al.*, 1969) and the epithelium of the yolk sac of the chick (Fujita, 1962).

The concept that differential mitosis (different mitotic rates in different tissues) plays a big rôle in shaping an embryo is such an obvious one that it is particularly instructive to consider when it does not.

(d) Cell movements

Gastrulation does not appear to be the direct result of regions of high proliferation (p. 85) but to depend rather on highly correlated cell movements.

Similarly, as we have seen (p. 218), elaborate migrations of tissues and organs also take place at later stages. Examples are: the migration of the neural crest cells (p. 227); the shifting of the primordium of the thyroid from the floor of the pharynx where it originates to its final position more caudally (p. 274); the movements of the gut, which herniates into the extra-embryonic coelom and there undergoes rotation through 90° before being withdrawn into the abdominal cavity; the descent of the testes in most mammals (such as man) into the developing scrotum; the movement of the heart from its place of origin in the neck to its final position in the thorax. During their migratory movements all these organs continue to grow and to develop their cellular specialisations, but it appears that their movements are not brought about as the result of localised proliferation.

(e) Cell death

As well as regions of high proliferation, there are also in the embryo areas that have a high level of necrosis. Probably in all embryonic tissues there is a continuous loss of cells through death, but in certain regions the death rate outstrips the proliferative rate. This means that just as some regions become enlarged by rapid cell division others become 'eroded' away. The dying cells undergo a process of breakdown and gradually become phagocytosed by other cells (Glücksmann, 1951; Bellairs, 1961b; Biggers, 1964). Probably the phagocytic cells then migrate to other parts of the body. Two interesting examples of the rôle of cell death in morphogenesis are the parts it plays in the shaping of the wing bud and of the toes of the developing chick.

We have seen (p. 245) that the limbs of the chick arise as paddle-shaped buds. As they elongate, a large necrotic patch appears in the posterior margin of the wing and a similar one has been found on the pre-axial margin of the leg bud (Saunders et al., 1962). In an eight-hour period as many as 1,000–2,000 cells may die in the posterior necrotic zone of the wing (Saunders, 1966). It has been suggested that these regions play an active rôle in shaping the developing limb. Saunders' interpretation is that these cells die because they have become committed to respond to their normal environment by dying; in a similar way most cells are committed to respond to their normal environment by developing.

If the necrotic zone of the wing bud is excised and grafted to a different region of the embryo, the cells still die at the same time as they would have done if they had remained in their normal location. Further, cells from other parts of the body grafted into the wing bud in place of the prospective necrotic zone develop normally. He therefore concluded that the necrotic zone cells do not die because of their position but because of a built-in programme for death which he describes as 'an internal death clock which ticks off necrosis at a pre-set time' (Saunders and Fallon, 1966). The position is further complicated by the fact that although the control of this death programme lies within the cell itself, nevertheless under certain experimental conditions this basic control can be overcome by the environment. Thus, if the post-necrotic zone is grafted into a much younger embryo the cells frequently fail to degenerate, and instead form extra cartilage, skin and feathers.

At the tips of the limb buds of most amniote embryos, there is another region of localised necrosis. This has been demonstrated in the chick (Saunders and Fallon, 1966), the mouse (Milaire, 1963; Menkes et al., 1965; Ballard and Holt, 1968), the rat and the human (Menkes et al., 1965). It appears to be important in sculpting out the fingers and toes. In the duck (Deleneau, 1965) necrotic patches of cells appear between the first and second toes, which are the only digits not connected by a web. Necrosis, is however, greatly reduced in the regions between the other toes, where webs subsequently do develop.

In the fowl mutant, 'talpid[3]', which includes in its syndrome of abnormalities shortened, polydactylous limbs, the massive patches of necrotic cells are absent (Hinchliffe and Ede, 1967) although small patches may be present at the tip of the limb bud.

The ultimate mechanism by which cell death is controlled is not understood; Saunders and Fallon (1966) have, however, discovered that it is possible to prevent necrosis from taking place in the posterior necrotic zone. This can be done by growing in tissue culture both the posterior necrotic zone and a large piece of mesoderm from the central part of the wing or leg. If the two tissues are separated by a Millipore filter with pores not less than 0·45 μ diameter, some factor appears to be able to pass through the filter and 'turn off the death clock'. Further work may be expected to throw light on the nature of these factors. Meanwhile, the system provides for us one more example of the importance of cell to cell communication in the embryo.

Ede and Agerbak (1968) have suggested that programmed cell death may not be a property of individual cells, but that like other forms of differentiation it depends on forms of contact between cell groups. Certainly in the mutant 'talpid³' in which the posterior zone is absent, the cell contacts are different from normal, for there is some evidence for an increased adhesiveness between the cells.

In most if not all, the known examples of cell death, lysosomes appear to play an important rôle. Lysosomes, which are usually densely-staining bodies, are present in most cells. They are characterised by a high content of acid hydrolases, especially acid phosphatases. There is, however, great variation in the structure and enzyme content of lysosomes. Each lysosome is bounded by a unit membrane and when this breaks down the enzymes contained within it are released. It is not known what causes the membrane to break down, nor whether this is the first stage of necrosis (discussions by de Duve, 1963, de Duve and Baudhuin, 1966; Dingle and Fell, 1970). Other cytological changes that occur include the distribution of lumps of chromatin around the nuclear membrane, vacuolation of the mitochondria and the production of patches of large, regularly-arranged granules in the cytoplasm (Bellairs, 1961b). These cells, or parts of them, are phagocytosed by healthy cells. In the wing bud stage the phagocytic cells are macrophages which converge on the degenerating region. We do not know how they are attracted to the necrotic zone. Various authors have suggested that they are drawn there by some sort of chemotaxis (but see p. 22).

Many other examples of cell death as a factor in morphogenesis are known. For example, the lumen of the oesophagus of the chick embryo becomes obliterated at about five days of incubation; this is because the epithelium of the roof collapses on to the floor and fuses with it (Allenspach and Hamilton, 1962). At about seven and a half to eight days of incubation the oesophagus reopens and this is largely due to the degeneration of the epithelial cells (Allenspach, 1966). Localised cell death is also important in the regression of the chick mesonephros (J. E. Morris, 1966).

(f) Cell size and changes in cell shape

One of the characteristics of development is the changes that take place in cell size and shape. We have already seen that cleavage is a period when little if any growth in mass takes place. Consequently, as rapid cell division occurs the cells become smaller and smaller (Fig. 20). The differentiation that takes place in the tissues after cleavage, however, is accompanied and, indeed, manifested by changes in cell shape and size. For instance, at the primitive streak stage the ectoderm is an epithelium, whereas the mesenchyme consists of loosely arranged spindle-shaped cells (Fig. 29). Subsequently, cells in the ectoderm of the neural plate become enlarged and individually taller than those in the non-neural ectoderm.

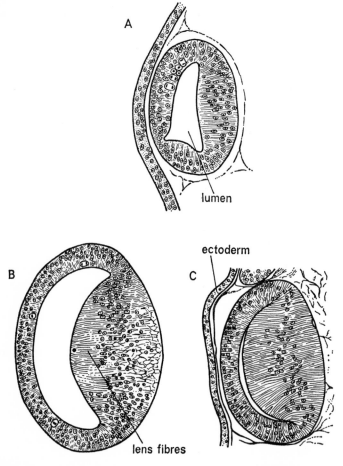

Figure 78. *Changes in the cell shape during lens formation in the chick embryo. A. The lens is at first a hollow sphere. B and C (not to scale). As the lens forms, some of the cells become more elongated than others. The cavity within the lens gradually becomes obliterated. (After Romanoff, 1960.)*

The final size of the organ is not controlled by the size of the individual cells, for if the cells are about twice the normal size, as sometimes happens in certain polyploid amphibians, the embryos themselves, and even the adults, are within the normal size range. Thus, in these animals fewer cells are present than normal (discussion by Goss, 1964).

It seems probable, however, that the final shape of many organs is affected by the shape of the individual cells. For instance, in its earliest stages, the lens consists of a hollow sphere, its walls being composed of columnar or cubical epithelium. With further development, these cells change shape and most of them become highly elongated. The amount of elongation varies, however (Fig. 78), those at the edge of the lens becoming less extended than those in the middle. The result of this elongation is correlated with a change in shape of the lens which now becomes 'lentoid'.

Similarly, the shape of the developing spinal cord is affected by the way in which the axons leave it, by the number and shape of the glial cells, and by the normality or otherwise of the surrounding vertebrae. The shape of the gut depends on the sizes and relationships of the different types of cells involved in it.

The shape of all organs is, however, modified by the inter-cellular matrix, by such extra-cellular materials as collagen, and by the shape of the surrounding organs.

(g) Amitosis

It is occasionally suggested that the cell may divide without undergoing mitosis. To most cytologists this seems unlikely. However, Bucher (1963), who has discussed the problem, has suggested that the nucleus, if not the whole cell, may divide in this way in certain circumstances. Similarly, Emanuelsson (1965) who has described multi-nucleated cells in the chick blastoderm at the cleavage stages has suggested that these may form by a process of amitosis.

B. CONTROL OF ORGAN SIZE IN THE EMBRYO

Perhaps understandably, scientists interested in growth have often been concerned to produce theories that explain not only the way in which the process is controlled in their own organ system but also in all other situations. It is often assumed that the same controls operate during embryonic growth, during the hypertrophy of organs under special circumstances, or during processes of regeneration. These are, however, by no means identical processes and it is questionable whether we are justified in attempting an all-embracing theory.

Several theories have been put forward to explain the way in which the organism controls the size of its individual parts. These have been critically examined in a stimulating book by Goss (1964) who recognises three main groups of theory. He points out that the basic requirement of any growth control theory is that it shall permit some sort of feedback mechanism which will 'inform' the organ when to stop growing. Inevitably, there must also be some sort of interaction between the feedback of all organs in the body.

(a) Control by tissue mass

This theory supposes that when an organ reaches a certain mass it ceases to grow (Weiss and Kavanau, 1957). The feedback mechanism is thought to be a chemical one, which has the merit of explaining the phenomenon of 'compensatory hypertrophy'. If one of a pair of organs is removed, the remaining organ frequently enlarges, and it could be argued that the chemical balance in the body has been disturbed by removing the one organ, but that this balance is restored by the remaining organ increasing its mass. Goss points out, however, that it is often difficult to distinguish in compensatory hypertrophy between the direct effects of tissue mass and of the physiological needs of the organism which require that the remaining tissue increases in size sufficiently to carry out the functions required of it. In an embryo we have relatively little information about the functional demand on the organs.

(b) Control by functional demand

This theory is based on the idea that if an organ is used it will continue to grow, though presumably there must be some feedback mechanism that stops it from growing when it has reached a certain size. The main value of this theory is in explaining compensatory growth. Thus, if one kidney is extirpated the other becomes enlarged, and since the main effect is that it now carries out all the excretory functions, it is not surprising that the hypertrophy is largely restricted to the most important parts from the point of view of the physiology of the organ. Hypertrophy of the general connective tissue is less apparent. In discussing this theory, Goss points out that the process of functional demand tends to be restricted to organs in some sort of physiological communication, for instance, those that share a common circulation. He concludes that although many experiments on adult organs have lent support to the theory of growth control by functional demand, it is not easy to apply this concept to embryos, as the factors that control their size seem to be in operation before many of the organs are functional. He suggests that embryonic differentiation is controlled more by direct gene action than by functional demand. But even this idea is a complex one. For instance, it might be thought that polyploid embryos possessing twice or more than twice the genetic material, and having correspondingly larger cells, might tend to produce larger animals. The careful studies of Fankhauser on amphibian polyploids, however, shows that this is not so, for the organs remain within their normal size-range even though this sometimes leads to mechanical difficulties; the larger red blood corpuscles may have trouble squeezing through capillaries that do not have a bore-size any larger than usual.

In considering the feedback mechanisms involved in controlling the size of the organism, it is clear that the cells and organs in different regions of the body are in communication.

(c) Control by growth-regulating substances

This theory is concerned with the mechanisms by which cells and tissues communicate with one another. In general they are based on the idea that chemical substances (specific molecules) are passed from one part of the body to another. A well known example is that of the substance named 'chalone' that is thought to act as a feedback mechanism inhibiting mitosis in mouse epidermis (Bullough, 1962). The evidence is based mainly on the fact that a specific substance can be extracted from mouse epidermis, and indeed from other tissues which will inhibit mitosis in mouse skin. Bullough has suggested that these substances prevent mitosis from proceeding at the same rate in the adult as in the embryo. Others have suggested that mitotic stimulations also play a rôle in control. S. Cohen (1965) has obtained an epidermal growth factor (EGF) from the sub-maxillary gland of mice and found that this has a specific effect on the epidermal skin of chick embryos growing in culture. It stimulates both proliferation and keratinisation in organ cultures of intact skin and in dermis-free epidermal sheets. Subsequently, Cohen and his colleagues have been able to demonstrate that EGF acts by stimulating the synthesis of protein at the level of the ribosomes. Hoober and Cohen (1967) showed that there was no increase in DNA, but that there was an increase in the ability of the ribosomes to incorporate labelled amino acids. This is apparently brought about by the conversion of the pre-existing ribosomal monomers into functional polysomal structures.

As we have seen (p. 246 *et seq.*), the presence of a 'maintenance factor' has been invoked in the development of the limb bud. One of the most fully investigated embryonic growth regulating substances, however, is that of the nerve growth factor (NGF) of Levi-Montalcini (1952; *et al.*, 1964). This substance, initially obtained from mouse sarcomas and later from mouse salivary gland, has a specific effect on both chick embryos and newborn and adult mammals in that they respond by hypertrophy of certain nerve ganglia. In chicks the effect is principally on the sensory and sympathetic ganglia but there is also a tremendous (and abnormal) innervation of the viscera. The nerve growth factor has been studied biochemically and found to be a protein with molecular weight of about 44,000 (Cohen, 1958, 1960); it possesses antigenic properties so that if it is injected into rabbits it can call forth specific antibodies. If these antibodies are injected into newborn mammals the sympathetic cells are totally destroyed. The way in which the NGF affects the cells has also received considerable attention and it has been demonstrated that the addition of even very small amounts of NGF to nerve cells growing in culture increases the rate of incorporation of amino acids; this suggests that NGF acts by stimulating the production of mRNA. This is supported by work using inhibitors of mRNA synthesis, such as actinomycin D (p. 186) (Levi-Montalcini and Angeletti, 1965).

Finally, several authors have reported the results of experiments that are difficult to explain, except by postulating that growth regulating substances are involved. Chaytor (1963), for example, found that if he parabiosed chick embryos of nine days incubation with partners of twelve days, the livers of the younger embryos grew at a slower rate. As a result, he suggested that a 'liver growth inhibitor' acting as a feed-back mechanism in the older embryo had affected the growth rate of the younger.

C. HORMONES IN DEVELOPMENT

Another type of growth regulating substance is embryonic hormones. Several authors have suggested that hormones might bring about their effects by interfering with DNA transcription or else by interfering with the RNA at the level of translation (discussions by Scarano and Augusti-Tocco, 1967; Tata, 1968).

Our knowledge of the rôle played by embryonic hormones in development is still very limited compared with the information we possess about adult hormones. This is largely due to the difficulties involved in dealing with materials that are initially produced only in minute quantities. It is especially difficult to know when the embryo begins to secrete a particular hormone, though we can be confident that this does not happen until the organ has become at least partially differentiated histologically. Thus, the thyroid promordium will not begin to secrete until the follicles have formed. On the other hand, since the morphological and histological differentiation of an endocrine organ does not in itself prove that hormones are being secreted, attempts have been made to demonstrate their activity by experimental means. These experiments are of the same type as are classically used by students of adult endocrinology.

(1) Organs are extirpated if possible (though this is often technically impossible in embryos). Astonishingly enough, extirpation of the pituitary is usually by the drastic expedient of decapitation! This horrific experiment is perhaps rather less crude than would at first sight appear, since its side effects can be controlled to some extent in the chick by additional experiments in which pituitary hormone is injected after decapitation.

(2) Additional pieces of the organ are taken from another embryo or adult and implanted into the embryo. In the chick, this can be done by grafting tissues on to the chorioallantoic membrane.

(3) Hormones are administered to the embryo. This experiment, which results in an abnormally high dose of hormone acting on the tissues, has the theoretical disadvantage that since supplies of embryonic hormone would be excessively tedious to collect, adult hormone is generally used and this is not necessarily identical.

(4) Sometimes the embryo can be treated with biological analogues of the hormone which, theoretically, have the effect of depriving the target organs of the real hormone. The disadvantages of biological analogues are discussed on p. 188.

(5) In some cases it has been possible to treat the embryo with suitable labels which are then taken up and incorporated into the hormones. For example, radio-active iodine is taken up by the thyroid just prior to colloid formation.

(6) The embryonic gland may be grown in tissue culture, possibly in combination with its target organ.

It is, perhaps, not over-idealistic to hope that not only will all these techniques soon be applied more widely to the embryonic endocrine glands but also we shall in addition know more of the molecular biology of the system. For instance, what are the instructions that initiate the onset of secretions and what are the feedback mechanisms that stimulate and control them?

The three endocrine systems about which most is understood are the thyroid, pituitary and gonads.

(a) The thyroid

This develops from an epithelial portion which invaginates from the floor of the pharynx and migrates caudally, and a mesenchymal portion which comes to surround it. The epithelium gives rise to the follicles, whereas the mesenchyme forms the general stroma and blood vessels. Interestingly, like many other organs of ectodermal-mesodermal origin (p. 238), there is probably an inductive relationship involved, for in the chick, at least, if the epithelium is deprived of its mesenchyme it fails to differentiate. Furthermore, it appears that the relationship is a highly specific one since, if the epithelium is presented with mesenchyme from another part of the body, it still cannot differentiate (Dossel, 1958). The interaction does not appear to be confined to the earliest stages, but continues to be necessary even after histological differentiation of the organ (Hilfer, 1968).

The ability of the thyroid to produce colloid appears to be dependent on stimulation from the pituitary, just as in the adult. In chicks, hypophysectomy (p. 273) leads to the production of small thyroids with reduced amounts of colloid (Fugo, 1940). In a further ingenious set of experiments, the same author found that thyroids from hypophysectomised chicks would develop normally if grafted on to the chorioallantoic membrane of normal embryos where they could come under the influence of blood-borne thyroid stimulating hormone (TSH). On the other hand, if they were grafted to the chorioallantoic membrane of a hypophysectomised host, they failed to develop properly. The thyroids of decapitated (i.e. hypophysectomised) slow worm (*Anguis*) embryos are also reduced in size (Raynaud, 1962b).

There is also indirect evidence that the thyroids of various amniotes (e.g. chick and sheep) are producing a hormone shortly after colloid first appears. A crucial test is to implant pieces of the organ into amphibian larvae and see if they bring about metamorphosis (discussion by Willier, 1955).

We have, unfortunately, little knowledge of the chemistry of thyroid hormones in the embryo and do not know if they are identical with those of the adult.

(b) The pituitary

This also has a dual origin, a downgrowth from the floor of the brain, the infundibulum fusing with an upgrowth from the roof of the mouth, Rathke's pouch. Once again, there is probably an inductive relationship between the two components. Both in amphibians and in birds, if the infundibulum is destroyed Rathke's pouch fails to differentiate properly (Hillemann, 1943). Moreover, if the chick pituitary rudiment is isolated from the body and grown either in tissue culture (Moscona and Moscona, 1952) or on the chorioallantoic membrane (Stein, 1929, 1933) it seems to be capable of proper differentiation only after the stage at which induction is presumed to have occurred.

We have seen that the pituitary has some effect on the development of the thyroid. Other effects are on the size of the embryo and on the gonads. Hypophysectomised chicks and mammals tend to be smaller than normal ones. This appears to be a direct effect of pituitary hormone deprivation, for if these hypophysectomised embryos are treated with extracts of anterior lobe of pituitary, they increase in size. Thus, it seems probable that in the late embryo, as in the infant and child, the pituitary exerts a controlling influence on the size of the entire organism and we can expect that in the foetus the pituitary will itself be controlled by the hormones produced by its target organs. Indeed, there is evidence suggestive of a feedback effect of the thyroid on the pituitary. Thus, in thyroidectomised amphibian embryos the anterior lobe pituitary becomes enlarged and contains more acidophils than in the control embryos (discussion by Willier, 1955).

The effect of the pituitary on the developing gonads has been shown in a variety of ways (discussion by Jost, 1969). For instance, if foetal pig pituitaries are grafted into immature juvenile mice there is a precocious development of sex organs (Smith and Dortzbach, 1929). The testes of hypophysectomised chicks and rabbits are smaller than normal and poor in interstitial cells.

So we can conclude that hormones are produced by the pituitary during development, and that although we cannot be certain that they are identical with the corresponding adult hormones, their effects on the target organs are comparable.

(c) The gonads

Although they do not themselves produce sexually mature germ cells until puberty, the gonads, nevertheless, appear to produce sex hormones

K

even in the foetal stages. In the early stages the gonad has both cortex and medulla and is capable of differentiating into either male or female. Similarly, although each organism is genetically determined, except in rare cases of chromosomal anomalies, to become either male or female, the initial stages of sex development are delicately balanced and can be swung one way or the other by environmental effects. The classic example is that of the development of free-martins in cattle. These are sterile females that develop when male and female twins share a common placenta, a situation which leads to a vascular connection between their circulations. Lillie (1917) was able to show that the interstitial cells of the testis differentiate earlier than the stroma of the ovary. The result is that the female gonad, which is genetically an ovary, becomes testicular, and Lillie concluded that this was because of the action of the male hormones. (A less orthodox view is that freemartins are the result of the migration of primordial germ cells from the male into the female.)

Further evidence comes from the fact that if a male rat foetus (Jost, 1965) is castrated before the sexual organs have differentiated properly, the Wolffian duct disappears and the urino-genital sinus and external genitalia become of the female type. The results are, however, less severe if castration does not take place till after the sex organs have differentiated.

The tissue culture experiments of Price and Pannabecker (1958) and Price and Ortiz (1965) are of particular interest in showing the effect of the gonads in development. They found that the entire genital tract of the male rat foetus at $17\frac{1}{2}$ days could differentiate well if grown in culture but that if the testes were removed the Wolffian duct regressed, there were no seminal vesicles and there was only a poor prostate. These omissions could, however, be corrected with the application of testosterone.

By contrast, the duct system of the female developed well whether or not the ovaries were present. Essentially, therefore, the testes appeared to produce a hormone for maleness. The story is probably more complex than that, however, for Price also found that the Müllerian ducts would only survive in tissue culture if the testes were absent. It seems likely that new light on the interrelationships of these organs will come from the studies on chimaeric (allophenic) mice (p. 136).

We can conclude, however, that sex hormones are produced as soon as the gonads begin to differentiate morphologically, for even at this stage castration causes immediate cessation of sexual development. We have as yet no evidence as to whether they are chemically identical with adult hormones, although they are undoubtedly similar for they produce comparable effects. Many aspects of the problem are discussed in Section VI of DeHaan and Ursprung (1965). Comparable experiments in reptiles are discussed by Raynaud (1967).

(d) Other hormones

It is outside the scope of this book to deal in detail with all the known hormonal effects during development, but reviews of special aspects of the subject are given by Moore (1950), Zwilling (1952) and various authors in the book edited by DeHaan and Ursprung (1965). The rôle of the placental hormones in mammalian pregnancy and embryogenesis are discussed by Simmer (1968), and Jost (1969).

Of especial interest have been the studies on the importance of insulin, especially in chick embryo, which results in disturbances in the carbohydrate cycle (Landauer, 1947; Zwilling, 1952).

D. CONTROL OF DEVELOPMENTAL PROCESSES BY THE NERVOUS SYSTEM

The rôle played by the presumptive notochord in inducing the nervous system has been discussed in Chapter 6. The immediate result is a thickening and rolling-up of the neural plate into a tube, though the mechanisms that bring about this rolling-up are not well understood. The experiments have been carried out mainly on amphibians, and these have indicated that changes in cell adhesiveness are important (Jacobson, 1968; see also discussion by Källen, 1965). The first signs of local differentiation in the central nervous system are in the head region, where forebrain, midbrain and hindbrain soon become visible as characteristic bulges. To some extent these bulges develop because of a high level of proliferation in each of those areas, but in addition they depend also on certain cell migrations taking place. There is some evidence that once the neural tube has been established the future topographic regions are already laid down. This has been illustrated by many experiments; for instance, pieces of neural tube have been extirpated and then replaced in a different position and yet each piece has developed according to its original fate rather than to its new situation (Birge, 1963). The position is not clear however since other work has shown that modifications of the fate can take place.   For example, Steding (1962) turned pieces of neural tube upside down and found that the development of individual cells could be modified by influence of the surrounding mesenchyme. Steding concluded that although the neural regions were determined to some extent there was a considerable degree of plasticity (see discussions in Chapter 1).

By about three days of development in the chick embryo, three distinct histological regions can be recognised in any transverse section through the central nervous system. These are the *ependyma*, the *mantle* layer and the *marginal* layer. As development proceeds the cord becomes larger in section and the mantle layer becomes a different shape, dorsal and ventral horns being visible. The ependyma is the part of the nervous system where proliferation of the cells takes place (see discussion on interkinetic migration on p. 264); the mantle layer is the region that contains neuroblasts which give rise to the neurons but it also possesses supporting cells; the marginal layer is made up principally of the axons that extend out from these neurons and pass either up or down the cord or leave the central nervous system and innervate other parts of the body.

Before considering the effect of the nervous system on the peripheral structures, we must first discuss the action of the end organs on the nervous system itself. In acting as a controller of developmental processes, the nervous system interacts with the organs it innervates. The effect of the central nervous system on the peripheral structures and the reciprocal action of the peripheral organs on the central nervous system have recently been reviewed in detail by Hughes (1968) and Hamburger (1968) so that only selected items will be discussed here. The effects of other factors on the central nervous system, such as hormones and temperature, are considered by Kollros (1968).

We have seen that there is a special growth promoting factor that particularly affects the nervous system (NGF, see p. 272) and that this is produced by other, non-neural, tissues. Perhaps the most elaborately investigated aspect of the problem, however, is the effect of the terminal connections upon the spinal ganglia. Initially these ganglia are derived from the neural crest (p. 227) but they come to lie at regular intervals down either side of the cord. The brachial and lumbrosacral horns rapidly become larger than the others, and this is due not only to a higher mitotic rate in these regions but also to the fact that fewer cells die than in other parts of the spinal cord (Hamburger and Levi-Montalcini, 1950).

If the limb of a chick embryo is extirpated, the nerve fibres that supply that limb are inevitably damaged. There is, however, also a striking reduction in the size of the ganglia from which those nerves emerge (Shorey, 1909; Hamburger, 1934). Conversely, if an additional limb is grafted to the chick, the ganglia enlarges. Hamburger and Levi-Montalcini concluded that the periphery controls the proliferation and initial differentiation of cells in both the sensory and motor centres. Similar experiments involving amputation of the limb have been carried out by others, including Prestige (1967) on *Xenopus* (review by Hughes,

1968). Prestige has interpreted his results as evidence of a chemical action of the peripheral organs on the central nervous system. Such a chemical would probably be related to the nerve growth factor. As Hughes points out, the alternative interpretation that they are due to impulses flowing along the afferent nerves to the central nervous system can be tested by interfering with the afferent nerves during development. Nevertheless, although this experiment has been performed by various workers the results are not clear-cut (discussion by Hughes, 1968).

The fact that the central nervous system can influence the development of the peripheral structures has been shown in many ways. In particular, most organs undergo some atrophy if their nerve supply is cut off (Singer, 1968). For instance, if the nerves supplying the lateral lines are cut there is a degeneration of the sensory cells. Conversely, if these nerves are then allowed to regenerate there is an increase in the numbers of these sensory cells (discussion by Hughes, 1968). Thus it appears that in some way the nerve fibres exert a trophic (or nourishing) action on the end organ.

This concept has been elaborately pursued by Singer and his collaborators, who have produced an impressive series of examples to show that nerves are required for regeneration. Their examples are largely drawn from invertebrates and from lower vertebrates, since regeneration is mainly limited in amniotes to structures such as skin and various soft tissues. The tails of lizards undergo regeneration, and there have been reports of occasional partial limb regeneration in this group (see A. d'A. Bellairs, 1970). Here regeneration is largely by the action of the spinal cord ependyma (Bryant, 1970).

Essentially, Singer suggests that a chemical is produced by the cell body of the nerve and that this chemical is then transmitted along the axon. It is produced primarily to maintain the axon itself. Similar chemicals may be produced by other tissues, but the nerve cells produce particularly large quantities. This is because of the special problems involved in maintaining axon cytoplasm which lies far from the cell body. The supposed chemical agent is not associated with conduction, however, since if conduction is blocked the action of the nerve on regeneration is not affected.

It is this trophic action of the nerve that Singer believes plays an essential rôle in stimulating regeneration, for without the nerves regeneration does not usually take place. A curious exception to the rule is that if during development the nerves are prevented from entering the limb, this limb is subsequently capable of regenerating without nerves.

Singer suggests that the ability of wounds to respond to the trophic

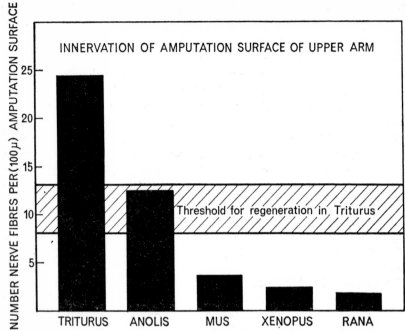

Figure 79. *Nerve fibres and regeneration in various vertebrates. In* Triturus, *regeneration occurs only when the number of fibres per unit area rises above the threshold. In many species the threshold value is never reached. In the lizard,* Anolis, *the limbs do not normally regenerate.* (*From Singer, 1968.*)

action of the nerves depends on a certain threshold of responsiveness, and that this threshold itself varies according to circumstances as well as to the animal. For instance, he found that if limb stumps were grafted to the back of the animal, they needed to be invaded by fewer nerves to enable them to regenerate than if they remained in their natural position. In comparing the ability of various species to regenerate limbs he found that this was correlated with the number of nerve fibres that entered the stump (Fig. 79). At first, it appeared that *Xenopus* might be an exception to this rule, since its limbs will regenerate even though only sparsely innervated. In this species, however, the cross-sectional area of each fibre was comparatively large. Thus it now appears that the ability to regenerate depends more on the amount of axoplasm entering the stump than upon the actual number of nerves (Singer *et al.*, 1967). Building on these results, he was able to test his theory that higher vertebrates do not normally regenerate limbs because their limb stumps are inadequately supplied with nerves. He doubled the nerve supply of a limb stump of a lizard by directing into it the sciatic nerve from unoperated side of the body. As a result he was able to achieve some degree of regeneration.

This elegant experiment, whilst supporting his ideas, still leaves us with many unanswered questions, such as the nature of the supposed trophic substance and the way it works on the regenerate. As a theory it can be criticised in that it is perhaps too flexible and too easily modified to accommodate all the facts; that it explains everything but is difficult to prove. But alternative theories, that regeneration is controlled by hormones or by influence from the epithelial cap over the wound, seem less satisfactory (discussions by Thornton, 1956, 1968; Hay, 1966). So, although the relationship is not understood, it is difficult not to be attracted by the idea that the nervous system affects the development and maintenance of some of, if not all, the organs that it innervates in the lower vertebrates; it follows that the situation may be similar in the amniotes as well.

We are at present ignorant of the interrelationships that exist between the action of the nerve growth factor of Levi-Montalcini and the trophic substance of Singer. At first sight it might seem unnecessary to look for a relationship, but it seems probable that to achieve a balance between the central nervous system and the end organs, each must produce a feed-back mechanism that informs the other. It is improbable that the two substances would be chemically identical, but perhaps not unlikely that they should be at opposite sides, as it were, of a continuous cycle of events.

E. SUMMARY

There is evidence that the biochemical changes that are the first sign of differentiation in a tissue may be related to a preceding burst of mitosis. Mitosis also plays a rôle in the shaping of organs and in the histological patterning of certain cells. Other controlling mechanisms in development that are discussed include various humoral agents.

ASSESSMENT

LET us now try to assess what we know about the development of higher vertebrates. In what ways does it resemble that of other organisms, and in what ways does it differ?

Our knowledge of many aspects of amniote development is poor compared with that of, say, amphibians. In this book, therefore, I have had to draw on information we possess about other groups. There can be little doubt that in many respects—at the cellular level, at least—the same type of biological laws apply throughout the animal and plant kingdoms. Life depends on such substances as nucleic acids, and on such processes as protein synthesis, whether we are dealing with bacteria, invertebrates or vertebrates.

The fertilized eggs of amniotes resemble those of other multi-cellular organisms in that they must normally pass through cleavage and gastrulation before the organs begin to form. The mechanisms of cell division and the behaviour of cells are much the same as in other groups. The organs form in a similar way to the comparable structures in lower vertebrates. Indeed, there is probably no developmental process that is unique to the amniotes and does not have its counterpart in other groups. As we have seen, even an 'amnion', which distinguishes the amniotes from all other vertebrates, develops in certain insect embryos.

These common processes vary, however, from one class of vertebrates to another, and it is by studying the variations that we begin to formulate general laws of development. Some of the ways in which variation of this type occurs are discussed in this book. They include such things as differences in the morphogenetic movements and variations in the time when the embryo begins to synthesise new ribosomes.

In its next phase, developmental biology will probably become more concerned with the way in which an embryo as a whole is co-ordinated. This is liable to be an even more complex problem than any that has yet been tackled. If we are to understand something of how this co-ordination and interaction takes place between different parts of an embryo, we need to be even more aware of the normal variations that occur.

The emphasis in this book has been on the amniotes. To some people this group has a strong emotional appeal since it includes our own class, the mammals. To others it is attractive because of the medical and veterinary aspects. But more important to the developmental biologist, perhaps, is the fact that it consists of three classes of vertebrates, each of which not only differs from the others, but also varies within itself. In the past our understanding of the developmental processes in amniotes has lagged behind that of amphibians. Let us hope that in the next decade the developmental processes of these higher vertebrates will not be neglected.

APPENDIX

NORMAL TABLES (TABLES OF NORMAL DEVELOPMENT)

A NORMAL table consists of a series of annotated illustrations of a particular species at successive stages of development. Usually only the external features are shown but descriptions of internal organs and their histology may also be given. It is useful to be able to describe the stage of development of an embryo in terms of a normal table, for not only does this eliminate the necessity for continual detailed accounts but it also makes it easy for the reader to decide whether different authors are referring to embryos of the same stage or not.

A. BIRDS

(a) Chick (Gallus)

Hamburger, V. and Hamilton, H. L. (1951) A series of normal stages in the development of the chick embryo. *J. Morph.* **88**, 48–92.

(b) Duck (Anas)

Von Koeke, H. U. (1958) Normalstadien der Embryonalentwicklung bei der Hausente (*Anas*). *Embryologia* **4**, 55–78.

(c) Japanese quail (Coturnix)

Padgett, C. S. and Ivey, W. D. (1960) The normal embryology of the *Coturnix* Quail. *Anat. Rec.* **137**, 1–11.

Zacchei, A. M. (1961). Lo sviluppo embrionale della quaglia giaponese (*Coturnix coturnix japonica*). *Archo. ital. Anat. Embriol.* **66**, 36–62.

(d) Turkey (Meleagris)

Phillips, R. E. and Williams, C. S. (1944) External morphology of the turkey during the incubation period. *Poultry Science* **23**, 270–277.

(e) Pheasant (Phasianus)

Fant, R. J. (1957) Criteria for aging pheasants. *J. Wildlife Management* **21**, 324–328.

(f) Penguin (Pygoscelis)

Herbert, C. (1967) A timed series of embryonic developmental stages of the Adelie penguin (*Pygoscelis adeliae*) from Signy Island, South Orkney Islands. *Br. Antarct. Surv. Bull.* **14**, 45–67.

B. MAMMALS

(a) Rat and mouse (Rattus and Mus)

Christie, G. (1964) Developmental stages in somite and post-somite rat embryo based on external appearance, and including some features of the macroscopic development of the oral cavity. *J. Morph.* **114**, 263–286.

Edwards, J. A. (1968) The external development of the rabbit and rat embryo. In *Advances in Teratology* 3, Ed. D. H. M. Woollam. Logos Press, London, pp. 239–263.

Grüneberg, H. (1943) Normal staging of the mouse embryo. *J. Hered.* **34**, 89–92.

Henneberg, B. (1937) Normentafeln zur Entwicklungsgeschichte der Wanderratte (*Rattus norvegicus* Erxleben). In Keibels *Normentafeln zur Entwicklungsgeschichte der Wirbeltiere* **15**. Fischer, Jena.

Rugh, R. (1967) *The Mouse, its Reproduction and Development*. Burgess, Minneapolis.

Snell, G. D. and Stevens, L. C. (1966) Early embryology. In *Biology of the Laboratory Mouse*, Ed. E. L. Green. McGraw-Hill, New York, pp. 205–245.

Witschi, E. (1962) Development: Rat. In *Growth VII. Prenatal Vertebrate Development*, Eds. D. O. Altman and O. D. Dittmer. Biological Handbooks of the Federation of American Societies for Experimental Biology, Washington.

(b) Rabbit (Lepus), deer (Cervus), pig (Sus), Tarsius, Loris, Homo

Tables of the normal development of these animals are given in Keibel's Normentafeln (see Henneberg above).

C. REPTILES

It must be emphasised that the stage of development reached at any moment of time will depend greatly on the temperatures to which the embryo has been subjected.

(a) Lizards (Sauria)

Dufaure, J. P. and Hubert, J. (1961) Table de développement du lézard vivipare. *Lacerta* (*Zootoca*) *vivipara* Jacquin. *Archs Anat. microsc. Morph. exp.* **50**, 309–327.

Lemus, D. A. (1967) Contribución al estudio de la embriología de reptiles chilenos. II. Tabla de desarrollo de la lagartija vivípara *Liolaemus gravenhorti* (Reptilia—Squamata—Iguanidae). *Biologica* **40**, 39–61.

Pasteels, J. J. (1956–7) Une table analytique du développement des reptiles. *Ann. Soc. Roy. Zool. Belg.* **87**, 217–241.

Peter, K. (1904) Normentafel zur Entwicklungsgeschichte der Zauneidechse (*Lacerta agilis*). In Keibel's *Normentafeln* (see Henneberg above).

(b) Snakes (Serpentes)

Hubert, J., Dufaure, J. P. and Collin, J. P. (1966) Matériaux pour une table de développement de *Vipera aspis* L. I. La periode d'organogenese. *Bull. Soc. zool. Fr.* **91**, 779–788.

Zehr, D. R. (1962) Stages in the normal development of the common garter snake, *Thamnophis sirtalis sirtalis. Copeia* No. 2, 322–329.

(c) Turtles (Chelonia)

Pasteels, J. J. (1956–7) See above.

Yntema, C. L. (1968) A series of stages in the embryonic development of *Chelydra serpentina. J. Morph.* **125**, 219–252.

(d) Crocodilia

Reese, A. M. (1915) *The Alligator and its Allies.* G. P. Putnam's Sons, New York and London.

Voeltzkow, A. (1902) Beiträge zur Entwicklungsgeschichte der Reptilien. I. Biologie und Entwicklung der ausseren Körperform von *Crocodilus madagascariensis* Grand. *Abh. Senkenberg. naturf. Ges.* **26**, 1–150. In *Wissenschaftliche Ergebnisse der Reisen in Madagascar und Ostafrika in den Jahren 1889–95.* This work contains other papers by Voeltzkow on reptilian embryology.

REFERENCES

Abbot, J. and Holtzer, H. (1965) Critical number of mitoses and differentiation of chondroblasts and myoblasts. *Anat. Rec.* **151**, 439.

Abercrombie, M. (1937) The behaviour of epiblast grafts beneath the primitive streak of the chick. *J. exp. Biol.* **14**, 302–318.

Abercrombie, M. (1939) Evocation in the chick. *Nature, Lond.* **144**, 1091.

Abercrombie, M. (1950) The effects of antero-posterior reversals of lengths of the primitive streak in the chick. *Phil. Trans. R. Soc. Ser. B.* **234**, 317–338.

Abercrombie, M. (1958) Exchanges between cells. In *The Chemical Basis of Development*, Eds. W. D. McElroy and B. Glass. Johns Hopkins Press, Baltimore, pp. 318–328.

Abercrombie, M. (1965) Cellular interactions in development. *Proc. Int. Cong. Zool.* **6**, 261–280.

Abercrombie, M. (1966) Contact inhibition: the phenomenon and its biological implications. *Nat. Cancer Inst. Monograph No.* 26, 249–277.

Abercrombie, M. (1967) General review of the nature of differentiation. In *CIBA Foundation Symposium on Cell Differentiation*, Eds. A. V. S. de Reuck and J. Knight. Churchill, London, pp. 3–12.

Abercrombie, M. and Ambrose, E. J. (1958) Interference microscope studies of cell contacts in tissue culture. *Exp. Cell Res.* **15**, 332–345.

Abercrombie, M. and Bellairs, R. (1954) The effects in chick blastoderms of replacing the primitive node by a graft of posterior primitive streak. *J. Embryol. exp. Morph.* **2**, 55–72.

Abercrombie, M. and Heaysman, J. E. M. (1953) Observations on the social behaviour of cells in tissue culture. I. Speed of movement of chick heart fibroblasts in relation to their mutual contacts. *Exp. Cell Res.* **5**, 111–131.

Abercrombie, M. and Heaysman, J. E. M. (1954) Observations on the social behaviour of cells in tissue culture. II. Monolayering of fibroblasts. *Exp. Cell Res.* **6**, 293–306.

Abercrombie, M. and Middleton, C. A. (1968) Epithelial-mesenchymal interactions affecting locomotion of cells in culture. In *Epithelial-Mesenchymal interactions*, Eds. R. Fleischmajer and R. E. Billingham. Williams and Wilkins, Baltimore, pp. 56–63.

Abercrombie, M. and Waddington, C. H. (1937) The behaviour of grafts of primitive streak beneath the primitive streak of the chick. *J. exp. Biol.* **14**, 302–318.

Adams, C. E. (1965) The influence of maternal environment on pre-implantation stages of pregnancy in the rabbit. In *Pre-implantation Stages of Pregnancy*, Eds. G. E. W. Wolstenholme and M. O'Connor. Churchill, London, pp. 3–22.

Adams, C. E. (1967) Ovarian control of early embryonic development within the uterus. In *Reproduction in the Female Mammal*, Eds. G. E. Lamming and E. C. Amoroso. Butterworths, London, pp. 532–550.

Adams, C. E., Hay, M. F. and Lutwak-Mann, C. (1961) The action of various agents upon the rabbit embryo. *J. Embryol. exp. Morph.* **9**, 468–491.

Afzelius, B. A. (1956) The ultrastructure of the cortical granules and their products in the sea urchin egg as studied with the electron microscope. *Exp. Cell Res.* **10,** 257–185.

Allenspach, A. L. (1966) The reopening process of the oesophagus in the normal chick and the crooked neck dwarf mutant. *J. Embryol. exp. Morph.* **15,** 67–76.

Allenspach, A. L. and Hamilton, H. L. (1962) Histochemistry of the oesophagus in the developing chick. *J. Morph.* **111,** 321–344.

Allenspach, A. L. and Roth, L. E. (1967) Structural variations during mitosis in the chick embryo. *J. Cell Biol.* **33,** 179–196.

Alliston, W., Howarth, B. and Ulberg, L. C. (1965) Embryonic mortality following culture *in vitro* of one and two-celled rabbit eggs at elevated temperatures. *J. Reprod. Fert.* **9,** 337–341.

Amoroso, E. C. (1952) Placentation. In *Physiology of Reproduction*, **2,** (3rd Edition), Ed. A. S. Parkes. Longmans, London, pp. 127–311.

Amoroso, E. C. (1959) The attachment cone of the guinea-pig blastocyst as observed under time-lapse phase contrast cinematography. In *Implanation of Ova*, Ed. P. Eckstein. Mem. Soc. Endocrinol., pp. 50–51.

Amos, H. and Kearns, K. E. (1962) Synthesis of 'bacterial' protein by cultured chick cells. *Nature, Lond.* **195,** 806–808.

Amprino, R. (1965) Aspects of limb morphogenesis in the chicken. In *Organogenesis*, Eds. R. L. DeHaan and H. Ursprung. Holt, Rinehart and Winston, New York, pp. 255–281.

Ancel, P. (1950) *La Chimiotératogenèse chez les Vertebrés.* Doin et Cie., Paris.

Anderson, J. M. (1965) Immunological inertia in pregnancy. *Nature, Lond.* **206,** 786–787.

Anderson, J. W. (1959) The placental barrier to gamma-globulins in the rat. *Am. J. Anat.* **104,** 403–430.

Apter, M. J. (1966) *Cybernetics and Development.* Pergamon Press, Oxford.

Asdell, S. A. (1965) *Patterns of Mammalion Reproduction.* Constable, London.

Assali, N. S., Dilts, P. V., Plentl, A. A., Kirschbaum, T. H. and Gross, S. J. (1968) Physiology of the Placenta. In *Biology of Gestation*, **1,** Ed. N. S. Assali. Academic Press, New York and London, pp. 186–289.

Attwood, H. D. and Park, W. W. (1961) Embolism to the lungs by trophoblast. *J. Obstet. Gynaec. Brit. Comm.* **68,** 611–617.

Auerbach, R. (1967) Development of immuno-competent cells. *Dev. Biol.* Suppl. 1, 254–263.

Auerbach, S. and Brinster, R. L. (1967) Lactate dehydrogenase isozymes in the early mouse embryo. *Exp. Cell Res.* **46,** 89–92.

Austin, C. R. (1956) Cortical granules in hamster eggs. *Expl. Cell Res.* **10,** 533–540.

Austin, C. R. (1961) *The Mammalian Egg.* Blackwell, Oxford.

Austin, C. R. (1965) *Fertilization.* Prentice-Hall, New Jersey.

Austin, C. R. and Bishop, M. W. H. (1958) Role of the rodent acrosome and perforatorium in fertilization. *Proc. Roy. Soc. London, Ser. B* **149,** 241–248.

Austin, C. R. and Braden, A. W. H. (1954) Induction and inhibition of the second polar division in the rat egg and subsequent fertilization. *Aust. J. biol. Sci.* **7,** 195–210.

Avery, G., Chow, M. and Holtzer, H. (1956) An experimental analysis of the development of the spinal column. V. Reactivity of chick somites. *J. exp. Zool.* **132**, 409–426.

Baker, C. M. A. (1968) The proteins of egg white. In *Egg Quality: a Study of the Hen's Egg*, Ed. T. C. Carter. Symposium No. 4 of the British Egg Marketing Board. Oliver and Boyd, Edinburgh, pp. 67–108.

Baker, T. G. and Franchi, L. (1966) Fine structure of the nucleus in the primordial oocyte of primates. *J. Anat.* **100**, 697–699.

Balinsky, B. I. (1925) Transplantation des Ohrbläschens bei Triton. *Arch. Entw-Mech. Org.* **105**, 718–731.

Balinsky, B. I. (1965) *An Introduction to Embryology* (2nd Edition). Saunders, Philadelphia.

Balinsky, B. I. and Walther, H. (1961) The immigration of presumptive mesoblast from the primitive streak in the chick as studied with the electron microscope. *Acta. Embryol. Morph. exp.* **4**, 261–283.

Ballard, K. J. and Holt, S. J. (1968) Cytological and cytochemical studies on cell death and digestion in the foetal rat foot: the role of macrophages and hydrolytic enzymes. *J. Cell Sci.* **3**, 245–262.

Ballowitz, E. (1901) Die gastrulation bei der Ringelnatter (*Tropidonotus natrix* Boie) bis zum Auftreten der Falterform der Embryonalanlage. *Z. Wiss. Zool.* **70**, 675–732.

Barack, B. M. (1968) Transport of spermatozoa from seminiferous tubules to epididymis in the mouse: a histological and quantitative study. *J. Reprod. Fert.* **16**, 35–48.

Barnett, S. A. (1965) Adaptation of mice to cold. *Biol. Rev.* **40**, 5–51.

Barros, C. and Franklin, L. E. (1968) Behaviour of the gamete membranes during sperm entry into the mammalian egg. *J. Cell Biol.* **37**, C13.

Barth, L. G. (1941) Neural differentiation without an organiser. *J. exp. Zool.* **87**, 371–382.

Bauchot, R. (1965) La placentation chez les reptiles. *Ann. Biol.* **4**, 547–575.

Bautzmann, H. (1926) Experimentelle Untersuchungen zur Abgrenzung des Organisations-zentrums bei *Triton taeniatus*. *Arch. Entw. Mech. Org.* **108**, 283–321.

Beatty, R. A. (1967) Parthenogenesis in vertebrates. In *Fertilization* 1, Eds. C. B. Metz and A. Monroy. Academic Press, New York and London, pp. 413–440.

Beaudoin, A. R. (1961) Teratogenic activity of several closely related disazo dyes on the developing chick embryo. *J. Embryol. exp. Morph.* **9**, 14–21.

Beck, F. and Lloyd, J. B. (1966) The teratogenic effect of azo dyes. In *Advances in Teratology* 1, Ed. D. H. M. Woollam. Logos Press, London, pp. 131–193.

Beck, F., Lloyd, J. B. and Griffiths, A. (1967) Lysosomal enzyme inhibition by trypan blue: a theory of teratogenesis. *Science, N.Y.* **137**, 1180–1182.

Bedford, J. M. (1964) Fine structure of the sperm head in ejaculate and uterine spermatozoa of the rabbit. *J. Reprod. Fert.* **7**, 221–228.

Bedford, J. M. (1967) The influence of the uterine environment upon rabbit spermatozoa. In *Reproduction in the Female Mammal*, Eds. G. E. Lamming and E. C. Amoroso. Butterworths, London, pp. 478–499.

Behrman, S. J. and Otani, Y. (1963) Transvaginal immunisation of the guinea pig with homologous testis and epididymal sperm. *Int. J. Fert.* **8**, 829–834.

Bekhtina, V. G. (1960) The early stages of cleavage in the chick embryo. *Arkiv. Anat., Gistol i. Embryol.* **38**, 77–85 (In Russian; translation no. 5050 of National Lending Library).

Bell, E. (1963) Protein synthesis in differentiating chick skin. *Nat. Cancer Inst. Monograph* **13**, 1–11.

Bell, E. (1965) The skin. In *Organogenesis*, Eds. R. L. DeHaan and H. Ursprung. Holt, Rinehart and Winston, New York, pp. 361–374.

Bellairs, A. d'A. (1970) *The Life of Reptiles.* Weidenfeld and Nicholson, London.

Bellairs, A. d'A. and Jenkin, C. R. (1960) The skeleton of birds. In *Biology and Comparative Physiology of Birds*, Ed. A. J. Marshall. Academic Press, New York and London, pp. 241–300.

Bellairs, R. (1951) Development of early reptile embryos *in vitro. Nature, Lond.* **167**, 687.

Bellairs, R. (1953) Studies on the development of the foregut in the chick embryo. II. The morphogenetic movements. *J. Embryol. exp. Morph.* **1**, 369–385.

Bellairs, R. (1954a) The effects of tetra-sodium 2-methyl 1:4-napthohydroquinone-diphosphate on early chick and amphibian embryos. *B. J. Cancer.* **8**, 685–692.

Bellairs, R. (1954b) The effects of folic acid antagonists on embryonic development. In *Chemistry and Biology of Pteridines*, Eds. G. E. W. Wolstenholme and M. P. Cameron. *CIBA Symp.*, Churchill, London, pp. 336–365.

Bellairs, R. (1957) Studies on the development of the foregut in the chick embryo. IV. Mesodermal induction and mitosis. *J. Embryol. exp. Morph.* **5**, 340–350.

Bellairs, R. (1958) The conversion of yolk into cytoplasm in the chick as shown by electron microscopy. *J. Embryol. exp. Morph.* **6**, 149–161.

Bellairs, R. (1959a) The yolk of the adder, *Vipera berus. Brit. J. Herpetol.* **2**, 155–158.

Bellairs, R. (1959b) The development of the nervous system in chick embryos, studied by electron microscopy. *J. Embryol. exp. Morphol.* **7**, 94–115.

Bellairs, R. (1960) Development of birds. In *Biology and Comparative Physiology of Birds*, Ed. A. J. Marshall. Academic Press, New York and London, pp. 127–189.

Bellairs, R. (1961a) The structure of the yolk of the hen's egg studied by electron microscopy. *J. biophys. biochem. Cytol.* **11**, 207–225.

Bellairs, R. (1961b) Cell death in chick embryos as studied by electron microscopy. *J. Anat.* **95**, 54–60.

Bellairs, R. (1963a) The yolk sac of the chick embryo studied by electron microscopy. *J. Embryol. exp. Morph.* **11**, 201–225.

Bellairs, R. (1963b) The development of somites in the chick embryo. *J. Embryol. exp. Morph.* **11**, 697–714.

Bellairs, R. (1964) Biological aspects of the yolk of the hen's egg. In *Advances in Morphogenesis*, **4**, Eds. M. Abercrombie and J. Brachet. Academic Press, London and New York, pp. 217–272.

Bellairs, R. (1965) The relationship between oöcyte and follicle in the hen's ovary as shown by electron microscopy. *J. Embrol. exp. Morph.* **13**, 215–233.

Bellairs, R. (1967) Aspects of the development of yolk spheres in the hen's oöcyte, studied by electron microscopy. *J. Embryol. exp. Morph.* **17**, 267–281.

Bellairs, R. (1969) Experimental twinning and multiple monsters in chick embryos. In *Teratology*, Ed. A. Bertelli. Excerpta Medica, Amsterdam, pp. 162–173.

Bellairs, R. and Boyde, A. (1969) Scanning electron microscopy of the shell membranes of the hen's egg. *Z. Zellforsch. mikrosk. Anat.* **96**, 237–249.

Bellairs, R., Boyde, A. and Heaysman, J. E. M. (1969) The relationship between the edge of the chick blastoderm and the vitelline membrane. *Arch. Entw Mech. Org.* **163**, 113–121.

Bellairs, R., Bromham, D. R. and Wylie, C. C. (1967) The influence of the area opaca on the development of the young chick embryo. *J. Embryol. exp. Morph.* **27**, 195–212.

Bellairs, R., Griffiths, I. and Bellairs, A. d'A. (1955) Placentation in the adder, *Vipera berus. Nature, Lond.* **176**, 657–658.

Bellairs, R., Harkness, M. and Harkness, R. D. (1963) The vitelline membrane of the hen's egg: a chemical and electron microscopical study. *J. Ultrastruct. Res.* **8**, 339–359.

Bellairs, R. and New, D. A. T. (1962) Phagocytosis in the chick blastoderm. *Exp. Cell Res.* **26**, 275–279.

Benoit, J., Leroy, P., Vendrely, C. and Vendrely, R. (1958) Phenotype du bec des canetons provenant de première et deuxième générations des Canard Pekin antérieurement traités a l'ADN de Canard Khaki Campbell. *C.R. Acad. Sci.,* **247**, 1049–1052.

Ben-or, S. and Bell, E. (1965) Skin antigens in the chick embryo in relation to other developmental events. *Dev. Biol.* **11**, 184–201.

Bernstein, M. H. (1966) Modifications of sperm structure in capacitation. *J. Cell Biol.* **31**, Abst. No. 20.

Berry, C. L. (1970) The effect of trypan blue on the growth of the rat embryo. *J. Embryol. exp. Morph.* **23**, 213–218.

Bhattacharya, D. R., Das, R. S. and Dutta, S. K. (1929) On the infiltration of Golgi bodies from the follicular epithelium to the egg. *Z. Zellforsch. mikrosk. Anat.* **8**, 566–577.

Biggers, J. D. (1964) Death of cells in normal multicellular organisms. In *Cellular Injury,* CIBA Foundation Symposium, Eds. A. V. S. de Reuck and J. Knight. Churchill, London, pp. 329–349.

Billett, F. S., Collini, R. and Hamilton, L. (1965) The effects of D- and L-threo-chloramphenical on the early development of the chick embryo. *J. Embryol. exp. Morph.* **13**, 341–356.

Billett, F. S., Hamilton, L. and Newth, D. R. (1964) A failure to transform metazoan cells by DNA. *Heredity* **19**, 259–269.

Billingham, R. E. (1964) Transplantation immunity and the maternal-foetal relation. *New Eng. J. Med.* **270**, 667–720; 720–725.

Billingham, R. E., Brent, L. and Medawar, P. B. (1953) Actively acquired tolerance of foreign cells. *Nature, Lond.* **172**, 603–606.

Billington, W. D., James, D. A. and Kirby, D. R. S. (1968) Some effects of genetic dissimilarity between mother and foetus. *J. Reprod. Fert. suppl.* **3**, 1–7.

Birge, W. J. (1962) A histochemical study of RNA in differentiating ependymal cells of the chick embryo. *Anat. Rec.* **143**, 147–152.

Birge, W. J. (1963) Mechanism of regulative development in the neural tube of the chick embryo. *Trans. Am. microsc. Soc.* **82**, 347–350.

Bishop, D. W. (1961) Biology of spermatozoa. In *Sex and Internal Secretions,* **2**, (3rd Edition), Ed. W. C. Young. Baillière, Tindall and Cox, London, pp. 707–796.

Bishop, D. W. and Tyler, A. (1956) Fertilizin of mammalian eggs. *J. exp. Zool.* **132**, 575–601.

Bland, K. P. and Donovan, B. T. (1966) Neural and hormonal stimuli from the uterus and the control of ovarian function. In the CIBA Foundation Study Group, No. 23 on *Egg Implantation*, pp. 29–37.

Blandau, R. J. (1949) Embryo endometrical interrelationship in the rat and guinea pig. *Anat. Rec.* **104**, 331–359.

Blandau, R. J., White, B. J. and Rumery, R. E. (1963) Observations on the movements of the living primordial germ cells in the mouse. *Fertil. Steril.* **14**, 482–489.

Bobr, L. W., Lorenz, F. W. and Ogasawara, F. X. (1964) Distribution of spermatozoa in the oviduct and fertility in domestic birds. I. Residence sites of spermatozoa in fowl oviducts. *J. Reprod. Fert.* **8**, 39–48.

Bobr, L. W., Ogasawara, F. X. and Lorenz, F. W. (1964) Distribution of spermatozoa in the oviduct and fertility in domestic birds. II. Transport of spermatozoa in the oviduct. *J. Reprod. Fert.* **8**, 49–58.

Bonner, J. T. (1965) *Size and Cycle*. University Press, Princeton.

Bonnet, R. (1882) *Die Uterinmilch und ihre Bedentung fur die Frucht*. Stuttgart. (Not seen; quoted by Boyd and Hamilton, 1952.)

Bons, J. (1963) Note preliminaire sur l'orientation de l'embryon dans l'oeuf chez le lézard *Agama bibroni* Dum. *Bull. Soc. Sci. Nat. Phys. Maroc.* **43**, 39–47.

Boveri, T. (1910) Die Potenzen der *Ascaris*-Blastomeren bei abgeänderter Furchung. Zugleich ein Beitrag Zur Frage qualitativungleicher Chromosomen-Teilung. *Festschrift z 60 Geburtstag R. Hertwigs*, **3**, Gustav Fischer, Jena.

Böving, B. G. (1968) Implantation in the rabbit. In *Annual Report of the Director of the Department of Embryology. Carnegie Inst. Wash. Year Bk.* (1966–1967), **66**, 54–6.

Boyd, J. D. and Hamilton, W. J. (1952) Cleavage, early development and implantation of the egg. In *Marshall's Physiology of Reproduction*, (3rd Edition), **2**, Ed. A. S. Parkes. Longmans, London, pp. 1–126.

Boyd, J. D. and Hamilton, W. J. (1966) Electron microscopic observations on the cytotrophoblast contribution to the syncytium in the human placenta. *J. Anat.* **100**, 535–548.

Boyd, M. M. M. (1942) The oviduct, foetal membranes, and placentation in *Hoplodactylus maculatus* Gray. *Proc. Zool. Soc. A* **112**, 65–104.

Boyden, E. A. (1927) Experimental obstruction of the mesonephric ducts. *Proc. Soc. exp. Biol. Med.* **24**, 572–576.

Boyer, C. C. (1950) Respiration of embryonic blood. *Proc. Soc. exp. Biol. Med.* **75**, 211–214.

Brachet, A. (1935) *Traité d'Embryologie des Vertébrés*. Masson et Cie, Paris.

Brachet, J. (1950) *Chemical Embryology* (Translated by L. G. Barth). Interscience, London.

Brachet, J. (1965) The history of chemical embryology. In *The Biochemistry of Animal Development*, **1**, Ed. R. Weber. Academic Press, New York and London, pp. 1–9.

Brachet, J. (1967) Behaviour of nucleic acids during early development. *Comprehensive Biochemistry* **28**, 23–54.

Brachet, J. and Denis, H. (1963) Effects of actinomycin D on morphogenesis. *Nature, Lond.* **198**, 205–206.

Braden, A. W. H. (1952) Properties of the membranes of rat and rabbit eggs. *Aust. J. sci. Res. Ser. B* **5**, 460–411.

Braden, A. W. H. and Austin, C. R. (1954) Reactions of unfertilized mouse eggs to some experimental stimuli. *Exp. Cell Res.* **7**, 277–280.

Bradley, T. R., Roosa, R. A. and Law, L. W. (1962) DNA transformation studies with mammalian cells in culture. *J. cell comp. Physiol.* **60**, 127–137.

Brambell, F. W. R. (1958) The passive immunity of the young mammal. *Biol. Rev.* **33**, 488–531.

Brambell, F. W. R., Hemmings, W. A. and Henderson, M. (1951) *Antibodies and Embryos*. Athlone Press, University of London, London.

Bremer, H. (1958a) Das Dottergefässystem beim Hühnchen als Beispiel einer Struktur-Entwicklung. *Arch. EntwMech. Org.* **150**, 702–748.

Bremer, von H. (1958b) Beobachtungen über die Wachstumsrichtungen der Gefässprosse am Dottergefässystem der Hühnerkeimscheibe. *Biol. Zbl.* **77**, 608–617.

Bremer, H. (1960) Untersuchungen über die Entwicklung des Dotterentoderms der Hühnerkeimscheibe. *Arch. EntwMech. Org.* **152**, 166–182.

Bremer, H. (1962) Untersuchungen über die Bildung von Differenzierungsmustern am Mesoderm der Hühnerkeimscheibe. *Arch. EntwMech. Org.* **154**, 103–123.

Brent, R. L. (1966) Immunologic aspects of developmental biology. In *Advances in Teratology*, **1**, Ed. D. H. M. Woollam. Logos Press, London, pp. 82–129.

Brierley, J. and Hemmings, W. A. (1956) The selective transport of antibodies from the yolk sac to the circulation of the chick. *J. Embryol. exp. Morph.* **4**, 34–41.

Bristow, D. A. and Deuchar, E. M. (1964) Changes in nucleic acid concentration during the development of *Xenopus laevis* embryos. *Exp. Cell Res.* **35**, 580–589.

Britt, L. G. and Herrmann, H. (1959) Protein accumulation in early chick embryos grown under different conditions of explantation. *J. Embryol. exp. Morph.* **7**, 66–72.

Britten, R. J. and Davidson, E. H. (1969) Gene regulation for higher cells: a theory. *Science* **165**, 349–357.

Britten, R. J. and Kohne, D. E. (1968) Repeated sequences in D.N.A. *Science* **161**, 529–540.

Brown, D. D. and Gurdon, J. B. (1966) Size distribution and stability of DNA-like RNA synthesised during development of anucleolate embryos of *Xenopus laevis. J. molec. Biol.* **19**, 399–422.

Brown, D. D. and Littna, E. (1964) RNA synthesis during the development of *Xenopus laevis*, the South African Clawed Toad. *J. molec. Biol.* **8**, 669–687.

Bryant, S. V. (1970) Regeneration in amphibians and reptiles. *Endeavour*, **29**, 12–17.

Bryant, S. V. and Bellairs, A. d'A. (1967) Amnio-allantoic constriction bands in lizard embryos and their effects on tail regeneration. *J. Zool. Lond.* **152**, 155–161.

Bucher, O. (1963) Le problème de l'amitose. In *Cell Growth and Cell Division*, **2**, Ed. R. J. C. Harris (Symp. No. 2 Int. Soc. Cell Biol.). Academic Press, New York and London, pp. 261–276.

Bullough, W. S. (1962) The control of mitotic activity in adult mammalian tissues. *Biol. Rev.* **37**, 307–342.

Bullough, W. S. (1965) Mitotic and functional homeostasis: a speculative review. *Cancer Res.* **25**, 1683–1727.

Burnett, F. M. and Fenner, F. (1949) *The Production of Antibodies*. Macmillan, New York.

Burns, R. K. (1942) The origin and differentiation of the epithelium of the urino-
genital sinus in the opossum, with a study of the modifications induced by
oestrogens. *Contr. Embryol.* **30,** 63–83.

Burns, R. K. (1956) Urinogenital system. In *Analysis of Development,* Eds. B. H.
Willier, P. A. Weiss and V. Hamburger. Saunders, Philadelphia, pp. 462–491.

Busch, H., Starbuck, W. C., Singh, E. J. and Ro, T. S. (1964) Chromosomal proteins. In *The Role of Chromosomes in Development,* Ed. M. Locke. (23rd Symp.
Soc. Study Dev. Growth). Academic Press, New York and London, pp.
51–82.

Butler, E. (1933) The capacity for differentiation of regions of the unincubated
chick blastoderm. *Anat. Rec.* **57** (suppl.), 71–72.

Butler, E. (1935) The developmental capacity of regions of the unincubated chick
blastoderm as tested in chorio-allantoic grafts. *J. exp. Zool.* **70,** 357–396.

Butros, J. (1960) Induction by desoxyribonucleic acids of axial structures in
post-nodal fragments of chick blastoderms. *J. exp. Zool.* **143,** 259–282.

Butros, J. (1962) Studies on the inductive action of the early chick axis on isolated
post-nodal fragments. *J. exp. Zool.* **149,** 1–19.

Butros, J. (1963) Differentiation of explanted fragments of early chick blastoderm.
II. Culture on protein deficient medium enriched with RNA. *J. exp. Zool.*
154, 125–134.

Butros, J. (1965) Action of heart and liver RNA on the differentiation of segments
of chick blastoderms. *J. Embryol. exp. Morph.* **13,** 119–128.

Cahn, R. and Lasher, R. (1967) Simultaneous synthesis of DNA and specialized
cellular products by differentiating cartilage cells *in vitro. Proc. nat. Acad. Sci.,
Wash.* **58,** 1131–1138.

Cairns, J. M. and Saunders, J. W. Jr. (1954) The influence of embryonic mesoderm
on the regional specification of epidermal derivatives in the chick. *J. exp.
Zool.* **127,** 221–248.

Cajal, S. R. y (1929) Études sur la neurogenèse de quelques vertébrés. In *Studies in
Vertebrate Neurogenesis,* Ed. Charles B. Thomas. Springfield, Illinois.

Calame, S. (1962) Contribution expérimentale à l'étude du développement du
systeme urogénital de l'embryon d'oiseau. *Arch. Anat.* **44,** suppl., 45–65.

Callan, H. G. and Lloyd, L. (1960) Lampbrush chromosomes of crested newts
Triturus cristatus (Laurenti). *Phil. Trans. R. Soc. Ser. B.* **243,** 135–219.

Camosso, M. and Racanelli, A. (1962) Comportamento della cresta apicale dell'ala
dell embrione di pollo in condizioni sperimentale. *Mon. Zool. Ital. (Suppl.* 70),
22, 35.

Carinci, P. and Manzoli-Guidotti, L. (1968) Albumen absorption during chick
embryogenesis. *J. Embryol. exp. Morph.* **20,** 107–118.

Cattaneo, L. (1963) Richerche microscopiche e submicroscopiche sull'area emato-
gena del blastoderma di pollo. *Z. Anat. EntGesch.* **123,** 397–419.

Chang, M. C. (1950) Cleavage of unfertilized ova in immature ferrets. *Anat. Rec.*
108, 31–43.

Chauhan, S. P. S. and Rao, K. V. (1970) Chemically stimulated differentiation of
post-nodal pieces of chick blastoderms. *J. Embryol. exp. Morph.* **23,** 71–78.

Chaytor, D. B. (1963) The control of growth of the chick embryo liver studied by
the method of parabiosis. *J. Embryol. exp. Morph.* **11,** 667–672.

Chen, B. K. (1932) The early development of the duck's egg, with special reference
to the origin of the primitive streak. *J. Morph.* **53,** 133–188.

Chen, P. S. (1967) Biochemistry of nucleo-cytoplasmic interactions in morphogenesis. In *Biochemistry of Animal Development*, 2, Ed. R. Weber. Academic Press, New York and London, pp. 115–192.

Child, C. M. (1928) The physiological gradients. *Protoplasma* 5, 447–476.

Chiquoine, A. D. (1954) The identification, origin and migration of the primordial germ cells in the mouse embryo. *Anat. Rec.* 118, 135–146.

Chrétien, C. H. (1966) Etude de l'origine, de la migration et de la multiplication des cellules germinales chez l'embryon de lapin. *J. Embryol. exp. Morph.* 16, 591–608.

CIBA Symposium (1965) *Preimplantation Stages of Pregnancy*, Eds. G. E. W. Wolstenholme and M. O'Connor. Churchill, London.

Clark, H., Florio, B. and Hurowitz, R. (1955) Embryonic growth of *Thamnophis s. sirtalis* in relation to fertilization date and placental function. *Copeia* 1, 9–13.

Clavert, J. (1963) Symmetrization of the egg of vertebrates. In *Advances in Morphogenesis*, 2, Eds. M. Abercrombie and J. Brachet. Academic Press, New York and London, pp. 27–60.

Clawson, R. C. and Domm, L. V. (1963) Developmental changes in glycogen content of primordial germ cells in chick embryo. *Proc. Soc. exp. Biol. Med.* 112, 533–537.

Clawson, R. C. and Domm, L. V. (1963) The glycogen content of primordial germ cells in the white Leghorn chick embryo. *Anat. Rec.* 145, 218–219.

Cohen, S. (1958) A nerve growth-promoting protein. In *The Chemical Basis of Development*, Eds. W. D. McElroy and B. Glass. Johns Hopkins Press, Baltimore, pp. 665–676.

Cohen, S. (1960) Purification of a nerve-growth promoting protein from the mouse salivary gland and its neuro-cytotoxic antiserum. *Proc. natn. Acad. Sci. Wash.* 46, 302–310.

Cohen, S. (1965) The stimulation of epidermal proliferation by a specific protein (E.G.F.). *Devl. Biol.* 12, 394–407.

Cole, R. J. (1967) Cinemicrographic observations on the trophoblast and zona pellucida of the mouse blastocyst. *J. Embryol. exp. Morph.* 17, 481–490.

Cole, R. J. and Paul, J. (1965) Properties of cultured preimplantation mouse and rabbit embryos and cell strains derived from them. In *Preimplantation Stages of Pregnancy*, Ed. S. E. W. Wolstenholme. *CIBA Symp.*, Churchill, London, pp. 60–75.

Cole, R. J., Edwards, R. G. and Paul, J. (1966) Cytodifferentiation and embryogenesis in cell colonies and tissue cultures derived from ova and blastocysts of the rabbit. *Devl. Biol.* 13, 385–407.

Coleman, J. R. and Moses, M. J. (1964) DNA and the fine structure of the synaptic chromosomes in the domestic rooster (*Gallus domesticus*). *J. Cell Biol.* 23, 63–78.

Colwin, L. H. and Colwin, A. L. (1967) Membrane fusion in relation to sperm-egg association. In *Fertilization*, 1, Eds. C. B. Metz and A. Monroy. Academic Press, New York and London, pp. 295–367.

Conklin, E. G. (1905) The organization and cell lineage of the ascidian egg. *J. Acad. nat. Sci. Philad.* 13, 1–119.

Coon, H. G. (1966) Clonal stability and phenotypic expression of chick cartilage cells *in vitro*. *Proc. Nat. Acad. Sci., Wash.* 55, 66–73.

Cooper, G. W. (1965) Induction of somite chondrogenesis by cartilage and noto-chord. A correlation between inductive activity and specific stages of cyto-differentiation. *Devl. Biol.* **12**, 185–212.

Corliss, C. (1953) A study of mitotic activity in the early rat embryo. *J. exp. Zool.* **122**, 193–227.

Cormack, D. H. (1966) Site of action of ribonuclease during its inhibition of egg cleavage. *Nature, Lond.* **209**, 1364–1365.

Coulombre, A. J., Coulombre, J. L. and Mehta, H. (1962) The skeleton of the eye. I. Conjunctival papillae and scleral ossicles. *Devl. Biol.* **5**, 381–401.

Cravens, W. W. (1952) Vitamin deficiencies and antagonists. *Ann. N.Y. Acad. Sci.* **55**, 188–195.

Croisille, Y. and Le Douarin, N. M. (1965) Development and regeneration of the liver. In *Organogenesis*, Eds. R. L. DeHaan and H. Ursprung. Holt, Rinehart and Winston, New York, pp. 421–466.

Cuellar, O. (1966) Oviducal anatomy and sperm storage structures in lizards. *J. Morph.* **119**, 7–20.

Cuellar, O. (1968) Additional evidence for true parthenogenesis in lizards of the genus *Cnemidophorus*. *Herpetologica* **24**, 146–150.

Cunningham, B. (1937) *Axial Bifurcation in Serpents*. Duke University Press, North Carolina.

Currie, G. A. (1968) Immunology of pregnancy: the foeto-maternal barrier. *Proc. Roy. Soc. Med.* **61**, 1206–1211.

Currie, G. A. (1969) The foetus as an allograft: the role of maternal unresponsive-ness to paternally derived foetal antigens. In *Foetal Autonomy*, Eds. G. E. W. Wolstenholm and M. O'Connor. *CIBA Symp.*, Churchill, London, pp. 37–56.

Curtis, A. S. G. (1960) Cortical grafting in *Xenopus laevis*. *J. Embryol. exp. Morph.* **8**, 163–173.

Curtis, A. S. G. (1965) Cortical inheritance in the amphibian *Xenophus laevis* preliminary result. *Archs. Biol., Liège* **76**, 523–546.

Curtis, A. S. G. (1967) *The Cell Surface: Its Molecular Role in Morphogenesis*. Logos Press, London.

Curtis, A. S. G. and Greaves, M. F. (1965) The inhibition of cell aggregation by a pure serum protein. *J. Embryol. exp. Morph.* **13**, 309–326.

Curzen, P. (1968) The antigenicity of human placenta. *J. Obstet. Gynaec. Brit. Cwlth.* **75**, 1128–1133.

Curzen, P. (1970) The antigenicity of the human placenta. *Proc. Roy. Soc. Med.* **63**, 65–66.

Dalcq, A. M. (1932) Études des localisations germinales dans l'oeuf vierge d'Ascidie par des expériences de mérogonie. *Arch. Anat. micr.* **28**, 224–333.

Dalcq, A. M. (1954) Fonctions cellulaires et cytochimie structurale dans l'oeuf de quelques Rongeurs. *C.R. Soc. Biol., Paris* **148**, 1332–1373.

Dalcq, A. M. (1955) Distribution et évolution de quelques métabolites dans l'oeuf des Rongeurs. *C.R. Ass. An. 42nd réun.* 409–419.

Dalcq, A. M. (1963) The relation to lysosomes of the *in vivo* metachromatic granules. In the CIBA Foundation Symposium on *Lysosomes*, Eds. A. V. S. de Reuck and M. Cameron. Churchill, London, pp. 226–263.

Daniel, J. C. Jr. (1964) Cleavage of mammalian ova inhibited by visible light. *Nature, Lond.* **201**, 316–317.

Daniel J. C. Jr. and Levy, J. D. (1964) Action of progesterone as a cleavage inhibitor of rabbit ova *in vitro*. *J. Reprod. Fert.* **7**, 323–329.

Daniel, J. C. Jr. and Olson, J. D. (1966) Cell movement, proliferation and death in the formation of the embryonic axis of the rabbit. *Anat. Rec.* **156**, 123–128.

Dantchakoff, V. (1934) La cellule germinale dans le dynamisme de l'ontogenèse. *Actual. Scient. ind.* Hermann, Paris.

Dareviski, I. S. and Kulikowa, W. N. (1961) Natürliche Parthenogenese in der polymorphen Gruppe der kaukasischen Felseidechse (*Lacerta saxicola* Eversmann). *Zool. Jahrb. Systematik.* **89**, 119–176.

Dareviski, I. S. and Kulikowa, V. N. (1964) Natural triploidy within a polymorphic group of *Lacerta saxicola* Eversmann, resulting from hybridization between bisexual and parthenogenetic subspecies of this species. *Dokl. Akad. Nauk. USSR.* **158**, 202 (In Russian, not seen).

Davidson, E. H. (1968) *Gene Activity in Early Development.* Academic Press, New York and London.

Davidson, E. H., Allfrey, V. G. and Mirsky, A. E. (1964) On the RNA synthesized during the lampbrush phase of amphibian oögenesis. *Proc. Nat. Acad. Sci., Wash.* **52**, 501–508.

Davidson, E. H., Crippa, M., Kramer, F. R. and Mirsky, A. E. (1966) Genomic function during the lampbrush chromosome stage of amphibian oogenesis. *Proc. Nat. Acad. Sci., Wash.* **56**, 856–863.

Davidson, E. H., Haslett, G. W., Finney, R. J., Allfrey, V. G. and Mirsky, A. E. (1965) Evidence for prelocalisation of cytoplasmic factors affecting gene activation in early embryogenesis. *Proc. Nat. Acad. Sci., Wash.,* **54**, 696–704.

Davidson, E. H. and Mirsky, A. E. (1965) Gene activity in oogenesis. In *Genetic Control of Differentiation,* 18th Brookhaven Symp. Brookhaven National Laboratory, New York, pp. 77–98.

Davidson, R. L., Ephrussi, B. and Yamamoto, K. (1966) Regulation of pigment synthesis in mammalian cells, as studied by somatic hybridization. *Proc. Nat. Acad. Sci., Wash.* **56**, 1437–1440.

Davies, J. (1960) *Survey of Research in Gestation and the Developmental Sciences.* Williams and Wilkins, Baltimore.

Davis, H. W. and Gunberg, D. L. (1968) Trypan blue in the rat embryo. *Teratology* **1**, 125–134.

Dawid, I. B. (1965) DNA in amphibian eggs. *J. molec. Biol.* **12**, 581–599.

De Bernardi, F., Cigada, M. L., Maci, R. and Ranzi, S. (1969) On protein synthesis during the development of lithium-treated embryos. *Experientia* **25** 211–213.

de Duve, C. (1963) The lysosome concept. In The CIBA Foundation Symposium on *Lysosomes,* Eds. A. V. S. Reuck and M. Cameron. Churchill, London, pp. 1–31.

de Duve, C. and Baudhuin, P. (1966) Peroxisomes (microbodies and related particles). *Physiol. Rev.* **46**, 323–357.

DeHaan, R. L. (1963) Migration patterns of the precardiac mesoderm in the early chick embryo. *Expl. Cell Res.* **29**, 544–560.

DeHaan, R. L. (1964) Cell interactions and oriented movements during development. *J. exp. Zool.* **157**, 127–138.

DeHaan, R. L. (1965) Morphogenesis of the vertebrate heart. In *Organogenesis*, Eds. R. L. De Haan and H. Ursprung. Holt, Rinehart and Winston, New York, pp. 377–420.

DeHaan, R. L. (1968) Emergence of form and function in the embryonic heart. *Devl. Biol. Suppl.* **2**, 208–250.

DeHaan, R. L. and Ursprung, H. (1965) *Organogenesis*. Holt, Rinehart and Winston, New York.

Delage, Y. (1901) Études expérimentales, sur la maturation cytoplasmique et sur la parthénogénèse artificielle chez les échinodermes. *Arch. Zool.* **9**, 285–326.

Deleanu, M. (1965) Toxic action upon physiological necrosis and macrophage reaction in the chick embryo leg. *Revue roum. Embryol. Cytol. ser. Embryol.* **2**, 45–56.

Delphia, J. M. and Elliott, J. (1965) The effect of high temperature incubation upon the myocardial glycogen in the chick embryo. *J. Embryol. exp. Morph.* **14**, 273–280.

Denis, H. (1964) Effets de l'actinomycine sur le développement embryonnaire. *Devl. Biol.* **9**, 435–483.

Denis, H. (1968) Role of messenger ribonucleic acid in embryonic development. In *Advances in Morphogenesis*, **7**, Eds. M. Abercrombie and J. Brachet. Academic Press, New York and London, pp. 115–150.

Deren, J. J., Padykula, H. A. and Wilson, T. H. (1966a) Development of structure and function in the mammalian yolk sac. II. Vitamin B^{12} uptake by rabbit yolk sacs. *Devl. Biol.* **13**, 349–369.

Deren, J. J., Padykula, H. A. and Wilson, T. H. (1966b) Development of structure and function in the mammalian yolk sac. III. The development of amino acid transport by rabbit yolk sac. *Devl. Biol.* **13**, 370–384.

Derrick, G. E. (1937) An analysis of the early development of the chick by means of the mitotic index. *J. Morph.* **61**, 257–284.

Deuchar, E. M. (1952) The effect of high temperature shock on early morphogenesis in the chick embryo. *J. Anat.* **86**, 443–458.

Deuchar, E. M. (1958) Experimental demonstration of tongue muscle origin in chick embryos. *J. Embryol. exp. Morph.* **6**, 527–529.

Deuchar, E. M. (1960) Adenosine triphosphatase activity in early somite tissue of the chick embryo. *J. Embryol. exp. Morph.* **8**, 251–258.

Deuchar, E. M. (1961) Amino-acid activation in embryonic tissues of *Xenopus laevis*. *Expl. Cell Res.* **25**, 364–373.

Deuchar, E. M. (1962) The roles of amino acids in animal embryogenesis. *Biol. Rev.* **37**, 378–421.

Deuchar, E. M. (1963a) Amino-acids and differentiation in animal embryos. *Cell Differentiation.* (Symp. Soc. exp. Biol. XVII). Cambridge University Press, pp. 58–73.

Deuchar, E. M. (1963b) Sites of earliest collagen-formation in the chick embryo, as indicated by uptake of tritiated proline. *Exp. Cell Res.* **30**, 528–540.

Deuchar, E. M. (1963c) Tracing amino-acids from yolk protein into tissue protein. I. Incorporation of tritiated leucine into oocytes and its distribution in the early embryo of *Xenopus laevis*. *Acta Embryol. Morph. Exp.* **6**, 311–323.

Deuchar, E. M. (1966) *Biochemical Aspects of Amphibian Development*. Methuen, London.

Deuchar, E. M. (1969) Effects of a mesoderm-inducing factor on early chick embryos. *J. Embryol. exp. Morph.* **22**, 295–304.

Deuchar, E. M. and Dryland, A. M. L. (1965) Inhibition by α-methyl-norvaline of valine and leucine uptake into protein in the chick embryo. *J. Embryol. exp. Morph.* **13**, 275–284.

Deuchar, E. M. and Herrmann, H. (1962) Uptake of amino acids into explanted chick embryos by epidermal and endodermal routes. *Acta. Emb. Morph. Exp.* **5**, 161–166.

Deysson, G. (1968) Antimitotic substances. *Int. Rev. Cytol.*, Eds. G. H. Bourne and J. F. Danielli. Academic Press, New York and London. **24**, 95–148.

Dickmann, Z. (1964) The passage of spermatozoa through and into the zona pellucida of the rabbit egg. *J. exp. Biol.* **41**, 177–182.

Dickman, Z. (1967) Shedding of the zona pellucida by the rat blastocyst. *J. exp. Zool.* **165**, 127–138.

Dickmann, Z. (1969) Shedding of the zona pellucida. In *Advances in Reproductive Physiology*, **4**, Ed. A. McLaren. Logos Press, London, pp. 187–206.

Dickmann, Z. and Dziuk, P. J. (1964) Sperm penetration of the zona pellucida of the pig egg. *J. exp. Biol.* **41**, 603–608.

Dietert, S. E. (1966) Fine structure of the formation and fate of the residual bodies of mouse spermatozoa with evidence for the participation of lysosomes. *J. Morph.* **120**, 317–346.

Dingle, J. T. and Fell, H. B. (1969) (Eds.) *Lysosomes in Biology and Pathology*. North Holland, Amsterdam.

Disse, J. (1878) Die Entwicklung des mittleren Keinblattes in Hühnerei. *Arch. mikr. Anat.* **15**, 67–94.

Diwan, B. A. (1966) A study of the effects of colchicine on the process of morphogesis and induction in chick embryos. *J. Embryol. exp. Morph.* **16**, 245–258.

Dmi'el, R. (1969) Circadien rhythm of oxygen consumption in snake embryos. *Life Sciences* **8**, 1333–1342.

Dodson, J. W. (1967) The differentiation of epidermis. II. Alternative pathways of differentiation of embryonic chicken epidermis in organ culture. *J. Embryol. exp. Morph.* **17**, 107–118.

Dondua, A. K., Efremov, V. I., Krichinskaya, E. B. and Nikolaeva, I. P. (1966) Mitotic index, duration of mitosis and proliferation activity in the early phases of the development of the chick embryo. *Acta. Biol. Acad. Sci., Hungary* **17**, 127–143.

Dorris, F. (1940) Behaviour of pigment cells from cultures of neural crest when grafted back into the embryo. *Proc. Soc. exp. Biol. Med.* **44**, 286–287.

Dossel, W. E. (1958) An experimental study of the locus of origin of the chick parathyroid. *Anat. Rec.* **132**, 555–562.

Driesch, H. (1891) Entwicklungsmechanische Studien. I Der Werth der beiden ersten Furchungszellen in der Echinodermenentwicklung Experimentelle Erzeugung von Theil- und Doppelbildungen. *Z. Zool.* **53**, 160–178.

Droller, M. J. and Roth, T. F. (1966) An electron microscope study of yolk formation during oogenesis in *Lebistes reticulatus* guppyi. *J. biol. Chem.* **28**, 209–232.

Dubois, R. (1964) Sur l'attraction des éléments germinaux de gonades indifférenciées par le jeune épithélium germinatif chez l'embryon de poulet, en culture, *in vitro*. *C.R. Acad. Sci., Paris* **258**, 5070–5072.

Dubois, R. (1965a) Sur les propriétés migratrices des cellules germinales de gonades embryonnaires différenciées, chez l'embryon de poulet, en culture *in vitro*. *C.R. Acad. Sci., Paris* **260**, 5108–5111.

Dubois, R. (1965b) Sur l'attraction exercée par le jeune épithélium germinatif sur les gonocytes primaires de l'embryon de poule, en culture *in vitro:* démonstration à l'aide de la thymidine tritiée. *C.R. Acad. Sci., Paris* **260**, 5885–5887.

Dubois, R. (1967) Localisation et migration des cellules germinales du blastoderme non incubé de Poulet, d'après les résultats de cultures *in vitro*. *Archs. Anat. micr. Morph. exp.* **56**, 245–264.

Dubois, R. (1968) La colonisation des ébauches gonadiques par les cellules germinales de l'embryon de Poulet, en culture *in vitro*. *J. Embryol. exp. Morph.* **20**, 189–213.

Dubois, R. (1969) Le méchanisme d'entrée des cellules germinales primordiales dans le réseau vasculaire, chez l'embryon de Poulet. *J. Embryol. exp. Morph.* **21**, 255–70.

Dulbecco, R. (1965) Interaction of viruses with the genetic material of the host cells. In *Reproduction: Molecular, Subcellular and Cellular*, Ed. M. Locke. (24th Symp. Soc. Dev. Biol.). Academic Press, London and New York, pp. 95–106.

Dufaure, J. P. (1964) Greffe allantoïdienne de la région cloacale d'embryons de Lézard vivipare (*Lacerta vivipara* Jacquin). *C.R. Soc. de Biol.* **158**, 2062.

Dunn, R. F. (1966) Studies on the retina of the gecko *coleonyx variegatus*. II. The rectilinear visual cell mosaic. *J. Ultrastruct. Res.* **16**, 672–684.

Ebert, J. D. (1953) An analysis of the synthesis and distribution of the contractile protein, myosin, in the development of the heart. *Proc. natn. Acad. Sci., Wash.* **39**, 333–344.

Ebert, J. D. (1958) Antigens as tracers of embryonic synthesis. In *Embryonic Nutrition*, Ed. D. Rudnick. University Press, Chicago, pp. 54–109.

Ebert, J. D. (1969) Annual report of the Director of the Department of Embryology. *Carnegie Inst. Wash. Year Bk.* **67**, 1967–1968.

Ebert, J. D. and DeLanney, L. E. (1959) Ontogenesis of the immune response. *Nat. Cancer Institute Monograph, No. 2*, 73–111.

Ebert, J. D. and Kaighn, M. E. (1966) The keys to change: factors regulating differentiation. In *Major Problems in Developmental Biology*, Ed. M. Locke. (25th Symp. Soc. Dev. Biol.). Academic Press, New York and London, pp. 29–84.

Ede, D. A. and Agerbak, G. S. (1968) Cell adhesion and movement in relation to the developing limb pattern in normal and talpid[3] mutant chick embryos. *J. Embryol. exp. Morph.* **20**, 81–100.

Edidin, M. (1964) Transplantation antigens in the mouse embryo. The fate of early embryo tissues transplanted to adult hosts. *J. Embryol. exp. Morph.* **12**, 309–316.

Edmonds, R. H. (1966) Electron microscopy of erythropoiesis in the avian yolk sac. *Anat. Rec.* **154**, 785–805.

Edström, J. F. (1960) Composition of ribonucleic acid from various parts of spider oocytes. *J. Biophys. Biochem. Cytol.* **8**, 47–51.

Edström, J. F. and Gall, J. C. (1963) The base composition of ribonucleic acid in lampbrush chromosomes, nucleoli, nuclear sap and cytoplasm of *Triturus* oocytes. *J. Cell Biol.* **19**, 279–284.

Edström, J. E. and Kawiak, J. (1961) Microchemical deoxyribonucleic acid determination in individual cells. *J. Biophys. Biochem. Cytol.* **9,** 619–626.

Ellem, K. A. O. and Gwatkin, R. B. L. (1968) Patterns of nucleic acid synthesis in the early mouse embryo. *Devl. Biol.* **18,** 311–330.

Ellison, M. L., Ambrose, E. J. and Easty, G. C. (1969a) Chondrogenesis in chick embryo somites *in vitro. J. Embryol. exp. Morph.* **21,** 331–340.

Ellison, M. L., Ambrose, E. J. and Easty, G. C. (1969b) Myogenesis in chick embryo somites *in vitro. J. Embryol. exp. Morph.* **21,** 341–346.

Emanuelsson, H. (1961) Mitotic activity in chick embryos at the primitive streak stage. *Acta Physiol. Scand.* **52,** 211–233.

Emanuelsson, H. (1962) Growth and nucleic acid mobilization in the early chick embryo Thesis of the University of Lund, pp. 1–34.

Emanuelsson, H. (1965) Cell multiplication in the chick blastoderm up to the time of laying. *Expl. Cell Res.* **39,** 386–399.

Emanuelsson, H. (1966) Incorporation of tritiated uridine in the early chick blastoderm. *Expl. Cell Res.* **42,** 537–542.

Emanuelsson, H. (1968) Ultrastructure of nuclei, yolk granules and mitochondria in the early chick blastoderm. *Ark. f. Zool. (ser.* 2) **20,** 513–531.

Emanuelsson, H. and Von Mecklenburg, C. (1968) Localization of extra-nuclear DNA in early chick blastoderm cells with electron microscopical autoradiography. *Zool.* **22,** 155–162.

Enders, A. C. (1963) Fine structural studies of implantation in the armadillo. In *Delayed Implantation,* Ed. A. C. Enders. University Press, Chicago, pp. 281–292.

Enders, A. C. (1964) Electron microscopy of an early implantation stage, with a postulated mechanism of implantation. *Devl. Biol.* **10,** 395–410.

Enders, A. C. (1965) A comparative study of the fine structure of the trophoblast in several haemochorial placentas. *Am. J. Anat.* **116,** 29–68.

Enders, A. C. and Schlafke, S. (1967) A morphological analysis of the early implantation stages in the rat. *Am. J. Anat.* **120,** 185–226.

Enders, A. C. and Schlafke, S. (1969) Cytological aspects of trophoblast-uterine interaction in early implantation. *Am. J. Anat.* **125,** 1–29.

Endo, Y. (1952) The role of cortical granules in the formation of the fertilization membrane in eggs from Japanese sea urchins. *Expl. Cell Res.* **3,** 406–418.

England, M. (1969) Millipore filters studied in isolation and *in vitro* by transmission electron microscopy and stereoscanning electron microscopy. *Expl. Cell Res.* **54,** 222–230.

Engström, K. (1963) Structure, organization and ultrastructure of the visual cells in the Teleost family Labridae. *Acta Zoologica* **44,** 1–41.

Everett, N. B. (1945) The present status of the germ-cell problem in vertebrates. *Biol. Rev.* **20,** 45–55.

Eyal-Giladi, H. and Spratt, N. T. Jr. (1965) The embryo-forming potencies of the young chick blastoderm. *J. Embryol. exp. Morph.* **13,** 267–274.

Faigle, W., Keberle, H., Riess, W. and Schmid, K. (1962) Metabolic fate of thalidomide. *Experientia* **18,** 389–397.

Fargeix, N. (1963) L'orientation dominante de l'embryon de la Caille domestique (*Coturnix coturnix Japonica*) et la règle de von Baer. *C.R. Soc. de Biol.* **157,** 1431–1434.

Fargeix, N. (1964) Répartition des cellules germinales chez des embryons jumeaux de Canard. *C.R. Seanc. Soc. Biol., Paris* **158,** 1507–1510.

Fargeix, N. (1966) Localisation des cellules germinales de l'embryon de Canard au stade des premières paires de somites. *C.R. Acad. Sci., Paris* **262**, 2259–2262.

Fawcett, D. W. (1965) The anatomy of the mammalian spermatozoan with particular reference to guinea pig. *Z. Zellforsch.* **67**, 279–296.

Fawcett, D. W. and Hollenberg, R. D. (1963) Changes in the acrosome of guinea pig spermatozoa during passage through the epididymis. *Z. Zellforsch.* **60**, 276–292.

Feldman, M. and Waddington, C. H. (1955) The uptake of methionine-[35] by the chick embryo and its inhibition by ethionine. *J. Embryol. exp. Morph.* **3**, 45–58.

Fell, H. B. and Mellanby, E. M. (1953) Metaplasia produced in cultures of chick ectoderm by high vitamin A. *J. Physiol. Lond.* **119**, 470–488.

Fell, H. B. and Robison, R. (1929) The growth, development and phosphatase activity of embryonic avian femora and limb-buds cultivated *in vitro*. *Biochem. J.* **23**, 767–784.

Fern, V. H. and Beaudoin, A. R. (1960) Absorptive phenomena in the explanted yolk sac placenta of the rat. *Anat. Rec.* **137**, 87–91.

Fisk, A. and Tribe, M. (1949) The development of the amnion and chorion of reptiles. *Proc. zool. Soc. Lond.* **119**, 83–114.

Flaxman, B. A., Lutzner, M. A. and van Scott, E. J. (1968) Ultrastructure of cell attachment to substratum *in vitro*. *J. Cell Biol.* **36**, 406–410.

Flaxman, B. A., Maderson, P. F. A., Szabo, G. and Roth, S. I. (1968) Control of cell differentiation in lizard epidermis *in vitro*. *Devl. Biol.* **18**, 354–374.

Fleischmajer, R. and Billingham, R. E. (1968) *Epitheliol-Mesenchymal Interactions*. Williams and Wilkins, Baltimore.

Flickinger, R. A. (1961) Formation, biochemical composition and utilization of amphibian egg yolk. In *Symposium on Germ Cells and Development*, Ed. S. Ranzi. Instut. Intern. d'Embryol. and Fondazione. A. Baselli, Milano, pp. 29–48.

Flynn, T. T. (1930) On the unsegmented ovum of Echidna (*Tachyglossus*). *Quart. J. micr. Sci.* **74**, 119–132.

Fox, W. (1956) Seminal receptacles of snakes. *Anat. Rec.* **124**, 519–533.

Franchi, L. L. and Mandl, A. (1962) The ultrastructure of oogonia and oocytes in the foetal and neonatal rat. *Proc. Roy. Soc. B* **157**, 99–114.

Fraser, R. C. (1960) Somite genesis in the chick. III. The role of induction. *J. exp. Zool.* **145**, 151–167.

Fridhandler, L. (1968) Gametogenesis to implantation. In *Biology of Gestation*, **1**, Ed. N. S. Assali. Academic Press, New York and London, pp. 67–92.

Fridhandler, L., Hafez, E. S. E. and Pincus, G. (1957) Developmental changes in the respiratory activity of rabbit ova. *Exp. Cell Res.* **13**, 132–139.

Fugo, N. W. (1940) Effects of hypophysectomy in the chick embryo. *J. exp. Zool.* **85**, 271–298.

Fujita, S. (1962) Kinetics of cellular proliferation. *Expl. Cell Res.* **28**, 52–60.

Fujita, S. (1964) Analysis of neuron differentiation in the central nervous system by tritiated thymidine autoradiography. *J. Comp. Neurol.* **122**, 311–327.

Furshpan, E. J. and Potter, D. D. (1969) Low-resistance junctions between cells in embryos and tissue culture. *Current Topics in Dev. Biol.* **3**, 95–128.

Galarco, P. G. and Moyer, F. H. (1966) Structural changes in the murine yolk-sac during gestation: cytochemical and electron microscope observations. *J. Morph.* **119**, 341–356.

Galassi, L. (1968) Autoradiographic study of the decidual cells reaction in the rat. *Devl. Biol.* **17**, 75–84.

Gall, J. G. (1954) Lampbrush chromosomes from oocyte nuclei of the newt. *J. Morph.* **94**, 283–351.

Gall, J. G. (1958) Chromosomal differentiation. In *The Chemical Basis of Development*, Eds. W. D. McElroy and B. Glass. Johns Hopkins Press, Baltimore, pp. 103–135.

Gallera, J. (1959) Le facteur 'temps' dans l'action inductrice du chordomésoblaste et l'âge de l'ectoblaste réagissent. *J. Embryol. exp. Morph.* **7**, 487–511.

Gallera, J. (1964) Exision et transplantation des différentes regions de la ligne primitive chez le poulet. *Bull. de l'Assoc. d'Anat.* **49**, 632–639.

Gallera, J. (1965) Quelle est la durée nécessaire pour déclencher des inductions neurales chez le poulet? *Experientia* **21**, 218–219.

Gallera, J. (1966) Le pouvoir inducteur de la chorde et du mésoblaste parachordal chez les oiseaux en fonction du facteur 'temps'. *Acta Anat. (Basel)* **63**, 388–397.

Gallera, J. (1968) Induction neurale chez les oiseaux. Rapport temporel entre la neurulation du blastoderme-hôte et l'apparition de l'ébauche neurale induite par un fragment de la ligne primitive. *Rev. Suisse Zool.* **75**, 227–234.

Gallera, J. (1969) Évolution intrinsèque de l'ectoblaste et induction neurale chez les oiseaux. *Acta Embryol. Exp.* **1**, 5–16.

Gallera, J. (1970) L'action de l'actinomycine D sur le pouvoir inducteur du noeud de Hensen et la compétence neurogène de l'ectoblaste de poulet. *J. Embryol. exp. Morph.* **23**, 473–489.

Gallera, J. and Nicolet, G. (1961) Quelques commentaires sur les méthodes de culture *in vitro* de jeunes blastodermes de poulet. *Experientia* **17**, 134–135.

Gallera, J. and Nicolet, G. (1969) Le pouvoir inducteur de l'endoblaste présomptif contenu dans la ligne primitive jeune de Poulet. *J. Embryol. exp. Morph.* **21**, 105–118.

Gallera, J., Nicolet, G. and Ballmann, M. (1968) Induction neurale chez les oiseaux à travers un filtre millipore: étude au microscope optique et électronique. *J. Embryol. exp. Morph.* **19**, 439–450.

Gallera, J. and Oprecht, E. (1948) Sur la distribution des substances basophiles dans le blastoderm de la poule. *Revue Suisse Zool.* **55**, 243–250.

Galton, M. (1962) DNA content of placental nuclei. *J. Cell Biol.* **13**, 183–191.

Garber, B. (1967) Aggregation *in vivo* of dissociated cells. II. Role of developmental age in tissue reconstruction. *J. exp. Zool.* **164**, 339–350.

Gardner, W. V. and Allen, E. (1942) Effects of hypophysectomy at mid-pregnancy in the mouse. *Anat. Rec.* **83**, 75–98.

Gascogne, C. le, Seki, M. and Shirasawa, H. (1966) Morphologic manifestation of physiologic activity in the endodermal cells of the rat yolk sac splanchnopleure. *6th Int. Cong. Elect. Micr., Japan* pp. 415–416.

Gatenby, J. B. and Hill, J. P. (1924) On an ovum of *Ornithorhynchus* exhibiting polar bodies and polyspermy. *Quart. J. micr. Sci.* **68**, 229–238.

Gelfant, S. (1963) A new theory on the mechanism of cell division. In *Cell Growth and Cell Division*, Ed. R. J. C. Harris. (2nd Symp. International Soc. Dev. Biol.), 229–259.

Ghiara, G. and Taddei, C. (1966) Cytological and ultrastructural data on a special type of basophil constituents of the cytoplasm of follicular cells and ovarian oocytes of reptiles (in Italian). *Boll. Soc. Ital. Biol. Sper.* **42,** 784–788.

Giacomini, E. (1891) Meteriali per la storia dello sviluppo del *Seps. chalcides Monit. Zool. Ital.* **2,** 179–198.

Giersberg, H. (1922) Untersuchungen über Physiologie und Histologie des Eileiters der Reptilien und Vogel. *Z. Wiss. Zool.* **120,** 1–97.

Gillman, J., Gilbert, C., Gillman, T. and Spence, I. (1948) Preliminary report on hydrocephalus, spina bifida and other congenital anomalies in the rat produced by trypan blue. *S. Afr. J. med. Sci.* **13,** 46–90.

Ginsburg, J. (1968) Breakdown in maternal protection: drugs. *Proc. roy. Soc. Med.* **61,** 1244–1247.

Gipouloux, J. B. (1964a) Les somites attirent les gonocytes primordiaux dans l'endoderme: demonstration expérimentale chez la grenouille verte *Rana esculenta. C.R. Acad. Sci., Paris* **258,** 1066–1068.

Gipouloux, J. D. (1964b) Influence de la corde dorsale et des uretères primaires sur l'édification des crêtes génitales et la migration des gonocytes primordiaux; démonstration expérimentale sur le crapand commun (*Bufo bufo* L.). *C.R. Acad. Sci., Paris* **257,** 1150–1152.

Glenister, T. W. (1961) Organ culture as a new method for studying the implantation of mammalian blastocysts. *Proc. Roy. Soc. B* **154,** 428–431.

Glenister, T. W. (1963) Observations on mammalian blastocysts implanting in organ culture. In *Delayed Implantation*, Ed. A. C. Enders. University Press, Chicago, pp. 171–182.

Glenister, T. W. (1965) The behaviour of trophoblast when blastocysts effect nidation in organ culture. In *Early Conceptus, Normal and Abnormal*, Ed. W. W. Park. Livingstone, Edinburgh and London, pp. 24–27.

Gluck, L. and Kulovich, M. V. (1964) Nucleic acid and protein changes during embryonic organ development in the chick. *Yale J. Biol. Med.* **36,** 361–378.

Glücksmann, A. (1951) Cell deaths in normal vertebrate ontogeny. *Biol. Rev.* **26,** 59–86.

Gluecksohn-Schoenheimer, S. (1949) Causal analysis of mouse development by the study of mutational effects. *Growth suppl.* **13,** 163–176.

Goldschmidt, R. B. (1924) Untersuchungen Zur Genetik der geographischen variation I. *Arch. mikr. Anat.* **101,** 92–337.

Good, R. A. and Papermaster, B. W. (1964) Ontogeny and Phylogeny of Adaptive Immunity. *Adv. in Immunol.* **4,** 1–116.

Goodwin, B. C. and Cohen, M. H. (1969) A phase-shift model for the spatial and temporal organisation of developing systems. *J. Theor. Biol.* **25,** 49–108.

Goodrich, E. S. (1930) *Studies on the Structure and Development of Vertebrates.* MacMillan, London.

Goss, R. J. (1964) *Adaptive Growth.* Logos Press, London.

Grabowski, C. T. (1956) The effects of the excision of Hensen's node on the early development of the chick embryo. *J. exp. Zool.* **133,** 301–344.

Grabowski, C. T. (1957) The induction of secondary embryos in the early chick blastoderm by grafts of Hensen's node. *Am. J. Anat.* **101,** 101–134.

Grabowski, C. T. (1962) Neural induction and notochord formation by mesoderm from the node area of the early chick blastoderm. *J. exp. Zool.* **150,** 233–245.

Grabowski, C. T. (1966) Physiological changes in the bloodstream of chick embryos exposed to teratogenic doses of hypoxia. *Devl. Biol.* **13,** 199–213.

Graham, C. F. (1970) Parthenogenetic mouse blastocysts. *Nature, Lond.* **226**, 165–167.

Grant, P. (1965) Informational molecules and embryonic development. In *The Biochemistry of Animal Development*, 1, Ed. R. Weber. Academic Press, New York and London, pp. 483–593.

Gräper, L. (1929) Die Methodik der stereokinematographischen Untersuchung der lebenden vitalgefärbten Hühnerembryos. *Arch. EntwMech. Org.* **115**, 523–543.

Grau, C. R. and Wilson, B. W. (1964) Avian oogenesis and yolk deposition. *Experientia* **20**, 1–3.

Greenfield, M. L. (1966) The oöcyte of the domestic chicken shortly after hatching, studied by electron microscopy. *J. Embryol. exp. Morph.* **16**, 297–316.

Greenwald, G. S. (1959) The comparative effectiveness of oestrogens in interrupting pregnancy in the rabbit. *Fert. Steril.* **10**, 155–161.

Greenwald, G. S. (1962) The role of the mucin layer in development of the rabbit blastocyst. *Anat. Rec.* **142**, 407–416.

Greenwald, G. S. (1968) Hormonal regulation of egg transport through the mammalian oviduct. In *Progress in Infertility*, Eds. S. J. Behrman and R. W. Kistner. Churchill, London, pp. 157–179.

Gregory, P. W. (1930) The early embryology of the rabbit. *Contr. Embryol.* **21**, 143–168.

Grobstein, C. (1953) Inductive epithelio-mesenchymal interaction in cultured organ rudiments. *Science* **118**, 52–55.

Grobstein, C. (1955) Inductive interaction in the development of the mouse metanephros. *J. exp. Zool.* **130**, 319–340.

Grobstein, C. (1966) What we do not know about differentiation? *Am. Zool.* **6**, 89–95.

Grobstein, C. (1967) Mechanisms of organogenetic tissue interaction. *Nat. Cancer Inst. Mon.* **26**, 279–299.

Grobstein, C. (1968) Developmental significance of interface materials in epithelio-mesenchymal interaction. In *Epithelial-Mesenchymal Interactions*, Eds. R. Fleischmajer and R. E. Billingham. Williams and Wilkins, Baltimore, pp. 173–176.

Grobstein, C. and Holtzer, H. (1955) *In vitro* studies of cartilage induction in mouse somite mesoderm. *J. exp. Zool.* **128**, 333–358.

Grodziński, Z. (1951) The yolk spheres of the hen's egg as osmometers. *Biol. Rev.* **26**, 253–264.

Grodziński, Z. (1953) The morphotic components in the yolk of the turtle *Geoclemys reevesi* Gray. *Bull. Acad. Polon. Sci. Lett. B* **II**, 68–116.

Gross, P. R. (1967) RNA metabolism in embryonic development and differentiation. *New England J. Med.* **276**, 1239–1246; 1297–1304.

Grosser, O. (1927) *Frühentwicklung Eihautbildung und Placentation des Menschen und der Säugetiere*. Deutsche Frauenheilkunde Bd. S. Bergmann, München.

Gruenwald, P. (1942) Experiments on distribution and activation of the nephrogenic potency in the embryonic mesenchyme. *Physiol. Zoöl.* **15**, 396–409.

Grüneberg, H. (1952) *The Genetics of the Mouse* (2nd Edition). Niyhoff, The Hague.

Grüneberg, H. (1969) Threshold phenomena versus cell heredity in the manifestation of sex-linked genes in mammals. *J. Embryol. exp. Morph.* **22**, 145–179.

Grünwald, P. (1937) Zur Entwicklungsmechanik des Urogenitalsystems keim Hühn. *Arch. EntwMech. Org.* **136**, 786–813.

Guraya, S. S. (1962) The structure and function of the so-called yolk nucleus in the oogenesis of birds. *Quart. J. micr. Sci.* **103,** 411–415.

Guraya, S. S. (1964) Histochemical studies on the yolk nucleus in the oogenesis of mammals. *Amer. J. Anat.* **114,** 283–292.

Gurdon, J. B. (1968a) Transplanted nuclei and cell differentiation. *Scient. Am.* **219,** 24–35.

Gurdon, J. B. (1968b) Nucleic acid synthesis in embryos and its bearing on cell differentiation. *Essays in Biochemistry,* **4,** 25–68.

Gurdon, J. B. and Woodland, H. R. (1968) The cytoplasmic control of nuclear activity in animal development. *Biol. Rev.* **43,** 233–267.

Guzsal, E. (1966) Histological studies on the mature and post-ovulation ovarian follicle of fowl. *Acta. Vet. Acad. Sci. Hung.* **16,** 37–44.

Hadek, R. (1963a) Submicroscopic changes in the penetrating spermatozoon of the rabbit. *J. Ultrastruct. Res.* **8,** 161–169.

Hadek, R. (1963b) Submicroscopic study on the cortical granules in the rabbit ovum. *J. Ultrastruct. Res.* **8,** 170–175.

Hadek, R. (1965) The structure of the mammalian egg. *Int. Rev. Cytol.* **18,** 29–72.

Hadek, R. (1969) *Mammalian Fertilization.* Academic Press, New York and London.

Hadorn, E. (1965) Problems of determination and transdetermination. *Brookhaven Symp.* **18,** 148–161.

Hafez, E. S. E. (1963) Physio-genetic interaction between mammalian blastocyst and endometrium. *J. exp. Zool.* **154,** 163–168.

Hafez, E. S. E. (1964) Transuterine migration and spacing of bovine embryos during gonadotrophin-induced multiple pregnancy. *Anat. Rec.* **148,** 203–208.

Hafez, E. S. E. and White, E. G. (1967) Endometrial and embryonic enzymes in relation to implantation of the rabbit blastocyst. *Anat. Rec.* **159,** 273–280.

Hahn, W. E. and Tinkle, D. W. (1965) Fat body cycling and experimental evidence for its adaptive significance to ovarian follicle development in the lizard *Utah stansburiana. J. exp. Zool.* **158,** 79–85.

Haines, T. P. (1940) Delayed fertilization in *Leptodeira annulata polysticta. Copeia* **2,** 116–118.

Hale, L. J. (1956) Mitotic activity during the early differentiation of the scleral bones in the chick. *Quart. J. micr. Sci.* **97,** 333–54.

Hall, E. K. (1937) Regional differences in the action of the organization center. *Arch. Entw. Mech. Org.* **135,** 671–689.

Hamburger, V. (1934) The effects of wing bud extirpation on the development of the central nervous system in chick embryos. *J. exp. Zool.* **68,** 449–494.

Hamburger, V. (1968) Origins of integrated behaviour. *Devl. Biol. suppl.* **2,** 251–271.

Hamburger, V. and Hamilton, H. L. (1951) A series of normal stages in the development of the chick embryo. *J. Morph.* **88,** 49–92.

Hamburger, V. and Levi-Montalcini, R. (1950) In *Genetic Neurology,* Ed. P. Weiss. University of Chicago Press, Chicago, pp. 128–160.

Hamilton, H. L. (1952a) *Lillie's Development of the Chick.* Henry Holt, New York.

Hamilton, H. L. (1952b) Sensitive periods during development. *Ann. N.Y. Acad. Sci.* **55,** 177–187.

Hamilton, W. J., Boyd, J. D. and Mossman, H. W. (1962) *Human Embryology* (3rd Edition). Heffer, Cambridge.

Hamilton, W. J., Harrison, R. J. and Young, B. A. (1960) Aspects of placentation in certain *Cervidae. J. Anat.* **94,** 1–33.

Hamlett, G. W. D. (1935) Delayed implantation and discontinuous development in the mammals. *Quart. Rev. Biol.* **10**, 432–447.

Hammond, W. S. and Yntema, C. L. (1964) Depletion of the pharyngeal arch cartilages following extirpation of cranial neural crest in the chick embryo. *Acta. Anat.* **56**, 21–34.

Hara, K. (1961) Regional neural differentiation induced by prechordal and presumptive chordal mesoderm in the chick embryo. Ph.D. thesis, University of Utrecht.

Hardy, M. H. (1968) Glandular metaplasia of hair follicles and other responses to vitamin A excess in cultures of rodent skin. *J. Embryol. exp. Morphol.* **19**, 157–80.

Harris, H. (1968) *Nucleus and Cytoplasm.* Oxford University Press.

Harris, H., Watkins, J. F., Ford, C. E. and Schoefl, G. I. (1966) Artificial heterokaryons of animal cells from different species. *J. Cell Sci.* **1**, 1–30.

Harrison, J. R. (1957) Morphogenesis of chick embryo *in vitro* after exposure to lowered temperature *in ovo*. *Physiol. Zoöl.* **30**, 187–197.

Harrison, L. and Weekes, H. C. (1925) On the occurrence of placentation in the Scincid lizard *Lygosoma entrecasteauxi*. *Proc. Linn. Soc. N.S.W.* **50**, 470–486.

Harrison, R. J. and Hamilton, W. J. (1952) The reproductive tract and the placenta and membranes of Père David's deer (*Elaphurus davidianus* Milne Edwards). *J. Anat.* **86**, 203–225.

Hashimoto, K. and Wilt, F. H. (1966) The heterogeneity of chicken hemoglobin. *Proc. natn. Acad. Sci. U.S.A.* **56**, 1477–1483.

Hay, E. D. (1966) *Regeneration.* Holt, Rinehart and Winston, New York.

Hayashi, Y. and Herrmann, H. (1959) Growth and glycine incorporation in chick embryo explants. *Devl. Biol.* **1**, 437–458.

Heape, W. (1890) Preliminary note on the transplantation and growth of mammalian ova within a uterine foster-mother. *Proc. Roy. Soc. B* **48**, 457–458.

Heilporn-Pohl, V. (1964) Effets de l'actinomycine D sur la morphogénèse et la métabolism des acides nucléiques chez l'embryon de poulet. *J. Embryol. exp. Morph.* **12**, 439–446.

Heimlich, E. M. and Heimlich, M. G. (1950) Uterine changes and placentation in the Yucca Night Lizard. *J. Entomol. Zool.* **42**, 5–12.

Hell, A. (1964) The initial synthesis of haemoglobin in de-embryonated chick blastoderms. II. The effect of metabolic inhibitors on the blastodisc cultured *in vitro*. *J. Embryol. exp. Morph.* **12**, 609–619.

Hemmings, W. A. (1956) Protein selection in the yolk-sac splanchnopleur of the rabbit: the distribution of isotope following injection of ^{131}I-labelled serum globulin into the uterine cavity. *Proc. Roy. Soc. B* **145**, 186–195.

Herbst, C. (1893) Experimentelle Untersuchungen über den Einfluss der veränderten chemischen Zusammensetzung des umgebenden Mediums auf die Entwicklung der Thiere. I. Theil. Versuche an Seeigeleiern. *Z. wiss. Zool.* **55**, 446–518.

Herrmann, H., Konigsberg, U. R. and Curry, M. (1955) A comparison of the effects of antagonists of leucine and methionine on the chick embryo. *J. exp. Zool.* **128**, 359–378.

Herrmann, H. and Marchok, A. (1963) Gain and loss of protein in explanted chick embryos. *Devl. Biol.* **7**, 207–217.

Herrmann, H. and Tootle, M. L. (1964) Specific and general aspects of the development of enzymes and metabolic pathways. *Physiol. Rev.* **44**, 289–371.

Heussner, A. and Zahnd, J. P. (1963) Etude de la consommation d'oxygène de l'embryon de poulet au cours du nycthémère. *C.R. Soc. Biol. (Paris)* **157**, 1498–1501.

Hicks, P. and D'Amato, C. J. (1966) Effects of ionizing radiations on mammalian development. In *Advances in Teratology*, **1**, Ed. D. H. M. Woollam. Logos Press, London, pp. 196–250.

Hilfer, S. R. (1968) Cellular interactions in the genesis and maintenance of thyroid characteristics. In *Epithelial-Mesenchymal Interactions*, Eds. R. Fleischmajer and R. E. Billingham. Williams and Wilkins, Baltimore, pp. 177–199.

Hill, J. P. (1910) The early development of the marsupialia, with special reference to the native cat (*Dasyurus viverrinus*). *Quart. J. micr. Sci.* **56**, 1–134.

Hill, J. P. (1918) Some observations on the early development of *Didelphys aurita*. *Quart. J. micr. Sci.* **63**, 91–137.

Hill, J. P. and Tribe, M. (1924) The early development of the cat (*Felis domestica*). *Quart. J. micr. Sci.* **68**, 513–602.

Hillemann, H. H. (1943) An experimental study of the development of the pituitary gland in chick embryos. *J. exp. Zool.* **93**, 347–374.

Hillman, N. W. and Niu, M. C. (1963) Chick cephalogenesis. I. The effect of RNA on early cephalic development. *Proc. Nat. Acad. Sci. Wash.* **50**, 486–93.

Hinchliffe, J. R. and Ede, D. A. (1967) Limb development in the polydactylous talpid mutant of the fowl. *J. Embryol. exp. Morph.* **17**, 385–404.

Hiromoto, Y. (1962) Microinjection of the live spermatozoa into sea urchin eggs. *Expl Cell Res.* **27**, 416–426.

His, W. (1876) Untersuchungen über die Entwicklung von Knochenfischen, besonders über diejenige der Salmens. *Z. Anat. Entwgsch.* **1**, 1–40.

Hoadley, L. (1926) The *in situ* development of sectioned chick blastoderms. *Arch. Biol., Paris* **36**, 225–308.

Holder, L. A. and Bellairs, A. d'A. (1962) The use of reptiles in experimental embryology. *Brit. J. Herpetol.* **3**, 54–61.

Holtfreter, J. (1943) A study of the mechanics of gastrulation. *J. exp. Zool.* **94**, 261–318; **95**, 171–212.

Holtzer, H. (1961) Aspects of chondrogenesis and myogenesis. In *Synthesis of Molecular and Cellular Structure*, Ed. D. Rudnick. Ronald Press, New York, pp. 35–87.

Holtzer, H. (1964) Control of chondrogenesis in the embryo. *Biophys. J. Suppl.* **4**, 239–250.

Holtzer, H. (1968) Induction of chondrogenesis: a concept in quest of mechanisms. In *Epithelial-Mesenchymal Interactions*, Eds. R. Fleischmajer and R. E. Billingham. Williams and Wilkins, Baltimore, pp. 152–164.

Holtzer, H., Marshall, J. M. and Finck, H. (1957) An analysis of myogenesis by the use of fluorescent antimyosin. *J. Biophys. Biochem. Cytol.* **3**, 705–724.

Hoober, J. K. and Cohen, S. (1967) Epidermal growth factor. I. The stimulation of protein and ribonucleic acid synthesis in chick embryo epidermis. *Bioch. Biophys. Acta.* **138**, 347–356.

Horne, H. W. and Thibault, J. P. (1962) Sperm migration through the human female genital tract. *Fert. Steril.* **13**, 135–139.

Hörstadius, S. (1950) *The Neural Crest.* Oxford University Press, London.

Howard, A. and Pelc, S. R. (1953) Synthesis of desoxyribonucleic acid in normal and irradiated cells and its relation to chromosome breakage. *Heredity Suppl.* **6**, 261–273.

Hoyes, A. D. (1969) The human foetal yolk sac. An ultrastructural study of four specimens. *Z. Zellforsch.* **99**, 469–490.

Hsu, T. S. (1962) Differential rate in RNA synthesis between euchromatin and heterochromatin. *Expl. Cell Res.* **27**, 332–334.

Hsu, T. C., Schmid, W. and Stubblefield, E. (1964) DNA replication sequences in higher animals. In *The Role of Chromosomes in Development*, Ed. M. Locke. (23rd Symp. Soc. Study Develop. Growth). Academic Press, New York and London, pp. 83–112.

Hubert, J. (1962) Étude histologique des jeunes stades du développement embryonnaire du Lézard vivipare (*Lacerta vivipara* Jacquin). *Arch. d'anat.* **51**, 11–26.

Hubert, J. (1963) Orientation et symétrisation de l'embryon de Lézard vivipare (*Lacerta vivipara* Jacquin). *C.R. Soc. Biol.* **157**, 2200.

Hubert, J. (1964) Essais de fissuration de l'oeuf de lézard vivipare (*Lacerta vivipara* Jacquin). *C.R. Soc. Biol.* **158**, 523.

Hubert, J. (1970) Localisation précoce et mode de migration des gonocytes primordiaux chez quelques reptiles. *Ann. Embryol. Morph.* **4**, 479–494.

Hubrecht, A. A. W. (1894) Spolia Nemoris. *Quart. J. micr. Sci.* **36**, 77–126.

Huehns, E. R. and Shooter, E. M. (1965) Human haemoglobins. *J. Med. Genetic.* **2**, 48–90.

Huggett, A. St. G. (1961) Carbohydrate metabolism in the placenta and foetus *Brit. Med. Bull.* **17**, 122–126.

Hughes, A. F. W. (1937) Studies on the area vasculosa of the embryo chick. II. The influence of the circulation on the diameter of the vessels. *J. Anat.* **72**, 1–17.

Hughes, A. F. W. (1968) *Aspects of Neural Ontogeny.* Logos Press, London.

Hulka, J. F., Hsu, K. C. and Beiser, S. M. (1961) Antibodies to trophoblasts during the post-partum period. *Nature, Lond.* **191**, 510–511.

Hunt, T. E. (1937) The development of the gut and its derivatives from the mesectoderm and mesentoderm of early chick blastoderms. *Anat. Rec.* **68**, 349–370.

Huxley, T. H. (1864) *Lectures on the Elements of Comparative Anatomy: On the Classification of Animals and the Vertebrate Skull.* Peene, London.

Hyman, L. H. (1927) The metabolic gradients of vertebrate embryos. III. Chick. *Biol. Bull.* **52**, 1–38.

Ikeda, A., Abbott, R. L. and Langman, J. (1968) Muscle proteins in the chick myotomes examined by the immunofluorescent method. *J. Embryol. exp. Morph.* **19**, 193–202.

Ishii, K. (1967) Adhesiveness and histogenesis of various tissues of ectodermal origin from chick embryos in culture. *Embryologia* **10**, 1–11.

Izawa, M., Allfrey, V. G. and Mirsky, A. E. (1963) Composition of the nucleus and chromosomes in the lampbrush stage of the newt oocyte. *Proc. Nat. Acad. Sci., Wash.* **50**, 811–817.

Izquierdo, L. and Roblero, L. (1965) The incorporation of labelled nucleosides by mouse morulae. *Experientia* **21**, 532–533.

Izquierdo, L. and Vial, J. D. (1962) Electron microscope observations in the early development of the rat. *Z. Zellforsch.* **56**, 157–179.

Jacob, F. and Monod, J. (1961) Genetic regulatory mechanisms in the synthesis of proteins. *J. molec. Biol.* **3**, 318.

Jacobson, C-O. (1968) Selective affinity as a working force in neurulation movements. *J. exp. Zool.* **168**, 125–136.

Jacobson, W. (1938a) The early development of the avian embryo. 1. Endoderm formation. *J. Morph.* **62**, 415–444.

Jacobson, W. (1938b) The early development of the avian embryo. II. Mesoderm formation and the distribution of presumptive embryonic material. *J. Morph.* **62**, 445–502.

Jacobson, W. (1954) The yellow pigment of the argentaffine cells of the mammalian gastro-intestinal tract. In CIBA Foundation Symposium on *Chemistry and Biology of Pteridines*, Eds. G. E. W. Wolstenholme and M. P. Cameron. Churchill, London, pp. 314–326.

Johnen, A. G. (1964) Experimentelle Untersuchungen über die Bedeutung des Zeitfaktors beim Vorgang der neuralen Induktion. II. *Arch. EntwMech. Org.* **155**, 302–313.

Johnson, E. M. and Spinuzzi, R. (1968) Enzymic differentiation of rat yolk-sac placenta as affected by a teratogenic agent. *J. Embryol. exp. Morph.* **16**, 271–288.

Johnston, M. C. (1966) A radioautographic study of the migration and fate of cranial neural crest cells in the chick embryo. *Anat. Rec.* **155**, 143–156.

Jollie, W. P. and Bencosme, A. (1965) Electron microscopic observations on primary decidua formation in the rat. *Amer. J. Anat.* **116**, 217–236.

Jost, A. (1965) Gonadal hormones in sex differentiation in mammals. In *Organogenesis*, Eds. R. L. DeHaan and H. Ursprung. Holt, Rinehart and Winston, New York, pp. 611–628.

Jost, A. (1969) The extent of foetal endocrine autonomy. In *Foetal Autonomy*, Eds. G. E. W. Wolstenholme and M. O'Connor. *CIBA Symp.*, Churchill, London, pp. 79–89.

Jurand, A. (1962) The development of notochord in chick embryos. *J. Embryol. exp. Morph.* **10**, 602–621.

Jurand, A. (1963) Anti-mesodermal activity of a nitrogen mustard derivative. *J. Embryol. exp. Morph.* **11**, 689–696.

Jurand, A. (1966) Early changes in the limb buds of chick embryos after thalidomide treatment. *J. Embryol. Exp. Morph.* **16**, 289–300.

Källén, B. (1965) Early morphogenesis and pattern formation in the central nervous system. In *Organogenesis*, Eds. R. L. DeHaan and H. Urpsrung. Holt, Rinehart and Winston, New York, pp. 107–128.

Källén, B. and Valmin, K. (1963) DNA synthesis in the embryonic chick central nervous system. *Z. Zellforsch. mikrosk. Anat.* **60**, 491–496.

Kallman, F. and Grobstein, C. (1966) Localization of glucosamine—incorporating materials at epithelial surfaces during salivary epithelio-mesenchymal interaction *in vitro*. *Devl. Biol.* **14**, 52–67.

Karnovsky, D. A., Patterson, P. A. and Ridgway, L. P. (1949) Effect of folic acid, '4-amino' folic acids and related substances on growth of chick embryo. *Proc. Soc. exp. Biol. N.Y.* **71**, 447–452.

Kernis, M. M. and Johnson, E. M. (1969) Effects of trypan blue and Niagara blue 2B on the *in vitro* absorption of ions by the rat visceral yolk sac. *J. Embryol. exp. Morph.* **22**, 115–125.

Kessel, R. G. (1968) Annulate lamellae. *J. Ultra. Res. Suppl.* **10**, 3–82.

Kihlman, B. A. (1966) *Actions of Chemicals on Dividing Cells.* Prentice-Hall, New Jersey.

Kionka, H. (1894) Die Furchung des Hühnereies. *Anat. Hefte* **3**, 391–445.

Kirby, D. R. S. (1962) The influence of the uterine environment on the development of mouse eggs. *J. Embryol. exp. Morph.* **10**, 496–506.

Kirby, D. R. S. (1963a) The development of mouse blastocysts transplanted to the scrotal and cryptorchid testis. *J. Anat.* **97**, 119–130.

Kirby, D. R. S. (1963b) Development of the mouse blastocyst transplanted to the spleen. *J. Reprod. Fert.* **5**, 1–12.

Kirby, D. R. S. (1965) The invasiveness of the trophoblast. In *The Early Conceptus, Normal and Abnormal,* Ed. W. W. Park. Livingston, Edinburgh and London.

Kirby, D. R. S. (1967) On the orientation of the implanting blastocyst. *J. Embryol. exp. Morph.* **17**, 527–532.

Kirby, D. R. S. (1968) Immunological aspects of pregnancy. In *Advances in Reproductive Physiology,* **3**, Ed. A. McLaren. Logos Press, London, pp. 33–80.

Kirby, D. R. S., Billington, W. D., Bradbury, S. and Goldstein, D. (1964) Antigen barrier of the mouse placenta. *Nature, Lond.* **204**, 548–549.

Kirby, D. R. S., Potts, D. M. and Wilson, I. B. (1967) On the orientation of the implanting blastocyst. *J. Embryol. exp. Morph.* **17**, 527–532.

Kischer, C. W., Gurley, L. R. and Shepherd, G. R. (1966) Nuclear histones and early embryogenesis of the chick. *Nature, Lond.* **212**, 304–306.

Kischer, C. W. and Hnilica, L. S. (1967) Analysis of histones during organogenesis. *Expl. Cell Res.* **48**, 424–430.

Klein, N. W., McConnell, E. and Buckingham, B. J. (1962) Growth of explanted chick embryos on a chemically defined medium and effects of specific amino acid deficiencies. *Devl. Biol.* **5**, 296–308.

Klein, N. W., McConnell, E. and Riquier, D. J. (1964) Enhanced growth and survival of explanted chick embryos cultured under high levels of oxygen. *Devl. Biol.* **10**, 17–44.

Kohne, D. (1965) A study of RNA synthesis during embryogenesis in the frog, *Rana pipiens.* Ph.D. thesis. Purdue University. (Not seen, quoted by Wilt, 1966.)

Köllicher, A. von (1879) *Die Entwicklungsgeschichte des Menchen und der höhren Wirbelthiere.* Engelmann, Leipzig.

Kollros, J. J. (1968) Order and control in neurogenesis (as exemplified by the lateral motor column). *Devl. Biol. Suppl.* **2**, 272–305.

Königsberg, I. R. (1965) Aspects of cytodifferentiation of skeletal muscle. In *Organogenesis,* Eds. R. L. DeHaan and H. Ursprung. Holt, Rinehart and Winston, New York, pp. 337–358.

Krehbiel, R. H. and Plagge, J. C. (1962) Distribution of ova in the rat uterus. *Anat. Rec.* **143**, 239–246.

Kury, G. and Craig, J. M. (1967) The effect of mitomycin C on developing chicken embryos. *J. Embryol. exp. Morph.* **17**, 229–238.

Kussäther, E., Drews, U. and Usadel, K. H. (1968) Histochemical demonstration of cholinesterase during the folding off process of the chick embryo. *Arch. EntwMech. Org.* **161**, 141–161.

Laird, K. (1966) Dynamics of embryonic growth. *Growth* **30**, 263–275.

Lakshmi, M. S. and Mulerkar, L. (1963) Paper chromatographic study of the role of sulphur—containing amino acids in the process of induction in the chick embryo. *Experientia* **19**, 155–156.

Lakshmi, M. S. and Sherbet, G. V. (1964) The effect of chloroacetophenone on reacting ectoderm in induction in the chick embryo. *Naturwissenschaften* **51**, 64–65.

Lallier, R. (1954) Chlorure de lithium et biochimie du développement de l'oeuf d'amphibien. *J. Embryol. exp. Morph.* **2**, 323–339.

Lambson, R. O. (1966) An electron microscopic visualization of transport across rat visceral yolk sac. *Am. J. Anat.* **118**, 21–52.

Landauer, W. (1947) Insulin-induced abnormalities of beak extremities and eyes in chickens. *J. exp. Zool.* **105**, 145–172.

Landauer, W. and Clark, E. M. (1964) Teratogenic risks of drug synergism. *Nature, Lond.* **203**, 527–528.

Langman, J. and Nelson, G. R. (1968) A radioautographic study of the development of the somite in the chick embryo. *J. Embryol. exp. Morph.* **19**, 217–226.

Lanman, J. T. (1969) The fetus as a homograft. In *The Foeto-Placental Unit*, Eds. A. Pecile and F. Finzi. *Excerpta Med. Foundation*, Amsterdam, pp. 43–48.

Lanot, R. (1963) Action of ribonuclease on the early development of the chick embryo. *C.R. Acad. Sci.* **257**, 3471–3474.

Lanzavecchia, G. and LaCoultre, A. (1958) Origine di mitocondri durante lo sviluppo embrionale di *Rana esculenta.* Studio al microscopio elettronico. *Archo ital. Anat. Embriol.* **63**, 445–458.

Larsen, J. F. (1961) Electron microscopy of the implantation site in the rabbit. *Am. J. Anat.* **109**, 319–314.

Larsen, J. F. (1962) Electron microscopy of the uterine epithelium in the rabbit. *J. Cell. Biol.* **14**, 49–64.

Larsen, J. F. (1963) Histology and fine structure of the avascular and vascular yolk-sac placenta and the obplacental giant cells in the rabbit. *Amer. J. Anat.* **112**, 269–283.

Lash, J. W. (1955) Studies on wound closure in Urodeles. *J. exp. Zool.* **128**, 13–28.

Lash, J. W. (1963) Tissue interaction and specific metabolic responses: Chondrogenic induction and differentiation. In *Cytodifferentiation and Macromolecular Synthesis*, Ed. M. Locke. Academic Press, New York and London, pp. 235–260.

Lash, J. W. (1967) Differential behaviour of anterior and posterior embryonic chick somites *in vitro. J. exp. Zool.* **165**, 47–56.

Lash, J. W. (1968) Somite mesenchyme and its response to cartilage induction. In *Epithelial-Mesenchymal Interactions*, Eds. R. Fleischmajer and R. E. Billingham. Williams and Wilkins, Baltimore, pp. 165–172.

Lash, J. W., Holtzer, S. and Holtzer, H. (1957) An experimental analysis of the development of the spinal column. *Expl. Cell Res.* **13**, 292–303.

Lash, J. W., Hommes, F. A. and Zilliken, F. (1962) Induction of cell differentiation. 1. The *in vitro* induction of vertebral cartilage with a low-molecular-weight tissue component. *Biochem. Biophys. Acta* **56**, 313–319.

Lecce, J. C., Morgan, D. O. and Matrone, G. (1964) Effect of feeding colostral and milk components on the cessation of intestinal absorption of large molecules (closure) in neonatal pigs. *J. Nutr.* **84**, 43–48.

Le Douarin, G., Obrecht, C. and Coraboeuf, E. (1966) Regional determinations in the presumptive cardiac region. Evidence from the embryo chick by the micro electrophysiologic method. *J. Embryol. exp. Morph.* **15**, 153–168.

Le Douarin, N. (1961) Radiodestructions partielles chez l'embryon de poulet aux stades jeunes et localisation des ébauches digestives. *J. Embryol. exp. Morph.* **9**, 1–8.

Le Douarin, N. (1962) Displacement of the endoderm during digestive organogenesis in the chick embryo. *C.R. Acad. Sci., Paris* **254**, 2075–2077.

Le Douarin, N. (1962) Experimental data on hepatic organogenesis in the chick embryo. *C.R. Acad. Sci. Paris.* **255**, 769–771.

Le Douarin, N. (1964) Isolement expérimental du mesenchyme propre du foie et rôle morphogène de la composante mésodermique dans l'organogenèse hépatique. *J. Embryol. exp. Morph.* **12**, 141–160.

Le Doux, L., LeClerk, J. and Brachet, J. (1955) Action de la ribonucléase sur le développement embryonnaire des Batraciens. *Expl. Cell Res.* **9**, 338–347.

Lerner, A. M., Bell, E. and Darnell, J. E. Jr. (1963) Ribosomal RNA in the developing chick embryo. *Science, N.Y.* **141**, 1187–1188.

Lesseps, R. S. (1963) Cell surface projections: their role in the aggregation of embryonic chick cells as revealed by electron microscopy. *J. exp. Zool.* **153**, 171–182.

Levak-Švajger, B. and Škreb, N. (1965) Intraocular differentiation of rat egg cylinders. *J. Embryol. exp. Morph.* **13**, 243–253.

Levey, R. H. (1964) The thymus hormone. *Scientific American* **211**, 66–77.

Levi-Montalcini, R. (1952) Effects of mouse tumor transplantation on the nervous system. *Ann. N.Y. Acad. Sci.* **55**, 330–343.

Levi-Montalcini, R. and Angeletti, P. V. (1965) In *Organogenesis*, Eds. R. L. DeHaan and H. Ursprung. Holt, Rinehart and Winston, New York, pp. 187–198.

Levi-Montalcini, R., Meyer, H. and Hamburger, V. (1954) *In vitro* experiments on the effects of mouse sarcomas 180 and 37 on the spinal and sympathetic ganglia of the chick embryo. *Cancer Res.* **14**, 49–57.

Lewis, E. B. (1964) Genetic control and regulation of developmental pathways. In *The Role of Chromosomes in Development*, Ed. M. Locke (23rd Symp. Soc. Dev. Growth). Academic Press, New York and London, pp. 231–251.

Lillie, F. R. (1917) The free-martin; a study of the action of sex hormones in the foetal life of cattle. *J. exp. Zool.* **23**, 371–452.

Lillie, F. R. and Wang, H. (1943) Physiology of development of the feather. VI. The production and analysis of feather-chimaerae in fowl. *Physiol. Zoöl.* **16**, 1–21.

Lindsay, D. T. (1964) Histones from developing tissues of the chicken: heterogeneity. *Science, N.Y.* **144**, 420–422.

Lister, V. (1969) Quoted by Curzen (1968).

Lloyd, J. B., Beck, F., Griffiths, A. and Perry, L. M. (1968) The mechanism of action of acid bisazo dyes. In the Biological Council Symposium on *The Interaction of Drugs and Subcellular Components in Animal Cells*, Ed. P. N. Campbell. Churchill, London, pp. 171–201.

Loewenstein, W. R. (1966) Permeability of membrane junctions. *Ann. N.Y. Acad. Sci.* **137**, 441–472.

Loewenstein, J. E. and Cohen, A. I. (1964) Dry mass, liquid content and protein content of the intact and zona-free mouse ovum. *J. Embryol. exp. Morph.* **12**, 113–121.

Lorenz, F. W. (1966) Behaviour of spermatozoa in the oviduct in relation to fertility. In *Physiology of the Domestic Fowl*, Eds. C. Horton-Smith and E. C. Amoroso. Oliver and Boyd, Edinburgh, pp. 39–43.

Lowe, C. H. and Wright, J. W. (1966) Evolution of parthenogenetic species of *Cnemidophorus* (Whiptail lizards) in Western North America. *J. Arizona Acad. Sci.* **4**, 81–87.

Lutwak-Mann, C. (1954) Some properties of the rabbit blastocyst. *J. Embryol. exp. Morph.* **2**, 1–13.

Lutwak-Mann, C. (1963) Uterine-blastocyst relationships at the time of implantation: biochemical aspects. In *Delayed Implantation*, Ed. A. C. Enders. University Press, Chicago, pp. 293–304.

Lutz, H. (1949) Sur la production expérimentale de la poly-embryonie et de la monstruosité double chez les oiseaux. *Archs Anat. microsc. morph. exp.* **38**, 79–144.

Lutz, H. (1955) Contribution expérimentale à l'étude de la formation de l'endoblaste chez les oiseaux. *J. Embryol. exp. Morph.* **3**, 59–76.

Lutz, H. (1962) Association de blastodermes d'oiseaux. *Anat. Anz.* **109**, 120–125.

Lutz, H. (1965) Symétrisation de l'oeuf d'oiseau. *Annls Fac. Sci. Marseille* **26**, 71–84.

Lutz, H., Departout, M., Hubert, J. and Pieau, C. (1963) Contribution à l'étude de la potentialité du blastoderm non incubé chez les oiseaux. *Devl. Biol.* **6**, 23–44.

Lyon, M. F. (1968) Chromosomal and subchromosomal inactivation. *Ann. Rev. Genet.* **2**, 31–52.

Lyser, K. M. (1968) Early differentiation of motor neuroblasts in the chick embryo as studied by electron microscopy. II. Microtubules and neurofilaments. *Devl. Biol.* **17**, 117–142.

Maderson, P. F. A. and Bellairs, A. d'A. (1962) Culture methods as an aid to experiment on reptile embryos. *Nature, Lond.* **195**, 401–402.

Malan, M. E. (1953) The elongation of the primitive streak and the localization of the presumptive chorda-mesoderm of the early chick blastoderm studied by means of coloured marks with Nile blue sulphate. *Archs. Biol., Paris* **64**, 149–182.

Malpoix, P. and Emelinckx, A. (1967) Effect of histones on morphogenesis and differentiation in chick embryos. *J. Embryol. exp. Morph.* **18**, 143–154.

Manasek, F. J. (1968) Embryonic development of the heart. I. A light and electron microscopic study of myocardial development in the early chick embryo. *J. Morph.* **125**, 329–366.

Manes, C. and Daniel, J. C. (1969) Quantitative and qualitative aspects of protein synthesis in the preimplantation rabbit embryo. *Expl. Cell. Res.* **55**, 261–268.

Markee, J. E. (1944) Intrauterine distribution of ova in the rabbit. *Anat. Rec.* **88**, 329–336.

Marks, P. and Kovach, K. (1966) Development of mammalian erythroid cells. In *Current Topics in Developmental Biology*, **1**, Eds. A. Moscona and A. Monroy. Academic Press, New York and London, pp. 213–252.

Martinovitch, P. N., Kanazir, D. T., Knezevitch, Z. A. and Simitch, M. M. (1962) Teratological changes in the offspring of chicken embryos treated with Tyrode or with Tyrode plus DNA. *J. Embryol. exp. Morph.* **10**, 167–177.

Mato, M., Aikawa, E., and Kirshi, K. (1964) Some observations on the interstice between mesoderm and endoderm in the area vasculosa of the chick. *Expl. Cell. Res.* **35**, 426–427.

Mayer, G. (1959) Recent studies on hormonal control of delayed implantation and superimplantation in the rat. *Mem. Soc. Endocrin.* **6**, 76–82.

Mazanec, K. and Dvořak, M. (1963) On the submicroscopical changes of the segmenting ovum in the albino rat. *C. S. Morfol.* **11**, 103–108.

Mazanec, K. (1965) Submikroskopische Veränderungen während der Furchung eines Saügetiereies. *Arch. biol., Liege* **76**, 69–85.

Mazia, D. (1961) Mitosis and the physiology of cell division. In *The Cell*, **3**, Eds. J. Brachet and A. E. Mirsky. Academic Press, New York and London, pp. 77–413.

McCabe, R. A. and Deutsch, H. F. (1952) The relationship of certain birds as indicated by their egg-white proteins. *Auk* **69**, 1–18.

McCallion, D. J. and Trott, J. C. (1964) Transient embryonic antigens in the chick. *J. Embryol. Exp. Morph.* **12**, 511–516.

McKenzie, J. and Ebert, J. D. (1960) The inhibitory action of antimycin A in the early chick embryo. *J. Embryol. exp. Morph.* **8**, 314–320.

McLaren, A. (1965) Maternal factors in nidation. In *Early Conceptus, Normal and Abnormal*, Ed. W. W. Park. Livingstone, Edinburgh and London, pp. 27–33.

McLaren, A. (1969) A note on the zona pellucida in mice. *Advances in Reproductive Physiology*, **4**, Ed. A. McLaren. Logos Press, London, pp. 207–210.

McLaren, A. (1970) The fate of the zona pellucida in mice. *J. Embryol. exp. Morph.* **23**, 1–19.

McLaren, A. and Michie, D. (1956) Studies on the transfer of fertilized mouse eggs to uterine foster-mothers. 1. Factors affecting the implantation and survival of native and transferred eggs. *J. exp. Biol.* **33**, 394–416.

McLaren, A. and Michie, D. (1959) Spacing of implantations in the mouse uterus. *Mem. Soc. Endoc. No.* **6**, 65–75.

McLary, D. C. and Fish, S. A. (1966) Fetal erythrocytes in the maternal circulation. *Amer. J. Obst. Gyn.* **95**, 824–830.

McLoughlin, C. B. (1961a) The importance of mesenchymal factors in the differentiation of chick epidermis. I. *J. Embryol. exp. Morph.* **9**, 370–384.

McLoughlin, C. B. (1961b) The importance of mesenchymal factors in the differentiation of chick epidermis. II. *J. Embryol. exp. Morph.* **9**, 385–409.

Menkes, B., Deleanu, M. and Ilies, A. (1965) Comparative study of some areas of physiological necrosis at the embryo of man, some laboratory-mammalians and fowl. *Revue roum. Embryol. Cytol. ser. Embryol.* **2**, 162–171.

Merbach, H. (1935) Beobachtungen an der Keimscheibe des Hühnchens vor dem Erscheinen des Primitivstreifens. *Z. Anat. EntwGesch.* **104**, 635–652.

Merker, H. J. (1961) Elektronenmikroskopische Untersuchungen uber die Bildung der *Zona pellucida* in der Follikeln des Karinchenovars. *Z. Zellforsch.* **54**, 677–678.

Metz, C. B. (1967) Gamete surface components and their role in fertilization. In *Fertilization*, **1**, Eds. C. B. Metz and A. Monroy. Academic Press, New York and London, pp. 163–236.

Metz, C. B. and Monroy, A. (1967) *Fertilization*, **1** and **2**. Academic Press, New York and London.

Meyer, A. W. (1939) *The Rise of Embryology*. Stanford University Press, California.

Midgely, A. R. and Pierce, G. B. (1963) Immunohistochemical analysis of basement membranes of the mouse. *Am. J. Path.* **43**, 929–943.

Milaire, J. (1963) Morphological and cytochemical study of development of the limbs of the mouse and mole. *Arch. Biol., Liège* **74**, 129–317.

Miller, J. F. A. P. (1964) The thymus and the development of immunologic responsiveness. *Science, N.Y.* **144**, 1544–1551.

Millonig, G. and Giudice, G. (1967) Electron microscopic study of the reaggregation of cells dissociated from sea urchin embryos. *Devl. Biol.* **15**, 91–101.

Mintz, B. (1957) Embryological development of primordial germ-cells in the mouse: influence of a new mutation Wj. *J. Embryol. exp. Morph.* **5**, 396–403.

Mintz, B. (1962) Experimental study of the developing mammalian egg: removal of the zona pellucida. *Science, N.Y.* **138**, 594–595.

Mintz, B. (1964a) Synthetic processes and early development in the mammalian egg. *J. exp. Zool.* **157**, 85–100.

Mintz, B. (1964b) Gene expression in the morula stage of mouse embryos, as observed during development of t^{12}/t^{12} lethal mutants *in vitro*. *J. exp. Zool.* **157**, 267–272.

Mintz, B. (1964c) Formation of genetically mosaic mouse embryos, and early development of 'lethal (t^{12}/t^{12}) normal' mosaics. *J. exp. Zool.* **157**, 273–291.

Mintz, B. (1965) Nucleic acid and protein synthesis in the developing mouse embryo. *Preimplantation Stages of Pregnancy*, Eds. G. E. W. Wolstenholme and M. O'Connor. (CIBA Foundation Symposium). Churchill, London, pp. 145–155.

Mintz, B. (1967) Gene control of mammalian pigmentary differentiation, 1. Clonal origin of melanocytes. *Proc. nat. Acad. Sci., Wash.* **58**, 344–351.

Mintz, B. (1968) Hermaphroditism, sex chromosomal mosaicism and germ cell selection in allophenic mice. *J. Animal Science* **27**, 51–60.

Mintz, B. and Baker, W. W. (1967) Normal mammalian muscle differentiation and gene control of isocitrate dehydrogenase synthesis. *Proc. nat. Acad. Sci., Wash.* **58**, 592–598.

Mintz, B. and Palm, J. (1965) Erythrocyte mosaicism and immunological tolerance in mice from aggregated eggs. *J. Cell Biol.* **27**, 66A.

Mintz, B. and Palm, S. (1969) Gene control of hemotopoiesis. I. Erythrocyte mosaicism and permanent immunological tolerance in allophenic mice. *J. exp. Med.* **129**, 1013–1027.

Mintz, B. and Russell, E. S. (1957) Gene-induced embryological modifications of primordial germ cells in the mouse. *J. exp. Zool.* **134**, 207–237.

Mintz, B. and Silvers, W. K. (1967) Intrinsic immunological tolerance in allophenic mice. *Science, N.Y.* **158**, 1484–1487.

Miura, Y. and Wilt, F. H. (1969) Tissue interaction and the formation of the first enythroblasts of the chick embryo. *Devl. Biol.* **19**, 201–211.

Modak, S. P. (1963) L'endoblaste se forme-t-il par invagination à travers la ligne primitive dans l'embryon de poulet? *International Congress Series No. 70*. Excerpta Medica Foundation, Amsterdam, London, pp. 122–124.

Modak, S. P. (1965) Sur l'origine de l'hypoblaste chez les oiseaux. *Experientia* **21**, 273–274.

Modak, S. P. (1966) Analyse expérimentale de l'origine de l'endoblaste embryonnaire chez les oiseaux. *Rev. Suisse Zool.* **73**, 877–908.

Modak, S. P., Morris, G. and Yamada, T. (1968) DNA synthesis and mitotic activity during early development of chick lens. *Dev. Biol.* **17**, 544–561.

Modlinski, J. (1970) The role of the zona pellucida in the development of mouse eggs *in vivo*. *J. Embryol. exp. Morph.* **23**, 539–547.

Monroy, A. (1965) *Chemistry and Physiology of Fertilization*, 1. Holt, Rinehart and Winston, New York.

Monroy, A. and Tyler, A. (1967) The activation of the egg. In *Fertilization*, **1**, Eds. C. B. Metz and A. Monroy, Academic Press, London and New York, pp. 369–412.

Moog, F. (1965) Enzyme development in relation to functional differentiation. In *The Biochemistry of Animal Development*, 1, Ed. R. Weber. Academic Press, New York and London, pp. 307–365.

Moore, B. C. (1959) Autoradiographic studies of H_3-thymidine incorporation in normal and hybrid frog embryos. *Anat. Rec.* **134**, 610–611.

Moore, B. C. (1963) Histones and differentiation. *Proc. nat. Acad. Sci., Wash.* **50**, 1018–1026.

Moore, C. R. (1950) The role of the fetal endocrine glands in development. *J. Clin. Endocr. Metab.* **10**, 942–985.

Moricard, R. (1960) Electron microscopical studies on changes in the acrosome during the penetration of the spermatozoan into the ovum of mammals. *C.R. Soc. Biol., Paris* **154**, 2187–2189.

Morita, S. (1936) Die künstliche Erzengung von Einzelmisbildungen, von Zwillingen, Drillingen und mehrlingen im Hühnerei. Vorläufige mitteilung. *Anat. Anz.* **82**, 81–102.

Morris, J. E. (1966) Chick mesonephric regression and its control. *Diss. Abs.* **26**, 4339.

Morris, V. B. (in press) Symmetry in the receptor mosaic demonstrated in the chick from the frequencies, spacing and arrangement of the types of retinal receptor, *J. Comp. Neurol.*

Morris, V. B. and Shorey, C. D. (1967) An electron microscope study of types of receptor in the chick retina. *J. Comp. Neurol.* **129**, 313–339.

Moscona, A. A. and Garber, B. B. (1968) Reconstruction of skin from single cells and integumental differentiation in cell aggregates. In *Epithelial-Mesenchymal Interactions*, Eds. R. Fleischmajer and R. E. Billingham. Williams and Wilkins, Baltimore, pp. 230–243.

Moscona, M. H. and Moscona, A. (1952) The development *in vitro* of the anterior lobe of the embryonic chick pituitary. *J. Anat.* **86**, 278–286.

Moscona, M. H. and Moscona, A. A. (1963) Inhibition of adhesiveness and aggregation of dissociated cells by inhibitors of protein and RNA synthesis. *Science, N.Y.* **142**, 1070–1071.

Moscona, M. H. and Moscona, A. A. (1965) Control of differentiation in aggregates of embryonic skin cells: suppression of feather morphogenesis by cells from other tissues. *Devl. Biol.* **3**, 402–423.

Moses, M. J. and Coleman, J. R. (1964) Structural patterns and the functional organization of chromosomes. In *The Role of Chromosomes in Development* (23rd Symp. Soc. Develop. Growth). Academic Press, New York and London, pp. 11–49.

Mossman, H. W. (1937) Comparative morphogenesis of the fetal membranes and accessory uterine structures. *Contr. Embryol.* **26**, 129–246.

Mounib, M. S. and Chang, M. C. (1964) Effect of *in utero* incubation on the metabolism of rabbit spermatozoa. *Nature, Lond.* **201**, 943.

Mukerjee, H., Ram, J. S. and Pierce, G. B. Jr. (1965) Basement membranes. V. Chemical composition of neoplastic basement membrane mucoprotein. *Amer. J. Path.* **46**, 49–57.

Mulherkar, L. (1958) Induction by regions lateral to the streak in the chick embryo. *J. Embryol. exp. Morph.* **6**, 1–14.

Mulherkar, L. (1960) The effects of trypan blue on chick embryos cultured *in vitro*. *J. Embryol. exp. Morph.* **8**, 1–5.

Mulherkar, L., Joshi, S. S., Diwan, B. A. and Joshi, P. N. (1967) Reversible effect of chloroacetophenone by sulphydryl groups on morphogenesis of chick embryos. *J. Embryol. exp. Morph.* **17**, 263–266.

Mulherkar, L., Rao, K. V. and Joshi, S. S. (1965) Studies on some aspects of the role of sulphydryl groups in morphogenesis. *J. Embryol. exp. Morph.* **14**, 129–135.

Müller, H. (1952) Bau und Wachstum der Netzhaut der Gupp (*Lebistes reticulatus*). *Zool. Jahrb.* **63**, 275–322.

Mulnard, J. (1955) Contribution à la connaissance des enzymes dans l'ontogénèse. Les phosphases acide et alcaline dans le développement du rat et de la souris. *Arch. Biol.* **66**, 525–685.

Mulnard, J. (1960) Problèmes de structure et d'organisation morphogénétique de l'oeuf de mammifère. In *Symposium on Germ Cells and Development.* Ed. S. Ranzi. Fondazione A. Baselli, Milan, pp. 639–688.

Mulnard, J. (1965a) Aspects cytochimiques de la régulation *in vitro* de l'oeuf de souris après destruction d'un des blastomères du stade II. I. La phosphomonoestérase acide. *Mem. Acad. Roy. Med. Belg.* **5**, 31–67.

Murray, P. D. F. (1932) The development *in vitro* of the blood of the early chick embryo. *Proc. R. Soc. B* **111**, 497–521.

Mystkowska, E. T. and Tarkowski, A. K. (1968) Observations on CBA-p/CBA-T6T6 mouse chimeras. *J. Embryol. exp. Morph.* **20**, 33–52.

Nakao, Y. (1939) Recherches sur les greffes chorioallantoidiennes des tissues embryonnaires du Lézard. *Zool. Mag., Tokyo* **51**, 683–697.

Nanney, D. L. (1958) Epigenetic control systems. *Proc. nat. Acad. Sci., Wash.* **44**, 712–717.

Nass, M. K. and Nass, S. (1963) Intramitochondrial fibers with DNA characteristics. 1. Fixations and electron dense staining. *J. Cell Biol.* **19**, 593–611.

Naz, J. F. and Rulon, O. (1946) Modification of development in the chick with LiCl and NaCNS. *Anat. Rec.* **96**, 555.

Needham, J. (1931) *Chemical Embryology.* Cambridge University Press.

Needham, J. (1950) *Biochemistry and Morphogenesis.* Cambridge University Press.

Needham, J. (1959) *A History of Embryology* (2nd Edition). Cambridge University Press.

Nelsen, O. E. (1953) *Comparative Embryology of the Vertebrates.* Blakiston, Toronto.

New, D. A. T. (1955) A new technique for the cultivation of the chick embryo *in vitro*. *J. Embryol. exp. Morph.* **3**, 326–331.

New, D. A. T. (1959) The adhesive properties and expansion of the chick blastoderm. *J. Embryol. exp. Morph.* **7**, 146–164.

New, D. A. T. (1966) *The Culture of Vertebrate Embryos.* Logos Press, London.

New, D. A. T. (1967) Development of explanted rat embryos in circulating medium. *J. Embryol. exp. Morph.* **17**, 513–526.

New, D. A. T. and Coppola, P. T. (1970) Effects of different oxygen concentrations on the development of rat embryos in culture. *J. Reprod. Fert.* **21**, 109–118.

New, D. A. T. and Stein, K. F. (1963) Cultivation of mouse embryos *in vitro. Nature, Lond.* **199**, 297–299.

New, D. A. T. and Stein, K. F. (1964) Cultivation of post implantation mouse and rat embryos on plasma clots. *J. Embryol. exp. Morph.* **12**, 101–112.

Nicholas, J. S. (1947) Experimental approaches to problems of early development in the rat. *Quart. Rev. Biol.* **22**, 179–195.

Nicholas, J. S. and Rudnick, D. (1933) The development of embryonic rat tissues upon the chick chorioallantois. *J. exp. Zool.* **66**, 193–261.

Nicolet, G. (1965a) Étude autoradiographique de la destination des cellules invaginées au niveau du noeud de Hensen de la ligne primitive achevée de l'embryon de poulet. *Acta. Emb. Morph.* **8**, 213–220.

Nicolet, G. (1965b) Action du chlorure de lithium sur la morphogénèse du jeune embryon de poulet. *Acta. Emb. Morph.* **8**, 32–85.

Nicolet, G. (1967) La chronologie d'invagination chez le poulet: Etude à l'aide de la thymidine tritiée. *Experientia* **23**, 576.

Nicolet, G. (1970) Analyse autoradiographique de la localisation des différentes ébouches présomptive dans la ligne primitive de l'embryon de poulet. *J. Embryol. exp. Morph.* **23**, 79–108.

Nicolet, G. and Gallera, J. (1963) Dans quelles conditions l'amnios de l'embryon de poulet peut-il se former en culture *in vitro*? *Experientia* **19**, 165–166.

Nieuwkoop, P. D. (1950) Causal analysis of the development of the primordial germ cells and the germ ridges in urodeles. *Archs Anat. microsc. Morph. exp.* **39**, 257–268.

Nieuwkoop, P. D. (1952) Activation and organization in the central nervous system in amphibians. *J. exp. Zool.* **120**, 1–108.

Nieuwkoop, P. D. (1967) Problems of embryonic induction and pattern formation in amphibians and birds. In *Morphological and Biochemical Aspects of Cytodifferentiation*, Eds. E. Hagen, W. Wechsler, P. Zilliken. Karger, Basel, pp. 22–36.

Nilsson, O. (1966) Estrogen-induced increase of adhesiveness in uterine epithelium of mouse and rat. *Expl. Cell Res.* **43**, 239–240.

Nisonoff, A. and Inman, F. P. (1965) Structural basis of the specificity of antibodies. In *Reproduction: Molecular, Subcellular and Cellular*. Ed. M. Locke (24th Symp. Soc. Dev. Biol.). Academic Press, London and New York, pp. 39–64.

Niu, M. C. and Twitty, V. (1953) The differentiation of gastrula ectoderm in medium controlled by axial mesoderm. *Proc. nat. Acad. Sci., Wash.* **39**, 985–989.

Nørrevang, A. (1968) Electron microscopic morphology of morphogenesis. *Int. Rev. Cytol.* Ed. G. H. Bourne and J. F. Danielli. Academic Press, New York and London, **23**, 114–186.

Noto, T. (1967) The effect of lithium chloride on the morphogenesis and differentiation in the early chick embryo. *Sci. Rep. Tohoku University (Biol.)* **33**, 51–57.

Noyes, R. W. (1968) Sperm transport. In *Progress in Infertility*, Eds. S. J. Behrman and R. W. Kistner. Churchill, London, pp. 181–194.

Obrecht, G., Coraboeuf, E. and Le Douarin, G. (1966) Inhibition poststimulative de l'automatisme de fragments de coeur embryonnaire de poulet cultivé *in vitro. J. Physiol., Paris* **58**, 577.

O'Connor, R. J. (1952) Growth and differentiation in the red blood cells of the chicken embryo. *J. Anat.* **86**, 320–325.

O'Dell, D. S. and McKenzie, J. (1963) The action of aminopterin on the explanted early chick embryo. *J. Embryol. exp. Morph.* **11**, 185–200.

Okazaki, K. and Holtzer, H. (1966) Myogenesis, fusion, myosin synthesis and the mitotic cycle. *Proc. nat. Acad. Sci., Wash.* **56**, 1484–1490.

Olenov, J. M. (1968) Transformationlike phenomena in somatic cells. *Int. Rev. Cytol.* Eds. G. H. Bourne and J. F. Danielli. Academic Press, New York and London, **23**, 1–24.

Olivo, O. M., Laschi, R. and Lucchi, M. L. (1964) Genesi delle miofibrille del cuore embrionale di pollo osservate al microscopio elettronico e inizio dell' attività contrattile. *Lo Sperimentale* **114**, 69–78.

Olsen, M. W. (1960) Nine year summary of parthenogenesis in turkeys. *Proc. Soc. exp. Biol. Med.* **105**, 279–281.

Olsen, M. W. (1962) Polyembryony in unfertilized turkey eggs. *J. Hered.* **53**, 125–129.

Olsen, M. W. (1965) Delayed development and atypical cellular organization in blastodiscs of unfertilized turkey eggs. *Devl. Biol.* **12**, 1–14.

Oppenheimer, J. M. (1967) *Essays in the History of Embryology and Biology.* Massachusetts Institute of Technology Press, Cambridge, U.S.A.

Orgebin-Crist, M. C. (1968) Maturation of spermatozoa in the rabbit epididymis delayed fertilization in does inseminated with epididymal spermatozoa. *J. Reprod. Fert.* **16**, 29–34.

Orts Llorca, F. (1963) Influence of the endoblast in the morphogenesis and late differentiation of the chick heart. *Acta. Anat., Basel* **52**, 202–214.

Orts Llorca, F. (1964) Influence of the ectoderm on heart differentiation and placement in the chicken embryo. *Arch. EntwMech. Org.* **155**, 162–180.

Orts Llorca, F. and Collado, J. J. (1967) Determination of heart polarity (anterio venous axis) in the chicken embryo. *Arch. EntwMech. Org.* **158**, 147–163.

Orts Llorca, F. and Gil, D. R. (1967) A causal analysis of the heart curvatures in the chicken embryo. *Arch. EntwMech. Org.* **158**, 52–63.

Overton, J. (1958) Effects of colchicine on the early chick blastoderm. *J. exp. Zool.* **139**, 329–348.

Overton, J. (1959) Studies on the mode of outgrowth of the amphibian pronephric duct. *J. Embryol. exp. Morph.* **7**, 86–93.

Overton, J. (1962) Desmosome development in normal and reassociating cells in the early chick blastoderm. *Devl. Biol.* **4**, 532–548.

Overton, J. (1968) The fate of desmosomes in trypsinized tissue. *J. exp. Zool.* **168**, 203–214.

Overton, J. (1969) A fibrillar intercellular material between reaggregating embryonic chick cells. *J. Cell Biol.* **40**, 136–143.

Oźdźeński, W. (1969) Fate of primordial germ cells in the transplanted hind gut of mouse embryos. *J. Embryol. exp. Morph.* **22**, 505–510.

Padykula, H. A. (1958) A histochemical and quantitative study of the rat's placenta. *J. Anat.* **92**, 118–129.

Padykula, H. A., Deren, J. J. and Wilson, T. H. (1966) Development of structure and function in the mammalian yolk sac. I. Developmental morphology and vitamin B12 uptake of rat yolk sac. *Devl. Biol.* **13**, 311–348.

Panigel, M. (1956) Contribution à l'étude de l'ovoviviparité chez les reptiles: gestation at parturition chez le lézard vivipare *Zootoca vivipara. Annls. Sci. nat. (Zool.)* **11**, 569–668.

Papaconstantinou, J. (1967) Metabolic control of growth and differentiation in vertebrate embryos. In *The Biochemistry of Animal Development*, **2**, Ed. R. Weber. Academic Press, New York and London, pp. 58–113.

Pasteels, J. J. (1937a) Etudes sur la gastrulation des Vertebrées méroblastiques. II. Reptiles. *Arch. Biol.* **48**, 105–184.

Pasteels, J. J. (1937b) Etudes sur la gastrulation des vertébrées méroblastiques. III. Oiseaux. IV. Conclusions générales. *Arch. Biol., Paris* **48**, 381–488.

Pasteels, J. J. (1940) Un aperçu comparatif de la gastrulation chez les chordés. *Biol. Rev.* **15**, 59–106.

Pasteels, J. J. (1945) On the formation of the primary entoderm of the duck (Anasdomestica) and on the significance of the bilaminar embryo in birds. *Anat. Rec.* **93**, 5–22.

Pasteels, J. J. (1953) Contribution à l'étude du développement de reptiles. 1. Origine et migration des gonocytes chez deux lacertiliens (*Mabuia megalura* et *Chamaeleo bitaeniatus*). *Arch. Biol., Paris* **64**, 227–244.

Pasteels, J. J. (1956) La formation de l'anmios chez les cameleons. *Ann. Soc. Roy. Zool. Belg.* **87**, 243–246.

Pasteels, J. J. (1957a) Une table analytique du développement des reptiles. I. Stades de gastrulation chez les chéloniens et les lacertiliens. *Ann. Soc. Roy. Zool. Belg.* **88**, 217–241.

Pasteels, J. J. (1957b) La formation de l'endophylle et de l'endoblaste vitellin chez les reptiles, cheloniens et lacertiliens. *Acta Anat.* **30**, 601–612.

Pasteels, J. J. (1962) La lignée germinale chez les reptiles et chez les mammifères. Séminaire. Hermann, Paris.

Pasteels, J. J. (1964) The morphogenetic role of the cortex of the amphibian egg. *Advances in Morphogenesis*, **3**, Eds. M. Abercrombie and J. Brachet. Academic Press, New York and London, pp. 363–388.

Pasteels, J. J. (1966) Les lysosomes existent—ils dans l'oeuf? *Bruxelles-Médical* **45**, 1065–1070.

Pasternak, L. and McCallion, D. J. (1962) Heterogeneous inductions in the chick embryo. *Canad. J. Zool.* **40**, 585–591.

Patten, B. M. (1953a) *Human Embryology* (2nd Edition). McGraw-Hill, New York.

Patten, B. M. (1953b) *Embryology of the Pig* (3rd Edition). Blakiston, Philadelphia, Toronto.

Patten, B. M. (1953c) *The Early Embryology of the Chick* (4th Edition). John Murray, London.

Patterson, J. T. (1909) Gastrulation in the pigeon's egg. A morphological and experimental study. *J. Morph.* **20**, 65–124.

Peebles, F. (1898) Some experiments on the primitive streak of the chick. *Arch. EntwMech. Org.* **7**, 405–429.

Peter, K. (1934) Die erste Entwicklung des Chamäleons *Chamaeleo vulgaris* verglichen mit der Eidechse. *Z. f. Anat. Gesch.* **103**, 147–188.

Peter, K. (1938) Die Entwicklung des Entoderms being Hühnchen. *Z. mikrosk-anat. Forsch.* **43**, 362–415.

Peters, H., Levy, E. and Crone, M. (1962) DNA synthesis in oocytes of mouse embryos. *Nature* **195**, 915–916.

Petersen, R. D. A., Fichtelius, K. E. and Liden, S. (1967) An inquiry into the possible reutilization of follicular cell DNA by the mouse oocyte. *Exp. Cell. Res.* **48**, 118–124.

Petry, G. and Kühnel, W. (1964) Beitrag zur Kenntnis des Baues von Basalmembranen. *Z. Zellforsch.* **64**, 533–540.

Pierce, G. B. Jr. and Midgley, A. R. Jr. (1964) The origin and function of human syncytiotrophoblastic giant cells. *Amer. J. Path.* **43**, 153–173.

Pierce, G. B. Jr., Midgley, A. R. Jr., Ram, J. S. and Feldman, J. D. (1962) Parietal yolk sac carcinoma: clue to the histogenesis of Reichert's membrane of the mouse embryo. *Amer. J. Path.* **41**, 549–566.

Pierro, L. J. (1961) Teratogenic action of actinomycin D in the embryo chick. II. Early development. *J. exp. Zool.* **148**, 241–250.

Pincus, G. (1939) The breeding of some rabbits produced by recipients of artificially activated ova. *Proc. natn. Acad. Sci., Wash.* **25**, 557–559.

Potts, D. W. and Wilson, I. B. (1967) The pre-implantation conceptus of the mouse at 90 hours post coitum. *J. Anat.* **102**, 1–11.

Potts, M. (1969) The ultrastructure of egg implantation. *Advances in Reproductive Physiology*, **4**, Ed. A. McLaren. Logos Press, London, pp. 241–267.

Prasad, M. R. N., Dass, C. M. S. and Mohla, S. (1968) Action of oestrogen on the blastocyst and uterus in delayed implantation—an autoradiographic study. *J. Reprod. Fert.* **16**, 97–104.

Prestige, M. C. (1967) The control of cell number in the lumbar spinal ganglia during the development of *Xenopus laevis* tadpoles. *J. Embryol. exp. Morph.* **17**, 453–472.

Price, D. and Ortiz, E. (1965) The role of foetal androgen in sex differentiation in mammals. In *Organogenesis*, Eds. R. L. DeHaan and H. Ursprung. Holt, Rinehart and Winston, New York, 629–652.

Price, D. and Pannabecker, R. (1958) A study of sex differentiation in the fetal rat. In *The Chemical Basis of Development*, Eds. W. D. McElroy and B. Glass. Johns Hopkins Press, Baltimore, pp. 774–777.

Psychoyos, A. (1966) Recent researches on egg implantation. In CIBA Foundation Study Group no. 23 on *Egg Implantation*, Eds. G. E. W. Wolstenholme and M. O'Connor. Churchill, London, pp. 4–15.

Ranzi, S. (1957) Early determination in development under normal and experimental conditions. In *The Beginnings of Embryonic Development*, Eds. A. Tyler, R. C. von Borstel and C. B. Metz. American Association for the Advancement of Science, Washington, pp. 291–318.

Ranzi, S. (1962) Investigations on molecular biology. *Pontif. Acad. Sc.: Scripta Varia* **22**, 255–267.

Ranzi, S. (1968) Considerations on transcription and translation in embryonic development. *Att. Acc. Naz. Lincei* **365**, 227–232.

Rauber, A. (1876) *Ueber die Stellung des Huhnchens in Entwicklungsplan*. Engelmann, Leipzig.

Raven, C. P. (1952) Lithium as a tool in the analysis of morphogenesis in *Limnea stagnalis*. *Experientia* **8**, 252–257.

Raven, C. P. (1961) *Oogenesis*. Pergamon Press, London.

Raven, C. R. (1954) *An Outline of Developmental Physiology*. Pergamon Press, London.

Rawles, M. E. (1936) A study in the localization of organ-forming areas in the chick blastoderm of the head—process stage. *J. exp. Zool.* **72**, 271–316.

Rawles, M. E. (1955) Skin and its derivatives. In *Analysis of Development*, Eds. B. H. Willier, P. A. Weiss, and V. Hamburger. Saunders, Philadelphia and London, pp. 499–519.

Rawles, M. E. (1960) The integumentary system. In *Biology and Comparative Physiology of Birds*, Ed. A. J. Marshall. Academic Press, London and New York, pp. 189–240.

Raynaud, A. (1959a) Une technique permettant d'obtenir le développement des oeufs d'Orvet. (*Anguis fragilis L.*) hors de l'organisme maternel. *C.R. Acad. Sci., Paris* **249**, 1715–1717.

Raynaud, A. (1959b) Développement et croissance des embryons d'Orvet (*Anguis fragilis L.*) dans l'oeuf incubé *in vitro*. *C.R. Acad. Sci., Paris* **249**, 1813–1815.

Raynaud, A. (1961) Quelques phases du développement des oeufs chez l'orvet (*Anguis fragilis L.*). *Bull. biol.* **95**, 365–382.

Raynaud, A. (1962a) Étude histologique de la structure des ébauches des membres de l'embryon d'orvet (*Anguis fragilis L.*) au cours de leur développement et de leur régression. *C.R. Acad. Sci., Paris* **254**, 4505–4507.

Raynaud, A. (1962b) Une technique de décapitation du jeune embryon d'Orvet (*Anguis fragilis L.*). *C.R. Acad. Sci., Paris* **255**, 3041–3043.

Raynaud, A. (1967) Effets d'une hormone oestrogène sur le développement de l'appareil génital de l'embryon de lézard vert (*Lacerta viridis* Laur.). *Arch. d'anat. mikr.* **56**, 63–122.

Reese, A. M. (1915) *The Alligator and its Allies*. Knickerbocker, New York.

Restall, B. J. (1967) The biochemical and physiological relationship between the gametes and the female reproductive tract. In *Advances in Reproductive Physiology*, **2**, Ed. A. McLaren. Logos Press, London, pp. 181–212.

Reynaud, G. (1969) Transfert de cellules germinales primordiales de Dindon à l'embryon de poulet par injection intravasculaire. *J. Embryol. exp. Morph.* **21**, 485–507.

Riddle, O. (1911) On the formation, significance and chemistry of the white and yellow yolk of ova. *J. Morph.* **22**, 455–492.

Robertson, J. D. (1963) Unit membranes: a review with recent new studies of experimental alterations and a new subunit structure in synaptic membranes. In *Cellular Membranes in Development*, Ed. M. Locke (22nd Symp. Soc. Study Dev. Growth). Academic Press, New York and London, pp. 1–82.

Rogers, K. T. (1963) Experimental production of perfect cyclopia in the chick by means of LiCl with a survey of the literature on cyclopia produced by experimental means. *Devl. Biol.* **8**, 129–150.

Rogers, K. T. (1964) Radioautographic analysis of the incorporation of protein and nucleic acid precursors into various tissues of early chick embryos cultured *in toto* on medium containing LiCl. *Devl. Biol.* **9**, 176–196.

Rogulska, T. (1968) Primordial germ cells in normal and transected duck blastoderms. *J. Embryol. exp. Morph.* **20**, 247–260.

Rogulska, T. and Komar, A. (1969) The relationship between the orientation of the early chick embryo and the shape of the egg shell. *Experientia* **25**, 990–991.

Romanoff, A. L. (1960) *The Avian Embryo.* Macmillan, New York.

Romanoff, A. L. (1967) *Biochemistry of the Avian Embryo.* Wiley, New York and London.

Romanoff, A. L. and Romanoff, A. J. (1949) *The Avian Egg.* Wiley, New York.

Romer, A. S. (1966) *Vertebrate Paleontology* (3rd Edition). University Press, Chicago.

Rosenquist, G. C. (1966) A radioautographic study of labeled grafts in the chick blastoderm: development from primitive streak stages to stage 12. *Contr. Embryol. Carneg. Instn.* **38**, 71–110.

Rosenquist, G. C. and DeHaan, R. L. (1966) Migration of precardiac cells in the chick embryo: a radioautographic study. *Contr. Embryol.* **38**, 111–121.

Ross, P. D. and Scruggs, R. L. (1964) Electrophoresis of DNA. II. Specific interactions of univalent and divalent cations with DNA. *Biopolymers* **2**, 79–89.

Roth, S. A. and Weston, J. A. (1967) The measurement of intercellular adhesion. *Proc. natn. Acad. Sci. U.S.A.* **58**, 974–980.

Roth, T. F. and Porter, K. R. (1964) Yolk protein uptakes in the oocyte of the mosquito *Aedes aegypti* L. *J. Cell Biol.* **20**, 313–332.

Rothfels, Ursula (1954) The effects of some amino acid analogues on the development of the chick embryo *in vitro.* *J. exp. Zool.* **125**, 17–18.

Rothschild, Lord (1956) *Fertilization.* Methuen, London.

Rubella Symposium (1965) Various authors. *Amer. J. Diseases Children* **110**, 345–476.

Rudnick, D. (1935) Regional restriction of potencies in the chick during embryogenesis. *J. exp. Zool.* **71**, 83–99.

Rudnick, D. (1948) Prospective areas and differentiation potencies in the chick blastoderm. *Annals. N.Y. Acad. Sci.* **49**, 761–772.

Ruggeri, A. (1966) Sull'ultrastruttura della linea primitiva nel blastoderma di pollo. *Arch. Ital. Anat. Emb.* **71**, 407–432.

Rugh, R. (1967) *The Mouse, its Reproduction and Development.* Burgess, Minneapolis.

Runnström, J. (1966) Considerations on the control of differentiation in the early sea urchin development. *Arch. Zool.* **51**, 239–272.

Rutter, W. J., Clark, W. R., Kemp, J. D., Bradshaw, W. S. Sanders, T. G. and Ball, W. D. (1968) Multiphasic regulation in cytodifferentiation. In *Epithelial-Mesenchymal Interactions,* Eds. R. Fleischmajer and R. E. Billingham. Williams and Wilkins, Baltimore, pp. 114–131.

Saake, R. G. and Almquist, J. O. (1964) Ultrastructure of bovine spermatozoa. I. The head of normal ejaculated sperm. *Am. J. Anat.* **115**, 143–162.

Sabin, F. R. (1917) Origin and development of the primitive vessels of the chick and of the pig. *Contr. Embryol.* **6**, 61–124.

Sacoman, F. M., Morgan, C. F. and Wells, L. J. (1967) Radioautographic studies of DNA synthesis in the developing extra-embryonic membranes of the mouse. *Anat. Rec.* **158**, 197–206.

Saint Girons, H. (1962) Présence de receptacles séminaux chez les Caméléons. *Beaufortia* **9**, 165–172.

Salzgeber, B. (1966) Production élective de la phocomélie sous l'influence d'yperite azotée, chez l'embryon de poulet. II. Étude histologique des bourgeons de membres au cours du développement. *J. Embryol. exp. Morph.* **16**, 339–354.

Samoshkina, N. A. (1965) Inclusion of H^3-thymidine into nuclei of mouse embryo cells during the pre-implantation and implantation period of development. In Russian: *Dok. Akad. Nauk SSSR* **161**, 1467–1470. (English abstract in *Excerpta Med.* section XXI (1966) **6**, no. 2857).

Sanyal, S. and Niu, M. C. (1966) Effects of RNA on the developmental potentiality of the posterior primitive streak of the chick blastoderm. *Proc. nat. Acad. Sci., Wash.* **55**, 743–750 .

Sauer, F. C. (1935) The cellular structure of the neural tube. *J. Comp. Neurol.* **63**, 13–24.

Sauer, F. C. (1936) The interkinetic migration of embryonic epithelial nuclei. *J. Morph.* **60**, 1–11.

Sauer, M. E. and Walker, B. E. (1959) Radioautographic study of interkinetic nuclear migration in the neural tube. *Proc. Soc. exp. Biol. Med.* **101**, 557–560.

Saunder, J. W. Jr. (1948) The proximo-distal sequence of origin on the parts of the chick wing and the role of the ectoderm. *J. exp. Zool.* **108**, 363–404.

Saunders, J. W. Jr. (1966) Death in embryonic systems. *Science, N. Y.* **154**, 604–612.

Saunders, J. W. Jr. and Fallon, J. F. (1966) Cell death in morphogenesis. In *Major Problems in Developmental Biology*, Ed. M. Locke. Academic Press, New York and London, pp. 289–316.

Saunders, J. W. Jr. and Gasseling, M. T. (1963) Trans-filter propagation of apical ectoderm maintenance factor in the chick embryo wing bud. *Devl. Biol.* **7**, 64–78.

Saunders, J. W. Jr., Gasseling, M. T. and Saunders, L. C. (1962) Cellular death in morphogenesis of the avian wing. *Dev. Biol.* **5**, 147–178.

Saunders, J. W. Jr. and Gasseling, M. T. (1968) Ectodermal-mesenchymal interactions in the origin of limb symmetry. In *Epithelial-Mesenchymal Interactions*, Eds. R. Fleischmajer and R. E. Billingham. Williams and Wilkins, Baltimore, pp. 78–97.

Saxén, L. (1961) Transfilter neural induction of amphibian ectoderm. *Devl. Biol.* **3**, 140–152.

Saxén, L. and Toivonen, S. (1962) *Primary Embryonic Induction*. Logos Press, London.

Scarano, E. and Augusti-Tocco, G. (1967) Biochemical pathways in embryos. *Comprehensive Biochemistry* **28**, 55–112.

Schindler, H., Ben-David, E., Hurwitz, S. and Kempenich, O. (1967) The relation of spermatozoa to the glandular tissue in the storage sites of the hen's oviduct. *Poultry Science* **46**, 1462–1471.

Schjeide, O. A., McCandless, R. G. and Munn, R. J. (1963) Possible participation of RNA in formation of mitochondria-like organelles. *Growth* **27**, 125–128.

Schjeide, O. A., McCandless, R. G. and Munn, R. J. (1964) Mitochondrial morphogenesis. *Nature, Lond.* **203**, 158–160.

Schlafke, S. and Enders, A. C. (1963) Observations of the fine structure of the rat blastocyst. *J. Anat.* **97**, 353–360.

Schuetz, A. W. (1969) Oogenesis: processes and their regulation. In *Advances in Reproductive Physiology*, **4**, Ed. A. McLaren. Logos Press, London, pp. 99–148.

Schultz, J. (1965) Genes, differentiation, and animal development. In *Genetic Control of Development*. Brookhaven Symp. No. 18, Brookhaven National Laboratory, New York, pp. 116–147.

Schultz, P. W. and Herrmann, H. (1958) Effect of a luecine analogue on incorporation of glycine into the proteins of explanted chick embryos. *J. Embryol. exp. Morph.* **6**, 262–269.

Schwartz, E. E., Shapiro, B. and Kollmann, G. (1964) Selective chemical protection against radiation in tumor-bearing mice. *Cancer Res.* **24,** 90–96.

Seidel, F. (1952) Die Entwicklungspotenzen einer isolierten Blastomere des Zweizellenstadiuums im Saügetiere. *Naturwiss.* **39,** 355.

Sengel, P. (1964) The determination of the differentiation of the skin and the cutaneous appendages of the chick embryo. In *The Epidermis,* Eds. W. Montagna and W. C. Lobitz. Academic Press, New York and London, pp. 15–34.

Seno, T. (1961) An experimental study on the formation of the body wall in the chick. *Acta. Anat.* **45,** 60–82.

Settle, G. W. (1954) Localization of the erythrocyte-forming areas in the early chick blastoderm cultivated *in vitro. Contr. Embryol.* **35,** 221–237.

Sharman, G. B. and Berger, P. J. (1969) Embryonic diapause in marsupials. In *Advances in Reproductive Physiology,* **4,** Ed. A. McLaren. Logos Press, London, pp. 211–240.

Shelesnyak, M. C. (1960) Nidation of the fertilized ovum. *Endeavour* **19,** 81–86.

Shelesnyak, M. C. (1963) Role of oestrogen in nidation. In *Delayed Implantation,* Ed. A. C. Enders. University Press, Chicago, pp. 265–280.

Shelley, H. J. (1961) Glycogen reserves and their changes at birth and in anoxia. *Brit. Med. Bull.* **17,** 137–143.

Sherbet, G. V. (1963) Studies on transplantations of amphibian anterior pituitary into chick embryo. Analysis of induction capacity by differential solubility and precipitation method. *J. Embryol. exp. Morph.* **11,** 227–253.

Sherbet, G. V. (1966) The inhibitory effects of calf thymus-histone fractions on the development of the chick embryo. *J. Embryol. exp. Morph.* **16,** 159–170.

Sherbet, G. V. and Mulherkar, L. (1963) The morphogenetic action of follicle-stimulating hormone on post-nodal fragments of early chick blastoderm. *Arch. EntwMech. Org.* **154,** 506–512.

Sherbet, G. V. and Mulherkar, L. (1965) A study of the inductive capacity of post-nodal primitive streak pieces after treatment with follicle stimulating hormone. *Arch. EntwMech. Org.* **155,** 701–708.

Sheridan, J. D. (1966) Electrophysiological evidence for low-resistance intercellular junctions in the early chick embryo. *J. Cell. Biol.* **37,** C1–C5.

Shettles, L. B. (1955) Morula stage of human ovum developed *in vitro. Fert. Steril.* **6,** 287–289.

Shoger, R. L. (1960) The regulative capacity of the node region. *J. exp. Zool.* **143,** 221–238.

Shorey, M. Louise (1909) The effect of the distruction of peripheral areas on the differentiation of the neuroblasts. *J. exp. Zool.* **7,** 25–64.

Simkiss, K. (1967) *Calcium in Reproductive Physiology, A Comparative Study of Vertebrates.* Chapman and Hall, London.

Simmer, H. H. (1968) Placental hormones. In *Biology of Gestation,* Ed. N. S. Assali. Academic Press, New York and London, pp. 290–354.

Simon, D. (1960) Contribution à l'étude de la circulation et du transport des gonocytes primaires dans les blastodermes oiseaux cultivés *in vitro. Arch. Anat. micr. Morph. exp.* **49,** 93–176.

Singer, M. (1968) Some quantitative aspects concerning the trophic role of the nerve cell. In *Systems Theory Biology,* Ed. M. D. Mesarovic. Springer-Verlag, Berlin, Heidelburg and New York, pp. 233–245.

Singer, M., Rzehak, K. and Maier, C. S. (1967) The relation between the caliber of the axon and the trophic activity of nerves in limb regeneration. *J. exp. Zool.* **166**, 89–97.

Sisken, J. E. and Kinosita, R. (1962) Timing of DNA synthesis in the mitotic cycle of mammalian cells *in vitro*. In *Biological Interactions in Normal and Neoplastic Growth*, Eds. M. J. Brennan and W. L. Simpson. Churchill, London, pp. 25–35.

Sissman, N. J. (1966) Cell multiplication rates during development of the primitive cardiac tube in the chick embryo. *Nature, Lond.* **210**, 504–507.

Škreb, N. and Frank, Z. (1963) Developmental abnormalities in the rat induced by heat shock. *J. Embryol. exp. Morph.* **11**, 445–447.

Smith, A. E. S. and Schechtman, A. M. (1962) Significance of the rabbit yolk sac. A study of the passage of heterologous proteins from mother to embryo. *Devl. Biol.* **4**, 339–360.

Smith, J. Maynard (1960) Continuous, quantized and modal variation. *Proc. Roy. Soc. B* **152**, 397–409.

Smith, L. D. (1966) The role of a germinal plasm in the formation of primordial germ cells in *Rana pipiens*. *Devl. Biol.* **14**, 330–347.

Smith, L. D. and Ecker, R. E. (1969) Role of the oocyte nucleus in physiological maturation in *Rana pipiens*. *Devl. Biol.* **19**, 281–309.

Smith, L. J. (1964) The effects of transection and extirpation on axis formation and elongation in the young mouse embryo. *J. Embryol. exp. Morph.* **12**, 787–804.

Smith, L. J. (1966) The changing pattern of basophilia in the mouse uterus from mating through implantation. *Am. J. Anat.* **119**, 1–14.

Smith, P. E. and Dortzbach, C. (1929) The first appearance in the anterior pituitary of the developing pig foetus of detectable amounts of the hormones stimulating ovarium maturity and general body growth. *Anat. Rec.* **43**, 277–299.

Smithells, R. W. (1966) Drugs and human malformations. In *Advances in Teratology*, **1**, Ed. D. H. M. Woollam. Logos Press, London, pp. 251–278.

Snell, G. D. and Stevens, L. C. (1966) Early embryology. In *Biology of the Laboratory Mouse*, Ed. E. L. Green. McGraw-Hill, New York, pp. 205–245.

Solomon, J. B. (1957a) Increase of deoxyribonucleic acid and cell number during morphogenesis of the early chick embryo. *Biochim. biophys. Acta.* **23**, 24–27.

Solomon, J. B. (1957b) Nucleic acid content of early chick embryos and the hen's egg. *Biochim. biophys. Acta.* **24**, 584–591.

Solomon, J. B. (1959) Changes in the distribution of glutamic, lactic and malic dehydrogenases in liver cell fractions during development of the chick embryo. *Devl. Biol.* **1**, 182–198.

Solomon, J. B. (1960) Constitutive enzymes of the developing chick embryo: adenosine deaminase. *Biochem. J.* **75**, 278–284.

Solomon, J. B. (1965) Development of nonenzymatic proteins in relation to functional differentiation. In *The Biochemistry of Animal Development*, **1**, Ed. R. Weber. Academic Press, London and New York, pp. 367–440.

Sorokin, S. (1965) Recent work on developing lungs. In *Organogenesis*, Eds. R. L. DeHaan and H. Ursprung. Holt, Rhinehart and Winston, New York, pp. 467–491.

Sorokin, S. P. and Padykula, H. A. (1964) Differentiation of rat's yolk sac in organ culture. *Am. J. Anat.* **114**, 457–479.

Soupart, P. and Noyes, R. W. (1964) Sialic acid as a component of the zona pellucida of the mammalian ovum. *J. Reprod. Fert.* **8**, 251–253.

Spemann, H. (1918) Ueber die Determination der ersten Organanlagen des Amphibien-embryo. *Arch. EntwMech. Org.* **43**, 448–555.

Spemann, H. (1938) *Embryonic Development and Induction.* Yale University Press, New Haven, Connecticut.

Spemann, H. and Mangold, H. (1924) Ueber Induktion von Embryonalanlagen durch Implantation artfremder Organisatoren. *Arch. EntwMech. Org.* **100**, 599–638.

Spirin, A. S. (1966) On masked forms of messenger RNA in early embryogenesis and in other differentiating systems. *Current Topics in Developmental Biology*, **1**, Ed. A. A. Moscona and A. Monroy. Academic Press, New York and London, pp. 2–38.

Spratt, N. T. Jr. (1946) Formation of the primitive streak in the explanted chick blastoderm marked with carbon particles. *J. exp. Zool.* **103**, 259–304.

Spratt, N. T. Jr. (1947) A simple method for explanting and cultivating early chick embryos *in vitro. Science, N.Y.* **106**, 452.

Spratt, N. T. Jr. (1949) Nutritional requirements of the early chick embryo. III. The metabolic basis of morphogenesis and differentiation as revealed by the use of inhibitors. *Anat. Rec.* **105**, 597–598.

Spratt, N. T. Jr. (1952) Localization of the prospective neural plate in the early chick blastoderm. *J. exp. Zool.* **120**, 109–130.

Spratt, N. T. Jr. (1954) Studies on the organizer center of the early chick embryo. In *Aspects of Synthesis and Order in Growth*, Ed. D. Rudnick. Princeton University Press, New Jersey, pp. 209–231.

Spratt, N. T. Jr. (1955) Analysis of the organizer center in the early chick embryo. I. Localization of prospective notochord and somite cells. *J. exp. Zool.* **128**, 121–164.

Spratt, N. T. Jr. (1957a) Analysis of the organizer center in the early chick embryo. II. Studies of the mechanics of notochord elongation and somite formation. *J. exp. Zool.* **134**, 577–612.

Spratt, N. T. Jr. (1957b) Analysis of the organizer center in the early chick embryo. III. Regulative properties of the chorda and somite centers. *J. exp. Zool.* **135**, 319–354.

Spratt, N. T. Jr. (1963) Role of the substratum, supracellular continuity and differential growth in morphogenetic cell movements. *Devl. Biol.* **7**, 51–63.

Spratt, N. T. Jr. (1966) Some problems and principles of development. *Am. Zool.* **6**, 9–19.

Spratt, N. T. Jr. and Haas, H. (1960a) Morphogenetic movements in the lower surface of the unincubated and early chick blastoderm. *J. exp. Zool.* **144**, 139–158.

Spratt, N. T. Jr. and Haas, H. (1960b) Importance of morphogenetic movements in the lower surface of the young chick blastoderm. *J. exp. Zool.* **144**, 257–276.

Spratt, N. T. Jr. and Haas, H. (1960c) Integrative mechanisms in development of the early chick blastoderm. I. Regulative potentiality of separated parts. *J. exp. Zool.* **145**, 97–137.

Spratt, N. T. Jr. and Haas, H. (1961a) Integrative mechanisms in development of the early chick blastoderm. II. Role of morphogenetic movements and regenerative growth in synthetic and topographically disarranged blastoderms. *J. exp. Zool.* **147**, 57–94.

Spratt, N. T. Jr. and Haas, H. (1961b) Integrative mechanisms in development of the early chick blastoderm. III. Role of cell population size and growth potentiality in synthetic systems larger than normal. *J. exp. Zool.* **147**, 271–293.

Spratt, N. T. Jr. and Haas, H. (1962) Integrative mechanisms in development of the early chick blastoderm. IV. Synthetic systems composed of parts of different development age. Synchronization of developmental rates. *J. ex. Zool.* **149**, 75–102.

Spratt, N. T. Jr. and Haas, H. (1965) Germ layer formation and the role of the primitive streak in the chick. I. Basic architecture and morphogenetic tissue movements. *J. exp. Zool.* **158**, 9–38.

Staehelin, M. (1965) Biochemical mechanism of information transfer. In *The Biochemistry of Animal Development*, **1**, Ed. R. Weber. Academic Press, New York and London, pp. 442–483.

Stalsberg, H. (1969) The origin of heart asymmetry: right and left contributions to the early chick embryo heart. *Devl. Biol.* **19**, 109–127.

Stalsberg, H. and DeHaan, R. L. (1968) Endodermal movements during foregut formation in the chick embryo. *Devl. Biol.* **18**, 198–215.

Stalsberg, H. and DeHaan, R. L. (1969) The precardiac areas and formation of the tubular heart in the chick embryo. *Devl. Biol.* **19**, 128–159.

Steding, G. (1962) Experiments on the morphogenesis of the spinal cord. Studies with chick embryos (*Gallus gallus*). *Acta Anat., Basel* **49**, 199–231.

Stedman, E. and Stedman, E. (1950) Cell specificity of histones. *Nature, Lond.* **166**, 780–781.

Stegner, H. E. and Wartenberg, H. (1961) Elektronenmikroskopische und histotopochemische Untersuchungen über Struktur und Bildung der Zona Pellucida menschlicher Eizellen. *Z. Zellforsch.* **53**, 702–713.

Stein, K. F. (1929) Early embryonic differentiation of the chick hypophysis as shown in chorio-allantoic grafts. *Anat. Rec.* **43**, 221–238.

Stein, K. F. (1933) The location and differentiation of the presumptive ectoderm of the forebrain and hypophysis as shown by chorioallantoic grafts. *Physiol. Zoöl.* **6**, 205–235.

Steinberg, M. S. (1964) The problem of adhesive selectivity in cellular interactions. In *Cellular Mechanisms in Development*, Ed. M. Locke (22nd Symp. Soc. Study Dev. Growth). Academic Press, New York and London, pp. 321–366.

Stéphan, F. (1949) Les suppléances obtenues expérimentalement, dans le systeme des arcs aortiques de l'embryon d'oiseau. *C. rend. assoc. anat.* **36**, 647–651.

Stéphan, F. and Sutter, B. (1961) Réaction de l'embryon de poulet au bleu trypan. *J. Embryol. exp. Morph.* **9**, 410–421.

Stephenson, N. G. (1966) Effects of temperature on reptilian and other cells. *J. Embryol. exp. Morph.* **16**, 455–467.

Sterzl, J. and Silverstein, A. M. (1967) Developmental aspects of immunity. *Advances in Morphogenesis*, **6**, Eds. M. Abercrombie and J. Brachet. Academic Press, New York and London, pp. 337–459.

Stevens, L. C. (1967) The biology of teratomas. In *Advances in Morphogenesis*, **6**, Eds. M. Abercrombie and J. Brachet. Academic Press, New York and London, pp. 1–31.

Stewart, J. A. and Papaconstantinou, J. (1967) A stabilization of RNA templates in lens cell differentiation. *Proc. nat. Acad. Sci. Wash.* **58**, 95–102.

Stockdale, F. E. and Topper, Y. J. (1966) The role of DNA synthesis and mitosis in hormone-dependent differentiation. *Proc. Nat. Acad. Sci., Wash.* **56**, 1283–1289.

Straus, W. L. Jr. and Rawles, M. E. (1953) An experimental study of the origin of the trunk musculature and ribs in the chick. *Am. J. Anat.* **92**, 471–509.

Strohl, J. (1925) Les serpents à deux têtes et les serpents doubles. *Ann. Des. Sc., Zool.* **9**, 105–132.

Strudel, G. (1955) Conséquences de l'excision de tronçons du tube nerveux sur la morphogenèse de l'embryon de poulet et sur la différenciation de ses organes; contribution à la genèse de l'orthosympathique. *Année Biol.* **31**, 669–675.

Strudel, G. (1962) Induction de cartilage *in vitro* par l'extrait de tube nerveux et de chorde de l'embryon de poulet. *Devl. Biol.* **4**, 67–86.

Strudel, G. (1963) Self-differentiation and induction of cartilage from chick somitic mesoderm cultured *in vitro*. *J. Embryol. exp. Morph.* **11**, 399–412.

Strudel, G. and Pinot, M. (1965) Différenciation en culture *in vitro* du mesonephros de l'embryon de poulet. *Devl. Biol.* **11**, 284–299.

Swanson, C. P. and Young, W. J. (1965) Chromosome reproduction in mitosis and meiosis. In *Reproduction: Molecular, Subcellular and Cellular*, Ed. M. Locke. (24th Symp. Soc. Dev. Biol.). Academic Press, New York and London, pp. 107–129.

Szollosi, D. (1962) Cortical granules: a general feature of mammalian eggs? *J. Reprod. Fert.* **4**, 223–224.

Szollosi, D. (1965) The fate of sperm middle piece mitochondria in the rat egg. *J. exp. Zool.* **159**, 367–378.

Szollosi, D. (1966) Time and duration of DNA synthesis in rabbit eggs after sperm penetration. *Anat. Rec.* **154**, 209–212.

Szollosi, D. (1967) Development of cortical granules and the cortical reaction in rat and hamster eggs. *Anat. Rec.* **159**, 431–446.

Szollosi, D. and Ris, H. (1961) Observations on sperm penetration in the rat. *J. bioch. biophys. Cytol.* **10**, 275–283.

Taderera, J. V. (1967) Control of lung differentiation *in vitro*. *Devl. Biol.* **16**, 489–512.

Takeda, A. (1966) Behaviour of spermatozoa in the genital tract of the hen. IV. Persistence of spermatozoa and their transport in the hen's oviduct following artificial insemination. *Jap. Poultry Sci.* **4**, 62–67.

Tampion, D. and Gibbons, R. A. (1963) Effect of mucus on the swimming rate of sperms. *Nature, Lond.* **196**, 290.

Tarkowski, A. K. (1959) Experiments on the development of isolated blastomeres of mouse eggs. *Nature, Lond.* **184**, 1286–1287.

Tarkowski, A. K. (1961) Mouse chimaeras developed from fused eggs. *Nature, Lond.* **190**, 857–860.

Tarkowski, A. K. (1964) True hermaphroditism in chimaeric mice. *J. Embryol exp. Morph.* **12**, 735–757.

Tarkowski, A. K. (1965) Embryonic and postnatal development of mouse chimeras. *Preimplantation Stages of Pregnancy*, Eds. G. E. W. Wolstenholme and M. O'Connor. *CIBA Symp.*, Churchill, London, pp. 183–193.

Tarkowski, A. K., Witkowska, A. and Nowicka, J. (1970) Experimental partheno-genesis in the mouse. *Nature, Lond.* **226**, 162–165.

Tarkowski, A. K. and Wroblewska, J. (1967) Development of blastomeres of mouse eggs isolated at the 4- and 8-celled stage. *J. Embryol. exp. Morph.* **18**, 155–180.

Tata, J. R. (1968) Hormonal regulation of growth and protein synthesis. *Nature, Lond.*, **219**, 331–337.

Taylor, L. W. and Kreutziger, G. O. (1966) The gaseous environment of the chick embryo in relation to its environment. 3. Effect of CO_2 and O_2 levels during the period of the ninth through the twelfth days of incubation. *Poultry Sci.* **45**, 867–884.

Terio, B. (1962) Ulteriori studi sul comportamento del nucleolo in oociti di rettili et di molluschi. *Atti. Soc. Pelorit.* **8**, 177–188.

Terzakis, J. A. (1963) The ultrastructure of normal human first trimester placenta. *J. Ultrastruct. Res.* **9**, 268–284.

Thibault, C. (1949) L'oeuf des mammifères: son développement parthénogénétique. *Annls. Sci. Nat. (Zool.)* **11**, 133–216.

Thomson, J. L. and Brinster, R. L. (1966) Glycogen content of pre-implantation mouse embryos. *Anat. Rec.* **155**, 97–102.

Thornton, C. S. (1956) (Ed.) *Regeneration in Vertebrates.* Chicago University Press, Chicago.

Thornton, C. S. (1968) Amphibian limb regeneration. In *Advances in Morphogenesis*, **7**, Eds. M. Abercrombie and J. Brachet. Academic Press, New York and London, pp. 205–247.

Tiedemann, H. (1966) The molecular basis of differentiation in early development of amphibian embryos. In *Current Topics in Developmental Biology*, Eds A. A. Moscona and A. Monroy. Academic Press, New York and London, pp. 85–112.

Tiedemann, H. (1967a) Inducers and inhibitors of embryonic differentiation: their chemical nature and mechanism of action. In *Morphological and Biochemical Aspects of Cytodifferentiation*, Eds. E. Hagen, W. Wechsler, and P. Zilliken. S. Karger, Basel, pp. 8–21.

Tiedemann, H. (1967b) Biochemical aspects of primary induction and determination. In *The Biochemistry of Animal Development*, **2**, Ed. R. Weber. Academic Press, New York and London, pp. 4–55.

Toivonen, S. (1967) Mechanisms of primary embryonic induction. In *Morphological and Biochemical Aspects of Cytodifferentiation*, Eds. E. Hagen, W. Wechsler, and P. Zilliken. S. Karger, Basel, pp. 1–7.

Torrey, T. W. (1965) Morphogenesis of the vertebrate kidney. In *Organogenesis*, Eds. R. L. DeHaan and H. Ursprung. Holt, Rinehart and Winston, New York, pp. 377–420.

Townes, P. L. and Holtfreter, J. (1955) Directed movements and selective adhesion of embryonic amphibian cells. *J. exp. Zool.* **128**, 53–120.

Trelstad, R. L., Hay, E. and Revel, J. P. (1967) Cell contact during early morpho-genesis in the chick embryo. *Devl. Biol.* **16**, 78–106.

Trinkaus, J. P. (1965) Mechanisms of morphogenetic movements. In *Organogenesis*, Eds. R. L. DeHaan and H. Ursprung. Holt, Rinehart and Winston, New York, pp. 55–104.

Trinkaus, J. P. (1966) Morphogenetic Cell Movements. In *Major Problems in Development Biology*, Ed. M. Locke. Academic Press, New York and London, pp. 125–176.

Tsung, S. D., Ning, I. L. and Shieh, S. P. (1965) Studies on the inductive action of the Hensen's node following its transplantation *in ovo* to the early chick blastoderm. II. Regionally specific induction of the node region of different ages. *Acta. Biol. Exp. Sinica* 10, 69–83. In Chinese; not seen. (English abstract in *Excerpta Medic.* section XXI (1966) 6, no. 2856).

Tuffrey, M., Bishun, N. P. and Barnes, R. D. (1969) Porosity of the mouse placenta to maternal cells. *Nature, Lond.* 221, 1029–1030.

Turkington, R. W. (1968) Hormone-dependent differentiation of mammary gland *in vitro. Current Topics in Developmental Biology*, 3, Eds. A. A. Moscona and A. Monroy. Academic Press, New York and London, pp. 199–218.

Twiesselmann, F. (1938) Expérience de scission précoce de l'aire embryongène chez le poulet. *Archs. Biol., Paris* 49, 285–367.

Tyler, A. (1963) The manipulation of macromolecular substances during fertilization and early development of animal eggs. *Am. Zool.* 3, 109–126.

Tyler, A. (1965) The biology and chemistry of fertilization. *Am. Nat.* 99, 309–334.

Tyler, A. (1967) Masked messenger RNA and cytoplasmic DNA in relation to protein synthesis and processes of fertilization and determination in embryonic development. *Devl. Biol.* suppl. 1, 170–226.

Tyler, C. (1965) A study of the eggshells of the Sphenisciformes. *J. Zool.* 147, 1–19.

Vakaet, L. (1960) A propos du raccourcissement de la ligne primitive du blastoderme du poulet. *J. Embryol. exp. Morph.* 8, 6–19.

Vakaet, L. (1962) Some new data concerning the formation of the definitive endoblast in the chick embryo. *J. Embryol. exp. Morph.* 10, 38–57.

Vakaet, L. (1964) Diversité fonctionelle de la ligne primitive du blastoderme de poulet. *C.R. Soc. Biol.* 158, 1964–1966.

Vakaet, L. (1965) Résultats de la greffe de noeuds de Hensen d'âge différent sur le blastoderme de poulet. *C.R. Soc. Biol., Paris* 159, 232–233.

Vakaet, L. (1967) Contribution a l'étude de la prégastrulation et de la gastrulation de l'embryon de poulet en culture *in vitro. Mem. Acad. Roy. Med. Biol.* 5, 235–237.

Vakaet, L. and Mareel, M. (1964) Quelques précisions sur la régénération de l'endoblaste du blastoderme de poulet. *C.R. Soc. Biol., Paris* 158, 902–903.

Vaughan, R. B. and Trinkaus, J. P. (1966) Movements of epithelial sheets *in vitro. J. Cell. Sci.* 1, 407–413.

Viswanath, J. R., Leikola, A. and Rostedt, I. (1968) Induction by killed Hensen's node in chick embryo ectoderm. *Annales Zool. Fennici.* 5, 384–388.

Voeltzkow, A. (1902) Beiträge zur Entwicklungsgeschichte der Reptilien I Biologie und Entwicklung der ausseren Korperform von *Crocodilus madagascariensis* Grand. *Abh. Senkenberg. naturf. Ges.* 26, 1–150.

von Baer, K. E. (1828) Ueber Entwickelungsgeschichte der Thiere. *Beobachtung und Reflexion.* I. Theil. Borntrager, Konigsberg.

von Hahn, H. P. and Herrmann, Heinz (1962) Effects of amino acid analogs on growth and catheptic activity of chick embryo explants. *Dev. Biol.* 5, 309–327.

Waddington, C. H. (1932) Experiments on the devlopement of chick and duck embryos, cultivated *in vitro. Phil. Trans. B* 221, 179–230.

Waddington, C. H. (1937) Experiments on determination in the rabbit embryo. *Arch. Biol.* **48**, 273–290.

Waddington, C. H. (1938a) Studies on the nature of the amphibian organisation centre. VII. Evocation by some further chemical compounds. *Proc. roy. Soc.* (B) **125**, 365–371.

Waddington, C. H. (1938b) The morphogenetic function of a vestigial organ in the chick. *J. exp. Biol.* **15**, 371–376.

Waddington, C. H. (1946) *How Animals Develop.* Allen and Unwin, London.

Waddington, C. H. (1952) *The Epigenetics of Birds.* Cambridge University Press.

Waddington, C. H. (1956) *Principles of Embryology.* Allen and Unwin, London.

Waddington, C. H. (1966) Fields and gradients. In *Major Problems in Developmental Biology*, Ed. M. Locke. (25th Symp. Soc. Dev. Biol.). Academic Press, New York and London, pp. 105–124.

Waddington, C. H., Feldman, M. and Perry, M. M. (1955) Some specific developmental effects of purine antagonists. *Exp. Cell. Res., Suppl.* **3**, 366–380.

Waddington, C. H. and Perry, M. (1958) Effects of some amino-acid and purine antagonists on chick embryos. *J. Embryol. exp. Morph.* **6**, 365–372.

Waddington, C. H. and Schmidt, G. A. (1933) Induction by heteroplastic grafts of the primitive streak in birds. *Arch. EntwMech. Org.* **128**, 522–563.

Waddington, C. H. and Taylor, J. (1937) Conversion of presumptive ectoderm to mesoderm in the chick. *J. exp. Biol.* **14**, 335–339.

Waddington, C. H. and Waterman, A. J. (1933) The development *in vitro* of young rabbit embryos. *J. Anat.* **67**, 355–370.

Wagley, P. F. and Morgan, H. R. (1948) Observations on the effect of folic acid antagonists, folic acid, liver extract and Vitamin B_{12} on embryonated eggs. *Bull. Johns Hopkins Hosp.* **83**, 275–278.

Wallace, R. A. (1964) Studies on amphibian yolk. VI. A protein kinase from the ovary of *Rana pipiens. Biochim. biophys. Acta.* **86**, 286–294.

Walter, H. and Mahler, H. R. (1958) Biochemical studies of the developing avian embryo. 1. Protein precursors *in vivo. J. biol. Chem.* **230**, 241–250.

Wang, T. Y., Hsieh, K. M. and Blumenthal, H. (1953) The effect of growth hormone on the nucleic acid content of the developing chick embryo. *Endocrinology* **53** 520–526.

Warburg, O. (1908) Beobachtungen über die Oxydationprosse im Seeigeler. *Zeit. Physiol. Chem.* **57**, 1–16.

Ward, R. T. (1962a) The origin of protein and fatty yolk in *Rana pipiens.* I. Phase microscopy. *J. Cell. Biol.* **14**, 303–308.

Ward, R. T. (1962b) The origin of protein and fatty yolk in *Rana pipiens.* II. Electron microscopical and cytochemical observations of young and mature oöcytes. *J. Cell. Biol.* **14**, 309–341.

Warner, N. L. and Szenberg, A. (1964) Immunologic studies on hormonally bursectomized and surgically thymectomized chickens: Dissociation of immunologic responsiveness. In *The Thymus in Immunobiology*, Eds. R. A. Good and A. E. Gabriellsen. Harper and Row, New York, pp. 395–410.

Warren, D. C. (1949) Formation of the hen's egg. In *Fertility and Hatchability of Chicken and Turkey Eggs*, Ed. L. W. Taylor. Wiley, New York, pp. 52–94.

Warren, K. B. (1968) (Ed.) *Differentiation and Immunology* (Symp. No. 7, *Int. Soc. Cell Biol.*). Academic Press, London and New York.

Warren, R. H. and Enders, A. C. (1964) An electron microscope study of the rat endometrium during delayed implantation. *Anat. Rec.* **148**, 177–195.

Waterman, A. J. (1936) Experiments on young chick embryos cultured *in vitro*. *Proc. natn. Acad. Sci., Wash.* **22**, 1–3.

Watson, J. D. (1965) *Molecular Biology of the Gene.* Benjamin, New York.

Watterson, R. L. (1965) Structure and mitotic behaviour of the early neural tube. In *Organogenesis*, Eds. R. L. DeHaan and H. Ursprung. Holt, Rinehart and Winston, New York, pp. 129–160.

Watterson, R. L., Fowler, I. and Fowler, B. J. (1954) The role of the neural tube and notochord in development of the axial skeleton of the chick. *Am. J. Anat.* **95**, 337–400.

Weakly, B. S. (1966) Electron microscopy of the oocyte and granulosa cells in the developing follicles of the golden hamster. *J. Anat.* **100**, 503–534.

Weber, R. (1965) (Ed.) *The Biochemistry of Animal Development.* Academic Press, New York and London.

Weber, R. (1969) The mechanism of tissue destruction during anuran metamorphosis. In *Teratology*, Eds. A. Bertelli and L. Donati. Excerpta Medica, Amsterdam, pp. 139–144.

Weber, R. and Boell, E. J. (1962) Enzyme patterns in isolated mitochondria from embryonic and larval tissues of *Xenopus*. *Devl. Biol.* **4**, 452–472.

Weekes, H. C. (1927) Placentation and other phenomena in the scincid lizard *Lygosoma* (*Hinulia*) *quoyi*. *Proc. Linn. Soc. New S. Wales* **1**, 499–554.

Weekes, H. C. (1935) A review of placentation among reptiles with particular regard to the function and evolution of the placenta. *Proc. zool. Soc. Lond.* **3**, 625–645.

Weiss, L. (1967) *The Cell Periphery Metastasis, and Other Contact Phenomena.* North Holland, Amsterdam.

Weiss, P. (1934) *In vitro* experiments on the factors determining the course of the outgrowing nerve fiber. *J. exp. Zool.* **68**, 393–448.

Weiss, P. (1939) *Principles of Development.* Holt, Rinehart and Winston, New York.

Weiss, P. (1945) Experiments on cell and axon orientation *in vitro:* the role of colloidal exudates in tissue organization. *J. exp. Zool.* **160**, 353–386.

Weiss, P. and Kavanu, J. L. (1957) A model of growth and growth control in mathematical terms. *J. Gen. Physiol.* **41**, 1–47.

Weitlauf, H. M. and Greenwald, G. S. (1967) A comparison of the *in vivo* incorporation of S_{35} methionine by 2-celled mouse eggs and blastocysts. *Anat. Rec.* **159**, 249–254.

Wessells, N. K. (1964) Tissue interactions and cytodifferentiation. *J. exp. Zool.* **157**, 139–152.

Wessells, N. K. (1965) Morphology and proliferation during early feather development. *Devl. Biol.* **12**, 131–153.

Wessells, N. K. (1968) Problems in the analysis of determination, mitosis and differentiation. In *Epithelial-Mesenchymal Interactions*, Eds. R. Fleischmajer and R. E. Billingham. Williams and Wilkins, Baltimore, pp. 132–151.

Weston, J. A. (1963) A radioautographic analysis of the migration and localization of trunk neural crest cells in the chick. *Dev. Biol.* **6**, 279–310.

Weston, J. A. and Butler, Sarah L. (1966) Temporal factors affecting localization of neural crest cells in the chicken embryo. *Devl. Biol.* **2**, 246–266.

Wetzel, R. (1929) Untersuchungen am Huhnchen. Die Entwicklung des Keins Wahrend der ersten beiden Bruttage. *Arch. EntwMech. Org.* **119**, 188–321.

Wetzel, R. (1936) Primitivstreifen und Urkörper nach Storungsversuchen am 1–2 Tage bebrüteten Hühnchen. *Arch. EntwMech. Org.* **134**, 357–465.

Wharton, L. R. Jr. (1949) Double ureters and associated renal anomalies in early human embryos. *Contr. Embryol.* **33**, 103–112.

Whitten, W. K. and Dagg, C. P. (1961) Influence of spermatozoa on the cleavage rate of mouse eggs. *J. exp. Zool.* **148**, 173–183.

Wild, A. E. (1970) Protein transmission across the rabbit foetal membranes. *J. Embryol. exp. Morph.* (in press).

Wilde, C. E. (1955) The urodele neuroepithelium. II. The relationship between phenylalanine metabolism and the differentiation of neural crest cells. *J. Morph.* **97**, 313–344.

Williams, J. (1965) Chemical constitution and metabolic activities of animal eggs. In *The Biochemistry of Animal Development*, 1, Ed. R. Weber. Academic Press, New York and London, pp. 14–71.

Williams, J. (1967) Yolk utilization. In *The Biochemistry of Animal Development*, 2, Ed. R. Weber. Academic Press, New York and London, pp. 343–382.

Willier, B. H. (1937) Experimentally produced sterile gonads and the problem of the origin of germ cells in the chick embryo. *Anat. Rec.* **70**, 89–112.

Willier, B. H. (1955) Ontogeny of endocrine correlation. In *Analysis of Development*, Eds. B. H. Willier, P. A. Weiss and V. Hamburger. Saunders, Philadelphia and London, pp. 574–619.

Willier, B. H. (1968) Glycogen synthesis, storage and transport mechanisms in the yolk sac membrane of the chick embryo. *Arch. EntwMech. Org.* **161**, 89–117.

Willier, B. H. and Rawles, M. E. (1935) Organ-forming areas of the early chick blastoderm. *Proc. Soc. exp. Biol.*, **32**, 1293–1296.

Wilson, I. B. (1963) A tumour tissue analogue of the implanting mouse embryo. *Proc. Zool. Soc., Lond.* **141**, 137–151.

Wilson, J. T. and Hill, J. P. (1907) Observations on the development of Ornithorhynchus. *Phil. Trans. B* **199**, 31–168.

Wilt, F. H. (1965) Erythropoiesis in the chick embryo: the role of the endoderm. *Science, N.Y.* **147**, 1588–1590.

Wilt, F. H. (1966) The concept of messenger RNA and cytodifferentiation. *Am. Zoöl.* **6**, 67–74.

Wilt, F. H. (1967) The control of embryonic hemoglobin synthesis. In *Advances in Morphogenesis*, 6, Eds. M. Abercrombie and J. Brachet. Academic Press, New York and London, pp. 89–125.

Wilt, F. H. and Wessells, N. K. (1968) (Eds.) *Methods in Developmental Biology*. Crowell, New York.

Wimber, D. E. (1963) Methods for studying cell proliferation with emphasis on DNA labels. In *Cell Proliferation*, Eds. L. F. Lamerton and R. J. M. Fry. Blackwell, Oxford, pp. 1–17.

Wimsatt, W. A., Krutzsch, P. H. and Napolitano, L. (1966) Studies on sperm survival mechanisms in the female reproductive tract of hibernating bats. I. Cytology and ultrastructure of intra-uterine spermatozoa in *Myotis lucifugis*. *Am. J. Anat.* **119**, 25–60.

Wischnitzer, S. (1967) The ultrastructure of the nucleus of the developing amphibian egg. *Advances in Morphogenesis*, **6**, Eds. M. Abercrombie and J. Brachet. Academic Press, New York and London, pp. 173–199.

Wislocki, G. B. (1955) Discussion. In *Gestation* (No. 1. Josiah Macy Jr. Foundation Conference). Corlies, Macy and Co., New York.

Wislocki, G. B., Deane, H. W. and Dempsey, E. W. (1946) The histochemistry of the rodent's placenta. *Am. J. Anat.* **78**, 281–347.

Witschi, E. (1962) Development: rat. In *Growth. VII. Prenatal Vertebrate Development*, Eds. D. O. Altman and O. D. Dittmer. Biological Handbooks of the Federation of American Societies for Experimental Biology.

Wolff, E. (1936) Les bases de la tératogénèse expérimentale des Vertébrés amniotes, d'après les résultats de méthodes directes. *Archs. Anat. Histol. Embryol.* **22**, 9–374.

Wolff, E. (1964) *Méthodes Nouvelles en Embryologie*, Ed. E. Wolff. Hermann, Paris.

Wolff, E. (1968) Specific interactions between tissues during organogenesis. *Current Topics in Developmental Biology*, **3**, Eds. A. A. Moscona and A. Monroy. Academic Press, New York and London, pp. 65–94.

Wolpert, L. (1969) Position information and the spatial pattern of cellular differentiation. *J. theor. Biol.* **25**, 1–48.

Wolpert, L. and Mercer, E. H. (1961) An electron microscope study of fertilization of the sea urchin egg. *Psammechinus miliaris. Exp. Cell. Res.* **22**, 45–55.

Woodruff, M. F. A. (1958) Transplantation immunity and the immunological problem of pregnancy. *Proc. roy. Soc. B* **148**, 68–75.

Woodside, G. L. (1937) The influence of host age on induction in the chick blastoderm. *J. exp. Zool.* **75**, 259–282.

WORLD HEALTH ORGANIZATION (1967) Principles for the testing of drugs for teratogenicity. Technical report series, No. 364.

Wyburn, G. M., Aitken, R. N. C. and Johnston, H. S. (1965) The ultrastructure of the zona radiata of the ovarian follicle of the domestic fowl. *J. Anat.* **99**, 469–484.

Wyburn, G. M., Johnston, H. S. and Aitken, R. N. (1966) Fate of the granulosa cells in the hen's follicle. *Z. Zellforsch.* **72**, 53–65.

Wylie, C. C. (1970) (Personal communication).

Wynn, R. M. (1968) Morphology of the placenta. In *Biology of Gestation*, Ed. N. S. Assali. Academic Press, New York and London, pp. 94–184.

Yamada, T. (1966) Control of tissue specificity: the pattern of cellular synthetic activities in tissue transformation. *Am. Zool.* **5**, 21–31.

Yamada, T. (1967) Factors of embryonic induction. *Comprehensive Biochemistry* **28** 113–143.

Yntema, C. L. (1960) Effects of various temperatures on the embryonic development of *Chelydra serpentina. Anat. Rec.* **136**, 305.

Yntema, C. L. (1964) Procurement and use of turtle embryos for experimental procedures. *Anat. Rec.* **149**, 577–586.

Yntema, C. L. and Hammond, W. S. (1955) Experiments on the origin and development of the sacral autonomic nerves in the chick embryo. *J. exp. Zool.* **129**, 375–413.

Young, J. D. (1950) The structure and some physical properties of the testudinian eggshell. *Proc. zool. Soc. Lond.* **120**, 455–469.

Young, W. C. (1931) A study of the function of the epididymis. III. Functional changes undergone by spermatozoa during their passage through the epididymis and vas deferens in the guinea-pig. *J. exp. Biol.* **8,** 151–162.

Zahnd, J. P. and Porte, A. (1963) Sur une structure particulière recontrée dans l'ovocyte de la Tortue terrestre: *Testudo ermania. C.R. Soc. Biol.* **157,** 1490–1491.

Zhinkin, L. N. and Samoshkina, N. A. (1967) DNA synthesis and cell proliferation during the formation of deciduomata in mice. *J. Embryol. exp. Morph.* **17,** 593–606.

Zwaan, J., Bryan, P. R. and Pearce, T. L. (1969) Interkinetic nuclear migration during the early stages of lens formation in the chicken embryo. *J. Embryol. exp. Morph.* **21,** 71–83.

Zwaan, J. and Ikeda, A. (1966) Studies of the differentiation of chicken eye lens with the fluorescent antibody technique. *Anat. Rec.* **154,** 447.

Zwilling, E. (1949) The role of epithelial components in the developmental origin of the 'wingless' syndrome of chick embryos. *J. exp. Zool.* **111,** 175–188.

Zwilling, E. (1952) The effects of some hormones on development. *Ann. N.Y. Acad. Sci.* **55,** 196–202.

Zwilling, E. (1956) Interaction between limb bud ectoderm and mesoderm in the chick embryo. IV. Experiments with a wingless mutant. *J. exp. Zool.* **132,** 241–253.

Zwilling, E. (1960a) Limb morphogenesis. In *Advances in Morphogenesis,* **1,** Eds. M. Abercrombie and J. Brachet. Academic Press, New York and London, pp. 301–328.

Zwilling, E. (1960b) Some aspects of differentiation: disaggregation and reaggregation of early chick embryos. *Nat. Cancer Inst. Monogr.* **2,** 19–39.

Zwilling, E. (1963) Survival and non-sorting of nodal cells following dissociation and reaggregation of definitive streak chick embryo. *Devl. Biol.* **7,** 642–652.

Zwilling, E. (1964a) Controlled degeneration during development. In *Cellular Injury.* CIBA Foundation Symposium, Eds. A. V. S. de Reuck and J. Knight. Churchill, London, pp. 352–362.

Zwilling, E. (1964b) Development of fragmented and of dissociated limb bud mesoderm. *Devl. Biol.* **9,** 20–37.

Zwilling, E. (1968) Morphogenetic phases in development. *Devl. Biol. Suppl.* **2,** 184–207.

Zwilling, E. and Hansborough, L. A. (1956) Interaction between limb bud extoderm and mesoderm in the chick embryo. III. Experiments with polydactylous limbs. *J. exp. Zool.* **132,** 219–240.

M

AUTHOR INDEX

SUBJECT INDEX